Environmental Management and Sustainable Development of Cities

A Case Study of Ibadan, Nigeria

By

Soji Oyeranmi

Environmental Management and Sustainable Development of Cities:
A Case Study of Ibadan, Nigeria

By Soji Oyeranmi

This book first published 2023

Ethics International Press Ltd, UK

British Library Cataloguing in Publication Data

A catalogue record for this book is available from the British Library

Print Book ISBN: 978-1-80441-072-1

eBook ISBN: 978-1-80441-073-8

Table of Contents

Abbreviations

AG	Associated Gas
CASSAD	Centre for African Settlement Studies and Development
CBA	Cost-Benefit Analysis
CBD	Convention of Biological Diversity
CBO	Community Based Organisations
CFCs	Chlorofluorocarbons
CO_2	Carbon dioxide
CRIN	Cocoa Research Institute
DDT	Dichlorodiphenyltrichloroethane
DFID	Department For International Development
ECOSOC	United Nations Economic and Social Council
EDC	Economically Developed Countries
EMAS	European Eco-Management and Audit Scheme
EMS	Environmental Management System
EPI	Environmental Performance Index
ESEH	European Society For Environmental History
FAO	Food and Agriculture Organisation
FDI	Foreign Direct Investment

FRIN	Forestry Research Institute of Nigeria
GETI	Global Education and Training Institute
IAAE	International Conference of Agricultural Economists
IDRC	International Research Development Centre
IFRA	Institut Francais de Recherche en Afrique
IITA	International Institute of Tropical Agriculture
INC	International Corporations
IPCC	Inter-Governmental Panel on Climate Change
ISO	International Standards Organisation
LCA	Life Cycle Assessment
LDC	Least Developed Countries
MDGs	Millennium Development Goals
MEA	Millennium Ecosystem Assessment
MNC	Multinational Corporations
MNOC	Multinational Oil Corporations
N_2O	Nitrous oxide
NAI	National Archive Ibadan
NEST	Nigerian Environmental study /Action Team
NGO	Non Governmental Organisations
NIHORT	National Horticultural Research Institute

NISER	Nigerian Institute of Social and Economic Research
NNPC	Nigeria National Petroleum Corporation
PAHO	Pan American Health Organization
PZ	Patterson and Zochonis
RA	Resilience Alliance
SCP	Sustainable City Project
SEA	Strategic Environmental Assessment
SIP	Sustainable Ibadan Project
SDGs	Sustainable Development Goals
TNC	Transnational Companies
UAC	United African Company
UCH	University College Hospital
UEM	Urban Environmental Management
UNCED	United Nations Conference on Environment and Development
UNDP	United Nations Development Programme
UNDRR	United Nations Office for Disaster Risk Reduction
UNEP	United Nations Environmental Programme
UNESCO	United Nations Educational Scientific and Cultural Organisation
UN-HABITAT	United Nations Human Settlements Programme

UNCHE	United Nations Stockholm Conference on the Human Environment
UNO	United Nations Organisation
UNOSSC	United Nations Office for South-South Cooperation
UNU	United Nations University
WCED	World Commission on Environment and Development
WHO	World Health Organisation

Preface

Primarily, this is a study in Environmental history of Nigeria from the pre-colonial to postcolonial era with a focus on the impact of urban environmental management on sustainable development of African Cities with Ibadan, Nigeria as a case study. Obviously, environmental unsustainability threatens all humanity's futures, but the threat appears to differ from place to place. In reality, it is the billion of world's poor (especially in Sub-Saharan Africa) that will bear the greatest loss. More so, the industrialized nations have been taking steps to implement sustainable development strategies and the poor ones (particularly those in Africa) are yet to have effective programmes or national action plans either to solve the existing environmental crises or avert the impending ecological cataclysm. Due to the complexities of the challenges, all the stakeholders need to fashion more collaborative ventures to respond to the virulent global/national environmental crises. This is one of the primary focuses of Environmental history which reflects its multidisciplinarity and the unbreakable link with the concept of sustainable development.

Environmental historians seek an understanding of the range of human interactions with the physical environment in the past, how this has shaped the present and what are possible implications for the future. But in performing this complex and arduous task; they must look beyond the discipline of history. Hence, as a pioneer study in Environmental history (urban environmental history) in Nigeria, this research work becomes unavoidably multidisciplinary.Consequently, it engaged and interacted with Scholars/disciplines such as: Economics, Geography, Development Studies,Urban Planning, Archeology, Anthropology, Agricultural Studies, Sociology and Natural sciences, such as biology and Ecology, to name a few.

Environmental history is a recent trend in historical research. It explores the long term interactions of man and environment. It started in the USA around 1970s and later spread to other parts of the World through Europe

(Worster, 1989). Environmental history also concerns how economy, technology, politics, social structures and value systems have affected the natural environment, the use of natural resources, and how the changes in natural environment have affected the communities. Most significantly, it is a problem solving field of study, which focuses on finding solution to practical problems such as soil erosion, air and water pollution, waste management, urban planning, sanitation and many more. But the history of cities, towns and industrial communities in sub-Saharan Africa (Nigeria especially) and the need to make their development plans sustainable by combining them with adequate environmental management strategies has not received much attention. This study is an effort at addressing this.

This study contends that adequate urban environmental management strategies are conterminous with sustainable development of cities in any country. Contrary to hitherto widely held belief, it demonstrates that cities as drivers of sustainable development could be positive forces in support of social equality, cultural vitality, economic prosperity and environmental sustainability. With a primary focus on Ibadan, the study laments that most cities in sub-Saharan Africa have become cynosure of disgrace and centres of filth: partly because the leaders and planners are yet to fully grasp the meaning of the concept of sustainable development and positions of cities as drivers of the concept; and majorly due to their utter failure to combine their bids to achieve economic development with qualitative urban environmental management strategies. Hence, development in Ibadan from pre-colonial to post-colonial eras has become convoluted and certainly unsustainable.

As a study of 'cautious optimism' it believes that Ibadan is still redeemable. While it is impossible to rebuild the city from the scratch, reformation and transformation are still very possible. Therefore, in order to transform Ibadan into a globally competitive and sustainable city, the study suggests the following: adoption of UN's Whole-Society and Urban Ecosystem Approaches by the city planners by revitalising the Sustainable Ibadan Project (SIP); revolutionising urban governance in Ibadan; establishing constructive and unbreakable mutually beneficial relationships among all the stakeholders in the management of Ibadan environment; making Ibadan economy sustainable by priotising and combining the welfare of the poor people and care for the environment with economic development programmes; entrenching environmental education and traditional environmental ethics into the Oyo State educational Curricular from primary to tertiary levels; proper revising the exiting environmental laws and introducing new ones to take care of the new and emerging environmental challenges; properly enforcing these laws; encouraging attitudinal changes in the people towards jettisoning their cultural practices that are inimical to environmental sustainability; inaugurating and coordinating street by street associations towards enhancing community driven care for the environment. In addition to the adoption of multidisciplinary approach in carrying out the research, the study makes use of the Participant-Observation method together with the use primary and secondary sources.

To further enunciate and expatiate the above suggestions, the study is divided into eight chapters. Chapter one provides a general overview for the study by discussing: background to the study, statement of the Problem, objectives, clarification of key terms, significance of the study, methodology, and brief history and relevance of the study area-Ibadan. Several literatures (with different philosophical postulations) have been advanced on environmental studies, particularly since the 1970s when environmental issues gained more global preponderance. It is based on this assertion that the conceptual framework and some relevant works were reviewed in chapter two. Chapter three essentially discussed: relationship between globalization and the environment, an overview of global environmental challenges, African Environmental Crisis, the Nigerian Dimension and the impact on Ibadan. In chapter four the indigenous environmental practices in Ibadan before the incursion of the British Colonialists were critically evaluated. This covered: the relationship between culture and nature (how they perceive land and taking care of it); indigenous environmental laws; sanitation and personal hygiene.

Ibadan actually witnessed a number of unprecedented urban expansion during the British rule so much that it could be concluded that modern Ibadan was a colonial creation. But the haphazard nature of the growth made it unsustainable as we later discovered that it set foundation for the convoluted urban development Ibadan witnessed in the postcolonial period (which still persist till date. Thus, chapter five examined: Environmental laws and enforcement as they affected: land procurement/distribution, Urban Planning, Waste Management, Sanitation and impact of Poverty in Ibadan during colonial rule.To demonstrate the fact that nothing much changed for Ibadan in terms of environmental mismanagement and unsustainable development in the post-colonial era; chapter six focused on Environmental Management and Sustainable Development in Postcolonial Ibadan which covered: City, Environment and Sustainable Development; Ibadan Economy and the Environment and The Stakeholders and the Environment. These were utilised to analyse various urban environmental problems that are confronting the city in post-colonial period.

As it was firmly established in the study, it will be an exercise in deep absurdity to assume that successive governments now or in the past did nothing to ameliorate the environmental decadence in Ibadan. So, it was the focus of chapter seven to evaluate the existing environmental management strategies in Ibadan; covering: Ibadan Town Planning Authority, Ibadan Solid Waste Management Board (ISWMB),Local Government Areas, Ministry of Environmen,t Sustainable Ibadan Project (SIP), Rural Water Supply Sanitation Agency (RUWASSA), The Ibadan Urban Flood Management Project, before coming up with some new paradigms towards ameliorating the nagging environmental issues. Expectedly, chapter eight that concluded the study, enunciated some possible new paradigms in regenerating, making the city resilient and achieves sustainable development in post-colonial Ibadan; that are also applicable to other cities in the Global South and particularly in sub-Saharan and Nigeria.

Dedication

To my Daughter and City Dwellers in Africa and the Global South

Chapter One
General Introduction

How we plan, build and manage our cities now will determine the outcome of our efforts to achieve a sustainable and harmonious development tomorrow. We believe that the battle for a more sustainable future will be won or lost in cities

(The United Nations, 2013)

Background to the Study

In all ages, man has always exploited his natural environment in order to make a living. This does not only reflect the unbreakable nexus between environmental sustainability and human development but also authenticates the fact that existence of Flora and Fauna has been central to human existence on earth.[1] However, human beings and the natural world are on a collision course.[2] This is largely because man's activities had contrived to destroy the natural equilibrium that had facilitated the co-existence of man and nature over the millennia. Consequently, our world continues to witness hard and often irreversible damage on the environment and critical resources. While global community has considered homicide, ethnocide and genocide as crimes against humanity, **ecocide** which is the deliberate destruction of environment is yet to attract the needed attention and sanctions. Some of the problems witnessed in this regard include: global warming induced climate change (which is regarded as the most catastrophic environmental crisis confronting the 21st Century humanity); pollution deforestation, desertification, flooding, hurricane, tsunami and so on.[3]

---

[1] Walter Rodney, *How Europe underdeveloped Africa*, (London: Bogle L Ouverture Publishers, 1986) 9-30

[2] Peter Okebukola, "Global Warming", in S.O. Otokiti and S.G. Odewunmi, eds. *Issues in Management and Development*, (Ibadan: Rex Charles Publication, 2001) 597

[3] This point has been constant in almost all global meetings on environment from Stockholm 1972, Rio de Janeiro 1992, World Commission on Environment and

The good news is that, the world is standing together more than ever before, brain-storming on how to ensure that human development strategies are environment friendly. A number of conferences have been held at: Stockholm in 1972, Kyoto in 1987, Rio de Janeiro in 1992, Johannesburg in 2002 and IPCC from 2007 to 2022.Others organised for this purpose include: several UN-World Urban Fora, Tokyo in 2008, Copenhagen in 2010, Rio in 2012, Conference of Parties (COPs) which began in march 1995 in Berlin, Germany with the latest one COP27 which took place in Egypt between 6 and 18 November 2022 ; several World Economic For a and many more. Till date, the most comprehensive and ambitious step taken by the global community though the United Nations is the Sustainable Development Goals (SDGs) adopted in 2015 by member states.[4] The universal resolution was vividly captured by Federico Major (Director-General, UNESCO) in 2002 when he said: "Our challenge is to device policies that will subordinate business and economy to the needs of the individuals based on a better balance, between mankind and the environment."[5] Since cities are the epic centres of human development, they present gravest challenges to environmental sustainability due mostly to inadequate environmental management strategies.[6]

The daily reality of people on this planet is increasingly an urban reality, with about 56% of the world's population (4.4 billion people) living in the cities.(footnote). According to the United Nations Environment Programme (UNEP), this proportion is expected to increase to 60% or 4.9 billion people by 2030.[7] Despite this, the situation of urban environmental issues within the internationally agreed goals and targets for sustainable development has remained marginal in the urban growth dynamics. The

---

Development 1987, Brundtland, 1987, World Urban forum 2003, World Bank 2007, 2008 UNDP, 2006, 2007, 2008 G8 Summit 2008 Copenhagen 2010, and many more

[4] For more on the UN Sustainable Development Goals (SDGS) please visit https://www.undp.org/content/undp/en/home/sustainable-development-goals/background/

[5] United Nations Environmental Programme (UNEP) *Global Environment Outlook 3: Past, Present and the Future Perspectives*, 2002.Available at: http://www.unep.org/geo/geo3/

[6] C. H. Nilon, A. R. Berkowitz, and K. S. Hollweg, "Editorial: Understanding Urban Ecosystems: A New Frontier for Science and Education", *Urban Ecosystems* 3, (1999) 3-4 and Josef Cougler and William F. Flanagan, *Urbanisation and Social Change in West Africa*, (Cambridge: Cambridge University Press, 1981)

[7] Available at: http://www.unep.org

United Nation through several international conventions and treaties has unsuccessfully tried to reverse the negative trends in urban development. For examples, while chapter seven of Agenda 21 sought to galvanise local level action to prioritise and develop human settlements, chapter twenty-one focuses on solid waste management and sewage infrastructure.[8] Goal number seven of the Millennium Development Goals (MDGs) also concerns the achievement of a significant improvement in the lives of at least, 100 million slum dwellers by 2020.In addition, it also seeks to reduce the proportion of people without sustainable access to safe drinking water by half by 2015.[9] Although these proposals are important, they are not enough. These targets represent part of the "brown agenda" for provision of basic needs that are needed to urgently address environmental challenges, which are also highly correlated to poverty. In its 2007 report, the Third World Urban Forum estimated that approximately 45 percent of the total urban populations in Africa, Latin America, Southeast and West Asia lack portable water in their homes.[10] Moreover, over 25 percent lack three or more basic services (in-house access to drinking water, functioning modern toilet and electricity.[11] While it is imperative that the world urgently address these challenges, they have not received enough attention.

However, while industrialised countries are already taking steps to implement sustainable development policies, underdeveloped or developing countries particularly those in Africa are yet to have effective programmes or national action plans to ensure meeting basic necessities of life and still maintain decent and healthy environment. For this reason, most sub Saharan African cities especially those that ought to be "epic centres of development" are bogged down with severe environmental problems. This has led many urban and environmental historians to tag most urban centres in Africa as "sick cities" overwhelmed by air pollution, noise, traffic, waste, racial tension, slum conditions,

---

[8] United Nations Conference on Environment and Development (UNCED) *Agenda 21*, 1992, available at: http://www.un.org/esa/sustdev/agenda21text.htm

[9] World urban forum 2007, www.unhabitat.org/worldurbanforum2007

[10] Available online at www.undp.org

[11] Available online at www.idrc.ca

maladministration and many other urban malaises.[12] Onibokun equally concluded that "today, the hearts of many cities in Africa are like islands of poverty in the seas of relative affluence."[13] The general consensus among other scholars who shared similar view about urban centres in Africa is that most people living in African cities today are facing great challenges such as: the deterioration of basic services, housing and environment, mass unemployment and underemployment, the virtual absence of State welfare and many more.[14] All of these culminated in what Ilda Lourenco Lindell called "urban crisis."[15] Although the impact of environmental problems in Africa's urban centres vary from place to place, no place in sub Saharan Africa seem to be immune against the tragedy. As one of the few countries in sub Saharan Africa, which had many large pre-industrial cities, Nigeria cannot possibly be an exception to the African urban environmental decay.[16]

Given its size, it is difficult if not impossible to do a nuanced study of environmental management and development of Nigeria. It is in order to ensure an in-depth study that this thesis focuses primarily on Ibadan-one of the largest cities in Nigeria. The choice of Ibadan is premised on the fact that city usually serve as microcosm of their societies.[17] Thus, Ibadan as a micro society within a larger entity called Nigeria could be said to visibly depict the convoluted developmental processes of the country. The general signs include but not limited to: often chaotic political situation, mass illiteracy, perpetual poverty, mixture of traditionalism and

---

[12] Herbert Werlin, *Governing an African City: A Study of Nairobi,* (New York: Africana Publishing Company, 1974)

[13] A.G. Onibokun, et. al, *Affordable Technology and Strategies for Waste Management in Africa: Lessons From Experiences* (Ibadan: Centre for African settlement – Studies and Development (CASSAD), 2000)

[14] I. L. Lindell, *Walking The Tight Rope: Informal Livelihood and Social Networks in a West African City* (Stockholm: Stockholm University Press, 2003); Anne V. Whyte, "Women, Environmental Perception and Participatory Research", in Eva M. Rathgeber ed., *Women's Role in natural Resource Management in Africa,* (Ottawa, International Research Development Centre (IDRC), 2004), 102-190; Hakeem Ibikunle Tijani (ed), *Nigerian Urban History Past and Present,* (Maryland: University Press of America, 2006)

[15] I. L Lindell, *Walking the tight Role*.....9

[16] Laurent Fourchard, *Urban Slums Reports: The Case of Ibadan, Nigeria,* (Ibadan: Institut Francais de Recherche en Afrique (IFRA), University of Ibadan, 2003)

[17] Justin Labinjoh, "Ibadan and the Phenomenon of Urbanism", in G. O.Ogunremi ed., *A Historical and Socio-Cultural Study of an African City,* (Ibadan: Oluyole Club, 1999) 238

modernism; porous nature of industrialisation, inadequate town planning and lack of adequate urban environment management. Ibadan no doubt has been a great cosmopolitan city in Nigeria and the history of its environmental decay could be the true representation of Nigeria's ecological tragedy. [18]

Nigeria, aptly described as a "country of a thousand villages" and Ibadan was described by P.C Llyod as a "city-village."[19] These descriptions truly reflect the frenetic mixture of traditionalism and modernism in most African "modern cities." Ibadan as the study shows continues to draw more people to her expansive land and over the years has developed a paradoxical character. Like other pre-industrial societies, Ibadan largely presents a sprawling agglomeration of buildings, spread out in numerous directions without any coherent order. At the same time in the words Osofisan "the city is showing the frantic pressure and restless energy associated with modern metropolis."[20] Unfortunately, the environmental disorderliness, which has been ravaging this huge urban space, has become dreadful. Yet, both government and the people are seemingly growing more apathetic about environmental issues. It is therefore, not surprising that Ibadan appears today as "a crippled city" in terms of environmental management and development. [21]

In a study of urban decay in forty Nigerian cities, Sylvester Abumere submitted that "most of these cities (Lagos, Ibadan, Benin, Kano, Onitsha etc.) are generally overcrowded and are with dirty and degraded environments"[22]. Several other studies have also corroborated this unenviable environmental conditions of most Nigerian urban centres and the common denominator is their dawning conclusions. According to

[18] Sylvester Abumere, "Urbanisation and Urban Decay in Nigeria", in A. G. Onibokun et. al, eds., *Urban Renewal in Nigeria*, (NISER, 1987). See also Labinjoh, 1999, Fourchard.... 2003

[19] Available online at www.unesco.org

[20] See "The Prologue", in Dapo Adelugba, ed., *Ibadan MESIOGO: A Celebration of a City, its History and People*, (Ibadan: Bookcraft Ltd., 2002) 2

[21] NISER 1987, Labinjoh, 1999 and Laurent Fourchard, *Urban Slums Reports: The Case of Ibadan, Nigeria.... 2003*

[22] Sylvester Abumere, "Urbanisation and Urban Decay in Nigeria" in A.G. Onibokun et. al eds, *Urban Renewal in Nigeria*, (Ibadan: NISER, 1987) 2-25

Mabogunje, most Nigerian cities are still pre-modern largely due to the fact that they have physical environment, which compares unfavourably with their modern counterparts.[23] Most importantly, these cities were not clearly divided into industrial and residential quarters, as cottage industries were located inside the dwelling places of the people. Consequently most of the so called "modern cities" in Nigeria remain "oversized villages."[24] Almost all have been overwhelmed by the culture of unregulated urban environmental management and perennial distortion of master plans.[25] This has continued to hinder development in most of these cities. Ibadan is no exception.

This study fully recognises the fact that a lush and healthy environment is not a luxury and that environmental sustainability is at the heart of meaningful development in cities. Unfortunately, failure of environmental management strategies has been a recurring decimal in Ibadan at both colonial and post colonial times; thereby exposing the city to constant environmental hazards.[26] Environment is defined as all the surrounding things, conditions and influences affecting the growth or development of living things.[27] This reflects the dependence of humanity on nature in their bid to develop. However, urban environments as largely artificial creations of man, required sophisticated management strategies to achieve the desired development. While the world is full of both urban environmental "successes" and "failures", most African cities are examples of urban decadence. Ibadan, which is the main focus of this study, is particularly notorious in this regard.

---

[23] A.L. Mabogunje, *Cities and African Development*, (Ibadan: Oxford University Press, 1976)

[24] Ayodeji Olukoju, "Historical Background of Nigerian Cities", in Toyin Falola and Steven J. Salm eds., *Nigerian Cities*, (New Jersey: African World Press Inc, 2004)10

[25] Toyin Falola and Steven J. Salm eds., *Nigerian Cities*, (New Jersey: African World Press Inc., 2004); and David Anderson and Richard Rathbone eds., *Africa's Urban Past*, (Oxford: James Currey, 2000)

[26] This point has been constant in almost all global meetings on environment from Stockholm 1972, Rio de Janeiro 1992, World Commission on Environment and Development 1987, Brundtland, 1987, World Urban forum 2003, World Bank 2007, 2008 UNDP, 2006, 2007, 2008 G8 Summit 2008 just to mention a few

[27] Robert K. Barnhart ed., *The World Book Dictionary*, Vol. One, (Chicago: World Book, Inc., 1995) 708

A glimpse at its history reveals that Ibadan (founded around 1829) is one of the few large pre-colonial cities in Africa. Indeed, according to Mabogunje "in a very real sense, it is regarded as the pinnacle of pre-European urbanism in Nigeria, the largest purely African City...which has attracted various epithets as 'Black Metropolis'.[28] He further describes Ibadan as 'the largest city in black Africa'....[29] However, despite the pre-colonial pre-eminence of Ibadan, the arrival of the British in 1893 opened a completely new epoch in the history of urban development in Ibadan. Although, the City witnessed a number of unprecedented urban development so much that it could be concluded that modern Ibadan was a colonial creation but the haphazard nature of the development completely negated the outcomes of the colonial urban development in Ibadan. The most debilitating effect was the evolution of a segregated society resulting into polarisation of Ibadan into two unequal worlds. While reflecting on the absolute inequalities that characterised the colonial cities in Africa, Frantz Fanon observed that:

> The colonial world is a world cut in to two-(the settler's zone-where Europeans and other foreigners lived and the native's zone, where indigenes resided; which is diametrically opposed to each other)....No conciliation is possible...The settler zone is strongly built, made of stone and steel; brightly lit; the streets are covered with asphalt.....The settler town is a well fed town, an easy going town; its belly is always full of good things....The part of the town belonging to the colonised people is a place of ill-fame, peopled by men of evil repute. They are born there, they die there....It is a world without spaciousness; men live there on top of each other....The native town is a hungry town, starved of bread, of meat, of shoes, of coal, of light. The native town is crouching village, a town on its kneels, a town wallowing in the mire....[30]

In order to promote a deliberate policy of segregation, the British created the Ibadan Township area with the promulgation of "The Townships

---

[28] A. L. Mabojunje, *Urbanization in Nigeria*, (London: University of London Press, 1968) 186

[29] A. L. Mabojunje.....186

[30] Frantz Fanon, *The Wretched of the Earth*, (New York: Penguin Books, 1983) 29-30

Ordinance of 1917".[31] This was at two levels. The first was between indigenes of a town and Nigerians from other places (the so-called alien natives).The second was to segregate Europeans from Nigerians, irrespective of the status of the latter. Europeans lived in reservations where they had access to the best medical attention, efficient security system, adequate water supply, good road and other social amenities. With this ugly arrangement, it is not surprising that colonialism created racial segregation and ethnic disaffection.[32] It also fostered class consciousness which was anchored on the concept of modernisation especially among the few educated elite. These "civilised" few felt that they were privileged to imitate western culture and those who also benefited from the colonial state began to see themselves as *olaju* (the civilised) to distinguish themselves from the *ara oko* (the rural, uncivilised).[33] As it has been demonstrated in this study, the dual personality created in Ibadan by colonial urbanism led to a lot of enduring environmental, social, economic and cultural changes which sharpened and dampened almost in the same manner the general character and patterns of growth and development of Ibadan from colonial era and subsist till post colonial period.

Statement of the Problem

It is axiomatic that adequate urban environmental management is essential for sustainable development in any country.[34] As a corollary, urban centres have become the most conspicuous environments in which economic capabilities are expanded or impeded; and social qualities of life are fulfilled or frustrated. However, urbanisation could be as dangerous as it is important for the development of most African countries. This is because the urban centres' crucial role has not been sufficiently realised, as the development planners are yet to grasp their position as the major media of development. In the struggle to create new capabilities, African leaders tend to be more pre-occupied with economics that they often lose

---

[31] Toyin Falola, *Politics and Economy in Ibadan,* (Lagos: Modelor Design Aids Ltd., 1989) 135

[32] Toyin Falola....135

[33] Toyin Falola....135

[34] See Herbert Werlin, *Governing an African City....*6

sight of the need for adequate/qualitative urban environmental management strategies for sustainable development. With the stimulated progress in education, transportation and communication, most African urban areas are growing more rapidly than the available opportunities. At the same time, the urban authorities are unable to cope with the housing, educational, healthy welfare and recreational needs of the population. The result is the evolution of what Blackwell called "Parasitic City."[35] Rather than providing the basis for sustained economic growth, cities have become serious impediments to sustainable development, especially in sub Saharan Africa.

In spite of the above identified crises, historical research on implications of adequate environmental management or lack of it for sustainable development has been relatively insignificant in Africa. For example, while Nigerian historians have documented much on issues like pre-colonial era, colonialism and responses of the people; intra and intergroup relations, political, economic and social history; they have done little or nothing about environmental history of the country. Hence, the reason for this research as a modest contribution towards providing literature and enhance the position of environmental history as a veritable field of study in Nigeria. Ibadan, one of the largest (and presumably, one of the dirtiest) cities in Africa is the primary focus of the study.

It is incontestable that Ibadan presents a terrible environmental picture. In fact, the city like most of the Nigerian urban areas, there is little evidence of any realistic physical planning. Visits to major streets and residential areas especially the core city centre, show a prevalence of uncontrolled heaps of refuse in open spaces and all pervasive stench from open sewers. The picture is generally that of urban disarray. The core of this problem is the prevalence of industrial and non-industrial pollutants. Factories and individuals are reckless in waste disposal resulting in environmental hazards inimical to human and animal health. There are also: infrastructural decay, deplorable roads, pitifully inadequate water supply, erratic electricity supply and acute shelter shortage[36]. The inadequacies of

---

[35] Jonathan M. Blackwell et. al, *Environment and Development in African Selected Case Studies*, (Washington D.C: The World Bank, 1992) 10

[36] Abumere, "Urbanisation and Urban Decay in Nigeria"...20-23

town planning and urban mismanagement are visible for everyone to see but no one seems to be showing expected serious constructive concern towards solving the problem.

Indeed, Ibadan has a long history of urban arbitrariness, which explains clearly the reasons for the present urban crisis in the so called "ancient city". For example a 1945 report revealed that though Ibadan was founded in 1829, it took the successive administrators a century to commence a comprehensive planning for the city.[37] Consequently, development had proceeded in Ibadan without the needed environmental management philosophy or policy that could result in standard street systems, parks, well structured buildings and so on. Ibadan therefore becomes what Labinjoh called "epitome of planlessness..."[38] Laurent Fourchard further corroborates this in a special report on Ibadan in 2003 that "after taken a deep historical search, what Ibadan reflects, is a near total absence of urban management and urban planning.[39]" While one may not totally agree with this assertion, it will be foolhardy to deny the enormous challenges the present generation is facing with the appalling environmental situation in Ibadan, which is a reflection of national urban failure in Nigeria.

Although every generation has to live and work in an inherited environment shaped in some cases by very distant predecessors[40]; this does not foreclose environmental rejuvenation by succeeding generations. This study attempts to document and put in proper historical perspective the impact of environment on the development in Ibadan over an identified period of time. This is with the view to mapping the relationship between successive policies on the environment on one hand and sustainable development (or lack of it) on the other. It is also meant to raise environmental consciousness of Nigerians in this direction. Although Ibadan is one of the largest cities in sub-Saharan Africa, yet, it

---

[37] Abumere....

[38] Justin Labinjoh, "Ibadan and the Phenomenon of Urbanism", in G. O.Ogunremi ed., *A Historical And Socio-Cultural Study of an African City*.....238

[39] Laurent Fourchard....37

[40] Gerald Burke, *Towns in the Making of London*, (London: Edward Arnold Publishers Ltd., 1975) 1

lacks: international airport, international markets, navigable river, standard rail system or centralised connecting bridges (fly-overs).Most of the major roads are single lanes. Most importantly: this study chronicles the evolution of Ibadan as an urban centre; analyses stages of urbanisation in the city; discusses the colonial urban environmental policies; evaluates its post-colonial urban environmental policies; establishes the nexus between urban environmental management strategies and development in colonial and postcolonial Ibadan; highlights its infrastructural problems and suggest how best to solve the problems.

This study covers the pre-colonial to post-colonial era in Ibadan .The choice of the pre-colonial era as the starting date is informed by the fact that the knowledge about where we are coming from will assist in understanding the current circumstances and guide us towards a better future. The significance of the colonial rule in the environmental history of Ibadan makes it mandatory to be understood by any research such as this. Ibadan actually witnessed a number of unprecedented urban expansion during the British rule so much that it could be concluded that the city was a colonial creation. But the haphazard nature of the growth made it unsustainable as we later discovered that it set foundation for the convoluted urban development Ibadan witnessed in the postcolonial period (which still persist till date). For example, with the promulgation of "Township ordinances" by the British colonial rulers[41]marked the institutionalisation of segregation in colonial urban policies across Nigerian Cities. The entrenchment of policy of segregation in Ibadan by colonial urbanism led to a lot of enduring environmental, social, economic and cultural changes which shaped the general character and patterns of growth and development of Ibadan from colonial era till date.

The creation of Oyo State ministry of Environment in year 2000 coincided with the end of military rule, it marked the beginning of a new democratic rule and equally heralded the 'magical' new millennium with great expectations of unprecedented development in every aspect of life. This included a hope for urban environmental rejuvenation. Ibadan with its national significance was expected to benefit from the 'monumental

---

[41] Toyin Falola, *Politics and Economy in Ibadan,* (Lagos: Modelor Design Aids Ltd., 1989) 135

millennium miracles' which has largely turned out as a mirage.[42]
Stretching the discussion to post-colonial offers a unique opportunity of
evaluating more than a hundred years of environmental history of
Ibadan. With this, we are able to draw important lessons from the pre-
colonial and colonial environmental practices to correct the current
environmental maladies confronting the city and its people

Objectives

Having accepted that a healthy and adequately managed environment is
a prerequisite for sustainable development[43], this study demonstrates that
a systematic and focused environmental management is imperative for
the regeneration and development of Nigerian cities as a whole and
Ibadan in particular. Apart from the broad concern of examining the
relationship among urban management, environment and sustainable
development of Ibadan during the peak of colonial rule and post colonial
era, the study has the following specific objectives:

1. establish the interdependence of environmental management and
    sustainable development and investigate the root of
    environmental crisis in Ibadan;
2. Identify major stakeholders and their roles in urban
    environmental management and development in Ibadan;
3. examine the efficacy of the governmental strategies in urban
    environmental management and development in Ibadan;
4. assess people's attitude and perception to personal hygiene,
    environmental management and development in Ibadan;
5. examine roles of industries in environmental management vis-à-
    vis development in Ibadan; and
6. generate ideas that will be useful for the people and
    environmental managers when planning for the future.

---

[42] Details of the Millennium Development Goals (MDGs) are online at
www.undp.org/MDGs
[43] W. I. Bell-Gam et. al eds., *Perspectives on the Human Environment*, (Port – Harcourt:
Amethyst & Co Publishers, 2004)

The main research questions are: What were the ways people in Ibadan preserved their environment before colonial era? What were the programs for urban management and development in Ibadan during the colonial period? Did colonial rule introduce any significant changes to the Ibadan environment and its management? To what extent did these changes enhance or impede the development of Ibadan? What groups were influential in the management of Ibadan environment and provision of infrastructure? What were the people's roles in the management of the environment? What were the cultural perspectives to the management of the environment in Ibadan? How is globalisation impacting Ibadan environment and development? How can Ibadan become a globally competitive city and achieve sustainable development?

Clarification of Terms

Before going into the details of this discourse, one must give it a clear direction by providing working explanations rather than definitions for the principal terms around which the study essentially revolved i.e. Environment, Environmental Management, Environmental History, and Development and Sustainable Development.

Environment

The word, "Environment" is from the Latin word *Environs* which is linguistically translated as one's surroundings[44].Generally, the term varies in meaning because of its complexities and scope which encompasses anything from biosphere to the habitat of the smallest creatures or organism. Consequently, environment as a concept has been variously described in the works of several scholars and organisations. For examples, Adeniyi, defines environment as the outer physical and biological systems in which man and other organisms live in a wholly albeit a complicated one with many interacting components.[45] On his part

---

[44] A. D.Nagam and M. Halle Runnalts, "Environment and Globalisation Understanding the Linkages", available online at www.google/environmental-2bisuses.band.com
[45] E. O. Adeniyi, *Environmental Management and Protection in Nigeria,* (Ibadan:NISER 1986) 9

Abrahams, sees environment as the sum total of all external conditions influencing the growth and development of an organism. These factors could be physical, biological, social and cultural.

The 1992 United Nations Stockholm Conference on the Human Environment (UNCHE) declaration referred to man's environment as "that which gives him physical sustenance and affords him the opportunity for intellectual, spiritual, moral and social growth..., both aspects of man's environment, the national and the man made are essential for his well being and enjoyment of basic human rights".[46] The Infoterra National Network, a global information exchange network views the environment as the "sum of all external conditions and influences affecting the life and development of organisms.[47] Princeton University simply describes environment as "the area in which something exists or lives; the country-the flat agricultural surroundings."[48] The World book Dictionary sees environment as all of the surrounding things, conditions and influences affecting the growth and development of living things. The environment is something you are very familiar with.[49] It's everything that makes up our surroundings and affects our ability to live on the earth—the air we breathe the water that covers most of the earth's surface, the plants and animals around us, and much more.

From the above definitions, one can rightly deduce that, the natural environment, commonly referred to simply as the environment, encompasses all living and non-living things occurring naturally on Earth or some region thereof. But in all, the study adopts the definition offered by The American Heritage Dictionary. The book defines environment as the "circumstances or conditions that surround one's surroundings... especially the combination of the external physical conditions that affect and influence the growth, development and survival of organisms."[50] The American Heritage Dictionary goes a little further by explaining that the environment could be all the biotic and abiotic factors that act on an

---

[46] www.google/environmental-2bisuses.band.com Access date-20/09/2010

[47] www.google.com/inforteranetwork

[48] Available on www.wordnetweb.princeton.edu/perl/webwn

[49] See Robert K. Barnhart.....708

[50] *Environment*, available online at www.americanheritage.com

organism, population or ecological community and influence its survival and developments. The Biotic factors include the organisms themselves, their food and interactions, while the abiotic factors include a number of factors including weather, climate and pollution. The environment of a particular location is made up of all the things and conditions found there. Non-living things and conditions like mountains and valleys, rivers and streams, rocks and soils, sunlight and heat, rain and snow make up the physical part of the environment. Living things, like plants, animals, fungi, and bacteria make up the biological part of the environment.

Environmental Management

From the various definitions and descriptions of environment, it becomes obvious that it is both critical and central to human survival and development. Hence, the need for man to device means of take adequate care of this inestimable gift through evolving sound environmental management strategies. The concept of Environmental Management was designed to perform the aforementioned role. Environmental Management offers: research and opinions on the use and conservation of natural resources; how to ensure sustainable cities and urban centres (through proper urban planning and excellent management strategies); protection of habitats and control of hazards, spanning the field of applied ecology without regard to traditional disciplinary boundaries[51]. It aims to improve communication, making ideas and results from any field available to practitioners from other backgrounds. As a multidisciplinary concept, Environmental Management draws knowledge from biology, botany, climatology, ecology, ecological economics, environmental engineering, fisheries, environmental law, environmental history, forest sciences, geology, information science, public affairs, zoology and more.

Environmental management is not, as the phrase ordinarily appears, the management of the environment as such, but rather the management of interaction by the modern human societies with, and impact upon the environment. The three main issues that affect managers are those

---

[51] Schaltegger Stefan et. al, *An Introduction to Corporate Environmental Management: Striving for Sustainability*, (Sheffield: Greenleaf, 2003) 9

involving politics (networking), programs (projects) and resources (money, facilities, etc.)[52]. The need for environmental management can be viewed from a variety of perspectives. A more common philosophy and impetus behind environmental management is the concept of carrying capacity. Simply put, carrying capacity refers to the maximum number of organisms a particular resource can sustain.[53]Environmental management is therefore not the conservation of the environment solely for the environment's sake, but rather the conservation of the environment for humankind's sake. This element of Environmental Management is referred to as sustainable exploitation.[54]

Environmental management also involves the management of all components of the bio-physical environment, both living (biotic) and non-living (abiotic). This is due to the interconnected and network of relationships amongst all living species and their habitats. Environmental management also involves the relationships of the human environment, such as the social, cultural and economic environment with the bio-physical environment. It is in the attempt to maintain positive symbiosm between the various components of environment that the concept of sustainable development was invented.

The goal of sustainable development has been incorporated in the agenda of many international organisations with necessary policies in order to ensure solid Environmental Management practice. The most obvious example is the European Union which has now integrated into the its policy objectives wider range of tools for environmental policy. In the recent 6th Community Environmental Action Programme, titled, "Environment 2010: Our Future, Our Choice", recognises this and aims to be a programme that "...completes and reinforces our body of environmental legislation where there are gaps and takes forward the implementation of our directives...does more in terms of mobilising

---

[52] G. Buchenrieder and A.R. Göltenboth, "Sustainable Freshwater Resource Management in the Tropics: The Myth of Effective indicators", A paper presented at the 25th International Conference of Agricultural Economists (IAAE) on "Reshaping Agriculture's Contributions to Society" in Durban, South Africa, 2003

[53] G. Buchenrieder and A.R. Göltenboth....

[54] Schaltegger Stefan et. al, *An Introduction to Corporate Environmental Management: Striving for Sustainability* 7

stakeholders for the environment and 'greening' the market."[55]On the need to ensure proper implementation, European Commissioner for the Environment at the time, Margot Wallström submitted that:

> ...we will not solve environmental problems by simply adding a few new directives every year to our existing 270 or so pieces of European environmental law, especially if we discover later on that these directives are not implemented by the member States...we need a broader range of instruments to tackle ever more diffuse sources of environmental pressures...We need instruments which: promote information, awareness and commitment with citizens and in the business community; give the right incentives for environmental improvements in the market place and; ensure the integration of the environment into other policies.[56]

As the principal user of nature, humanity should be responsible for ensuring that its environmental impacts are benign rather than catastrophic. In order to help in achieving this, Environmental Management presents the work of academic researchers and professionals outside universities, including those in businesses, governments, research establishments, and public interest groups. Hence, the accumulation of a wide spectrum of viewpoints and approaches will greatly assist humanity towards ensuring environmental sustainability.

Environmental Management at both people and organisational level revolves round Environmental Management System. An Environmental Management System (EMS) is a problem identification and problem solving tool that provides organisations with a method to systematically manage their environmental activities, products and services and helps to achieve their environmental obligations and performance goals. International Standards Organisation (ISO) defines an EMS as "the part of the overall management system that includes organisational structure, planning activities, responsibilities, practices, procedures, processes and resources for developing, implementing, achieving, reviewing and

---

[55] http://www.economics.noaa.gov/?goal=ecosystems&file=users/
[56] Schaltegger Stefan et. Al....

maintaining the environmental policy."[57] Like all management functions, effective management tools, standards and systems are required. An 'environmental management standard or system or protocol attempts to reduce environmental impact as measured by some objective criteria. The ISO 14001 standard is the most widely used standard for environmental risk management and is closely aligned to the European Eco-Management and Audit Scheme (EMAS). [58]

Environmental History

Environmental history is a relatively new branch of historical scholarship. Although the field officially came into existence since the formation of the American Society for Environmental History in the early 1970s; historians have been researching interactions between humans and the environment since early times.[59] Historians at the early period have tended to treat nature — when they treat it at all — as the setting for history, rather than a participant or active agent of change.[60] Conversely, Environmental historians maintain that as nature is a key influence on human affairs then it is both an appropriate and necessary subject for historical analysis through the study of human interaction with the natural world over time. Environmental history is history written with the acknowledgment that we shape our environment and it shapes us.[61] Consequently, irreversible interdependence of man and nature becomes the central focus of Environmental historians. As Thomas H. Huxley once asserted, "the question of questions for mankind – the problem which underlies all others" – was to ascertain "the place which man occupies in nature ....What are the limits of our power over nature and of nature's power

---

[57] www.afrol.com/Categories/Environment/env046_management.htm

[58] Schaltegger Stefan et. Al.....

[59] Ian G. Simmons, *Environmental History: A Concise Introduction*, (Oxford: Oxford University Press, 1993)

[60] Michael Bess, "Artificialization and its Discontents", in *Environmental History*, Vol. 10, No. 1, (2005)

[61] Carolyn, The *Columbia Guide to American Environmental History*, (New York: Columbia University Press, 2002) 10

over us?"[62] And Environmental history primarily desire and design to provide answer to such questions.[63]

Environmental historians seek an understanding of the range of human interactions with the physical environment in the past, how this has shaped the present and what are possible implications for the future. But in performing this complex and arduous task; they must look beyond the discipline of history. Hence, Environmental history becomes unavoidably multidisciplinary. Scholars engaging in environmental history must interact with disciplines such as: economics, historical geography, landscape architecture, urban planning, archeology, anthropology, agricultural studies, sociology and natural sciences, such as biology and ecology, to name a few.[64]

Useful guidance on the process of doing environmental history has been given by Donald Worster, Carolyn Merchant, William Cronon and Ian Simmons.[65]Worster's three core subject areas (the environment itself, human impacts on the environment, and human thought about the environment) are generally taken as a starting point for the student as they encompass many of the different skills required. The tools are those of both history and science with a requirement for fluency in the language of natural science and especially ecology.[66] In fact, methodologies and insights from a range of physical and social sciences is required time to time in order for environmental historians to accomplish their tasks.

---

[62] K. Jan Oosthoek, "Environmental History - Between Science and Philosophy", online at http://science.jrank.org/pages/7662/Environmental-History.html#ixzz0eOxOibWY
[63] Donald Worster, "The Two Cultures: Environmental History and the Environmental Sciences", *Environment and History*, 2, (1996). 3-14; Donald Worster, *The Wealth of Nature: Environmental History and the Ecological Imagination*, (Oxford: Oxford University Press, 1993) 12
[64] Donald Worster, *The Wealth of Nature. Environmental History and the Ecological Imagination*....
[65] Donald Worster ed., *The Ends of the Earth: Perspectives on Modern Environmental History*, (Cambridge: Cambridge University Press.1988); Ian G. Simmons, "Environmental History: A Concise Introduction"....; Merchant Carolyn, *The Columbia Guide to American Environmental History*, (New York: Columbia University Press, 2002).
[66] Donald Worster ed.,*The Ends of the Earth*.....

There is no universally accepted definition of environmental history. But in simple terms, it is a history that tries to explain why our environment is like it is and how humanity has influenced its current configuration, as well as elucidating the problems and opportunities of tomorrow.[67]Like many of such terminologies many reputable scholars have offered their perceptions in form of definitions of Environmental history. In 1988, Donald Worster offered a widely quoted definition that "Environmental history is the interaction between human cultures and the environment in the past."[68] In 1995, J. Donald Hughes defined the subject as "The study of human relationships through time with the natural communities of which they are a part in order to explain the processes of change that affect that relationship"[69] and, in 2006, as "... history that seeks understanding of human beings as they have lived, worked and thought in relationship to the rest of nature through the changes brought by time...."[70] "As a method, environmental history is the use of ecological analysis as a means of understanding human history ... an account of changes in human societies as they relate to changes in the natural environment."[71] Environmental historians are also "interested in what people think about nature, and how they have expressed those ideas in folk religions, popular culture, literature and art."[72] In 2000, McNeill suggested that environmental history was "... the history of the mutual relations between humankind and the rest of nature".[73]

With McNeil's observation, it has become obvious that no one can separate culture from nature or vice versa. According to K. Jan Oosthoek "the separation of nature from culture obscures the fact that culture is influenced by the nature surrounding it. But it is not a one-way street

---

[67] Beinart, William & Coates, Peter, *Environment and History: The Taming of Nature in the USA and South Africa*, (London, 1995) 1

[68] Donald Worster ed., *The Ends of the Earth: Perspectives on Modern Environmental History*....

[69] Donald J. Hughes, "Ecology and Development as Narrative Themes of World History", *Environmental History Review*, 19, (Spring 1995) 1-16

[70] Available online at www.google.com/enviromentalhistory/hughes.htm

[71] Donald Worster, *The Wealth of Nature. Environmental History*......

[72] Donald Worster, *The Wealth of Nature. Environmental History*......

[73] John R. McNeill, *Something New Under the Sun: An Environmental History of the Twentieth Century*, (2000)

because culture is also asserting its influence on the natural world".[74] Beinart and Coates included this ambivalent character into their definition of environmental history: "Environmental history deals with the various dialogues over time between people and the rest of nature, focusing on reciprocal impacts."[75] To understand these reciprocal impacts we must try to bridge the gap between culture and nature, between science and history. Environmental history is an attempt to unite the two worlds of science and history. Donald Worster described the essence of environmental history as follows: Its essential purpose is to put nature back into historical studies, or, to explore the ways in which the biophysical world has influenced the course of human history and the ways in which people have thought about and tried to transform their surroundings.[76]

Environmental history prides itself in bridging the gap between the arts and natural sciences although to date the scales weigh on the side of science. A definitive list of related subjects would be lengthy indeed and singling out those for special mention a difficult task. However, those frequently quoted include historical geography, the history and philosophy of science, history of technology and climate science. On the biological side there is, above all, ecology and historical ecology, but also forestry and especially forest history, archaeology and anthropology.[77] With increasing globalisation and the impact of global trade on resource distribution, concern over never-ending economic growth and the many human inequities, environmental history is now gaining allies in the fields of ecological and environmental economics.[78]

As established earlier, Environmental history evolved in the United States with the emergence of the American Society for Environmental History founded in 1975 and later spread to other parts of the world. While the first institute devoted specifically to environmental history in Europe was

---

[74] K. Jan Oosthoek, "Environmental History - Between Science and Philosophy...."1
[75] William Beinard and Peter Coates, *Environment and History*...1
[76] Donald Worster ed.,*The Ends of the Earth: Perspectives on Modern Environmental History*....
[77] Donald Worster ed.....
[78] John R. McNeill, *Something New Under the Sun*....

established in 1991, based at the University of St. Andrews in Scotland. In 1986, the Dutch foundation for the history of environment and hygiene *Net Werk* was founded and publishes four newsletters per year.[79] In the UK the White Horse Press in Cambridge has, since 1995 published the *Journal of Environment and History* which aims to bring scholars in the humanities and biological sciences closer together in constructing long and well-founded perspectives on present day environmental problems.[80] A similar publication *Tijdschrift voor Ecologische Geschiedenis* (Journal for Environmental History) is a combined Flemish-Dutch initiative mainly dealing with topics in the Netherlands and Belgium although it also has an interest in European environmental history. Each issue contains abstracts in English, French and German. In 1999 the Journal was converted into a yearbook for environmental history.[81]

Communication between European nations is restricted by language difficulties. In April 1999 a meeting was held in Germany to overcome these problems and to co-ordinate environmental history in Europe. This meeting resulted in the creation of the European Society for Environmental History in 1999.[82] Only two years after its establishment, ESEH held its first international conference in St. Andrews, Scotland. Around 120 scholars attended the meeting and 105 papers were presented on topics covering the whole spectrum of environmental history. The conference showed that environmental history is a viable and lively field in Europe and since then ESEH has expanded to over 400 members and continues to grow and attracted international conferences in 2003, 2005 and 2011. In 1999 the Centre for Environmental History was established at the University of Stirling. Some history departments at European universities are now offering introductory courses in environmental history and postgraduate courses in Environmental history have been established at the Universities of Nottingham, Stirling and Dundee and

---

[79] Michael Bess, Mark Cioc, and James Sievert, "Environmental History Writing in Southern Europe", *Environmental History*, 5, (2000) 545-56

[80] Michael Bess et.-al, "Anniversary Forum: What Next for Environmental History?"*Environmental History*, Vol. 10, No. 1, (2005) 30–109

[81] Michael Bess et.-al

[82] Michael Bess et.-al

more recently a Graduierten Kolleg was created at the University of Göttingen in Germany.[83]

Development

Development is a complex issue, with many different and sometimes contentious definitions.[84] Arguably, development at all levels (personal or national) in human society is a multi-faceted process. At the level of the individual, it implies multiplied skill and capacity, greater freedom, creativity, self-discipline, responsibility and material well-being. It must, however, be noted that the achievement of any aspect of personal development is strong tied to the state of the society as a whole. At the national level, development will naturally mean the pulling together of the above-stated personal virtues for the benefit and well-being of people within such a nation. More often than not, as Walter Rodney once contended, development is often used in an exclusive economic sense the justification being that the type of economy is itself an index of other social features.[85] This view basically equates development with economic growth. A society develops economically as its members increase jointly their capacity for dealing with the environment, which of course depends on the extent to which they understand laws of nature (science), on the extent to which they put that understanding into practice by devising tools (technology), and on the manner in which work is organised.[86]

But development has gone far beyond mere economic growth. According to the United Nations Development Programme "development is to lead long and healthy lives, to be knowledgeable, to have access to the resources needed for a decent standard of living and to be able to participate in the life of the community."[87] Another perspective of human development sees development as possibility of freeing people from

---

[83] Michael Bess et.-al

[84] //www.volunteeringoptions.org/VolunteeringDevelopment/WhatisDevelopment/tabid/78/Default.aspx

[85] For Full Details on Discourse on Development and Underdevelopment in Africa see Walter Rodney, *How Europe Underdeveloped Africa*, (London: Bogle L'Ouverture Publishers, 1986) 9-39

[86] Walter Rodney, *How Europe Underdeveloped Africa*....

[87] Available at www.undp.org

obstacles that affect their ability to develop their own lives and communities. In this regard, development is empowerment: it is about local people taking control of their own lives, expressing their own demands and finding their own solutions to their problems.[88]

According to the 1996 Human Development Report, published by the United Nations Development Programme, "human development is the end—economic growth a means."[89] It is true that economic growth, by increasing a nation's total wealth, also enhances its potential for reducing poverty and solving other social problems. But history offers a number of examples where economic growth was not followed by similar progress in human development. Instead growth was achieved at the cost of greater inequality, higher unemployment, weakened democracy, loss of cultural identity, or overconsumption of natural resources needed by future generations. As the links between economic growth and social and environmental issues are better understood, experts including economists tend to agree that this kind of growth is inevitably unsustainable—that is, it cannot continue along the same lines for long. [90]First, if environmental and social/human losses resulting from economic growth turn out to be higher than economic benefits (additional incomes earned by the majority of the population), the overall result for people's wellbeing becomes negative. Thus such economic growth becomes difficult to sustain politically. Second, economic growth itself inevitably depends on its natural and social/human conditions. To be sustainable, it must rely on a certain amount of natural resources and services provided by nature, such as pollution absorption and resource regeneration.[91] This is why this study sees development from strictly a sustainable sense.

Sustainable Development is a term widely used by people (politicians, scholars, environmentalists, activists etc) all over the world, even as the notion is still rather new and lacks a uniform interpretation. Important as

---

[88] www.volunteeringoptions.org

[89] www.undp.org

[90] Available at http://www.un.org/millennium/declaration/ares552e.pdf,

[91] This point has been constant in almost all global meetings on environment and development from Stockholm 1972, Rio de Janeiro 1992, World Commission on Environment and Development 1987, Brundtland, 1987, World Urban forum 2003, World Bank 2007, 2008 UNDP, 2006, 2007, 2008 G8 Summit 2008 just to mention a few

it is, the concept of sustainable development is still being developed and the definition of the term is constantly being revised and extended. Although the term may still largely means different things to different people, but the most frequently quoted definition is from the report *Our Common Future* (also known as the Brundtland Report): Sustainable development is development that meets the needs of the present without compromising the ability of future generations to meet their own needs."[92]Sustainable development in this sense could probably be otherwise called "equitable and balanced", meaning that, in order for development to continue indefinitely; it should balance the interests of all strata of the societies in focus.

It is usually understood that this "intergenerational" equity would be impossible to achieve with the present-day global social inequity and inequality. The situation is mostly exemplifies by the jeopardisation of the well-being of people living in poorer parts of the world by the economic activities of the most developed parts of the globe. Imagine, for example, that emissions of greenhouse gases, generated mainly by highly industrialised countries, lead to global warming and flooding of certain low-lying islands—resulting in the death, displacement and impoverishment of entire island nations; or consider the situation when higher profits of pharmaceutical companies are earned at the cost of millions of poor people being unable to afford medications needed for treating their life-threatening diseases.[93]

As varied as the views and thought about the concept of sustainable development may appear, they all have one central focus of urging humanity to see the world as a system—a system that connects space; and a system that connects time.[94] When we think of the world as a system

---

[92] World Commission on Environment and Development (WCED), *Our Common Future*. (Oxford: Oxford University Press, 1987) 43

[93] I. M. Mintzer, ed., *Confronting Climate Change: Risks Implications and Responses*, (Cambridge: Cambridge University Press, 1992). 163 -170; A *Time International Magazine*, 2000 Special edition an Earth Day, UNEP 2007 Reports on Climate Change. The 2008 G8 Summit in Japan also focused on global environmental crisis, so also was the 2008 Environmental Performance Index (EPI) by Yale University and University of Columbia

[94] Available at http://www.un.org/esa/sustdev/agenda21text.htm

over space, we will be quick to realise that an injury to one is injury to all by understanding that air pollution from North America affects air quality in Africa and Asia; that emissions from Chinese firms could contaminate air all over the world and that pesticides sprayed in Argentina could harm fish stocks off the coast of Australia. And when we think of the world as a system over time, we will begin start to grasp that the decisions earlier generations made about how to farm the land continue to affect agricultural practice today; and the economic policies we endorse today will have an impact on urban poverty when our children became adults.[95] Cities and other urban spheres must be made environment friendly for sustainability and equitability of global development drive.

This drive to ensure that the sporadic urbanisation which is a revelation of rapidity of human civilisation is sustainable has led to the creation of the concept of "Sustainable Cities."[96] As usual the Sustainable Cities Project (SCP) has impacted positively in cities in developed parts of the World while vast majority of cities in the Third World still largely remain huge squalors and mostly unsustainable. Thus, most First World Cities have become ecological havens through: taking in to account economic and environmental costs of urbanisation; self-reliance in terms of resource production and waste absorption; cities become compact and energy sufficient; and the needs and rights of all are well balanced.[97] According to Darshini Mahadevia, cities in the underdeveloped parts of the World can only become sustainable if: there is environmental sustainability, social equity, economic growth and redistribution of wealth, and the empowerment of the disempowered.[98]

---

[95] http://www.un.org/esa/sustdev/agenda21text.htm

[96] Darshini Mahadevia, "Sustainable Urban Development in India, An inclusive Perspective", in David Westerndorff ed., *From Unstainable to Inclusive Cities*, (Geneva: An UNRISD Publication in Collaboration with Swiss Agency for Development Cooperation, October 2002) 2-19

[97] Darshini Mahadevia, "Sustainable Urban Development in India, An inclusive Perspective...."4

[98] Darshini Mahadevia, "Sustainable Urban Development in India, An inclusive Perspective...." 4

Although, the concept of sustainable development or people-centred development has been heavily criticised,[99] no one has denied that our world (in the cities or rural areas) today is confronting complex and serious problems; which require new paradigms-a shift away from the existing traditional norms. This is why Sustainable Development has gained global currency as it is rooted in the vortex of systems thinking. It also helps us understand ourselves and our world. Though we may not be able to address the enormous problems that we are currently facing the way we created them; but we can address them more effective if only we genuinely allow the core principles of sustainable development to always guide our conducts.

Significance of the Study

Environmental history studies how economy, technology, politics, social structures and value systems have affected the natural environment the use of natural resources; and how the changes in natural environment have affected the communities. Most significantly, it is a problem solving field of study which focuses on solving practical environmental problems such as: soil erosion, air and water pollution, waste management, urban crisis and many more.[100] As Richards once observed "the great task for environmental historians is to record and analyse the effects of man's recently achieved control over the natural world."[101] He stressed further that "what is needed is a longer-term global, comparative, historical perspective that treats the environment as a meaningful variable."[102] Sadly, history of cities and industrial communities in Nigeria -especially the need to combine development plans with adequate environmental management strategies has not received much attention. This study has made efforts to address this with a primary focus on Ibadan.

---

[99] Darshini Mahadevia, "Sustainable Urban Development in India, An inclusive Perspective...." 3
[100] www.un.org/millennium/declaration/ares552e.pdf
[101] John Richards, "Documenting Environmental History: Global Patterns of Land Conversion," in *Environment*, 26 (9). 1984
[102] John Richards.....

The significance of the study also lies in the newness of the research field and in the ability to produce new, important knowledge on the management of urban environment; evaluate Nigerians' perception of it and analyse the implications for development in Ibadan in particular and Nigeria as a whole. At the same time, the study was embarked upon to underscore the relevance of environmental history in Nigeria. It is also important to recognise the different priorities that urban dwellers in Ibadan attach to management of the environment.

Research Methodology

As established earlier, the strength of urban and environmental research lies in its multidisciplinary or interdisciplinary approach.[103] Thus, this study which primarily belongs to the field of Environmental history have benefited tremendously from related fields such as Geography, History of technology, Philosophy, Environmental Economics, Environmental Psychology, Historical Sociology and the likes. However, the study primarily depends on qualitative analysis through critical examination of data derived from written and oral sources. The study benefitted maximally from these documentary materials which were supplemented and corroborated by oral evidences.

The written sources were divided into Secondary and Primary Sources as shown in the bibliography. For the Secondary Sources, a thorough and extensive search was undergone to unravel relevant materials to the study in several libraries such as the Kenneth Dike Library, University of Ibadan, the Central Library of Obafemi Awolowo University, Ile Ife, Faculty of Social Science library, University of Ibadan, Department of History library, University of Ibadan, IFRA library, Institute of African Studies, University of Ibadan, Personal libraries of Professors J.F.Ade Ajayi and Bolanle Awe both situated at Bodija area, Ibadan, libraries of Ministries of Environment, Urban Regional and Development, Sustainable Ibadan Project (SIP), NISER and many more. Quite considerable Volumes of secondary materials are in form of published

---

[103] K. A. Owolabi, *Because of our Future: The Imperative for an Environmental Ethic for Africa*, (Ibadan: IFRA, 1996) 122

textbooks, unpublished theses/Journal articles, Newspapers, Magazines, Government records and many more. The largest relevant primary sources for this study came from the National Archive, Ibadan. The information here came from files, personal notes/diaries/correspondences, minutes of meetings, reports, record of programmes, eye witness accounts of key public figures, colonial and post colonial administrators within and outside Ibadan.

The above investigation in libraries, Archives and the Internet were complemented with the gathering of oral evidence through direct/personal interviews. Two sets of structured questions were used. The first set was for selected households to elicit information from the people about their living conditions, the environmental situation and demographic characteristics, political awareness, the role of government in urban environmental management and the link with development; their perception on environmental hazards and personal hygiene; what constitute urban problems; their potentials to improve their living conditions; how they believe environmental issues should handled and how poverty can be effectively alleviated especially in the urban centers. Fifty people interviewed cut across political, social and economic groups such as Civil servants, private sector workers, students, artisans, farmers, women, apprentices and the unemployed.

The second set was administered on: Environmental/Health Officers, Sanitary inspectors, Local, State, and Federal government officials, traditional chiefs, community leaders, NGOs and professionals. This was done to gain their views on: what they perceived as the environmental problems of the city and the causes of such problems; the link between environment and development; the past and present efforts and programmes at solving the problems; the expected roles of the people, governments, industries, international organisations, NGOs in environmental management; how effective are these efforts; what could be done to ameliorate the situation.

These interviews were guided by ethics of oral interview provided by the University of South Africa, which I have studied carefully. This has equipped me with the spirit of critical inquiry and social responsibility governing the conduct of oral interview. Some of the most fundamental of these rules include: respect for human dignity, respect for free and informed consent by disclosing my identity (through a letter from my Supervisor and my official identification card etc) and explaining in details, the purposes of the interviews in order to obtain their voluntary and committed responses. Others are: respect for privacy and confidentiality, respect for vulnerable persons and respect for justice and inclusiveness.

As a resident of Ibadan, which is the study area for the research, I did not experience much difficulty in obtaining the required information from my informants. In the first place, I utilised my long earned experience as a researcher to convince and persuade my informants to extract necessary information from them. Secondly, issues about the environment in Ibadan have become a source of great worry to many (both experts and non experts) most of whom readily offer information on the causes and how best they think the problems could be carefully resolved. Thirdly, the Participant-Observation method was also extensively employed, which took the researcher to all the eleven Local Government Areas which constituted the study area for this study.

Furthermore, in order to prepare for the challenges of studying Environmental History at Doctoral level and learn more about current environmental management strategies, I have attended both local and international conferences about the environment through which I have met and maintained fruitful relationships with many authorities and experts on environment (government and non-governmental) many of whom would serve as informants for this study. Examples of such conferences include: "Building Resilience Against Climate and Environmental Shocks" (BRACE) Program (Hybrid) organised by the African Food Change-makers (AFC), Lagos and Abuja, Nigeria, from June to September, 2023; the UNOSSC, UNDRR GETI, PAHO/WHO, WHO Joint Certificate Training Program 20237, (Online Course) June 7, 14, 21, 27, 2023 -8:00 New York | 14:00 Geneva | 19:00 Bangkok | 21:00 |13:00 Nigeria; One day Conference (hybrid) on the 50th Celebration of World Environment day with the central theme of "Solutions for Plastic Pollution" organized by Green Hub Africa and United Nations Office in

Nigeria, Abuja on 6th June 2023; Africa Adapt (A Non Governmental Organisation "curbing the deadly impact of Climate Change in African Cities," conference held at Draper's hall, Institute of African Studies, University of Ibadan on Friday, the 30th November, 2012. Others include: the 5th Annual Lagos Studies Association Conference in Lagos, Nigeria from June 25-27, 2021, where I presented paper on "Urbanism and Poverty in the 21st Century African Cities- Ibadan example"; the Second Conference of School of Arts and Social Sciences, The College of Education, Lanlate, Oyo State, Nigeria from 9th to 13th September, 2019, where I co-presented a paper on "Climate Change, Economy and Poverty in the 21st Century Nigeria"; the 2011, Oyo State Environmental Summit; the 2009 African Conference on "Science, Technology and the Environment in Africa" held at the University of Texas at Houston from March 27-29, 2009 (where I presented two papers). I was also given a full scholarship to attend the 2009 Bergen Research Summer School on "Climate, Energy and the Environment, at University of Bergen, Bergen, Norway, from June 22nd to July 7th, 2009. The summer school did not only enhance my understanding of global environmental challenges and solutions but also introduced me to methods of conducting environmental research. It equally lifted me drastically from the "poverty of modern historical methodology" which most recently trained Nigerian historians are exposed to.

Most importantly, the data and information for the study came from: known experts in the field of urban environmental management and other related fields in Nigeria, Institutions, organisations and centres that are engaging in training, study and research in environment, urban management and development studies. In all, over one hundred published and unpublished secondary materials; over a hundred relevant archival materials; more than fifty respondents have be collated and utilised for the writing of the thesis.

Method of Data Analysis

The data assembled were collated, classified and analysed according to their format (i.e. Books, Journals, Monographs, articles, Reports, theses and so on). Methods also included analysing relevant historical source materials such as council records, newspapers, journals and statistical data. Archival sources were interpreted critically by taking to account the prevailing socio-economic atmosphere. Special efforts was also be made to avoid anachronistic interpretations by corroborating available facts with oral evidence derived from the interviews conducted.

The Study Area

Ibadan, which is the primary focus of this study, has a robust metropolitan tendency. As Toyin Falola observed, Ibadan "has always been great with an outstanding history."[104] Ibadan is reputed as one of the largest African cities. Located along the edge of a thickly wooded forest belt, the name was coined from Eba-Odan, meaning a town at the edge of the forest. Ibadan, a city in Southwestern Nigeria is the capital of Oyo State. It is located about 110 km (about 70 miles) Northeast of Lagos and is a major transit point between the coast and the northern parts of Nigeria. The city is on the railroad line linking Lagos with Kano and is well connected by road to other cities in the country. It also has a local airport. Ibadan is the center of trade for a farming area producing cocoa, palm oil, yams, cassava, corn, and fruit. Industries include agro-allied industries, brewery, vehicle assembly plant, and a tobacco industry. The University of Ibadan (founded as a college of the University of London in 1948 and as an autonomous university in 1962) and Ibadan Polytechnic (founded in 1970) are located in the city. The city is the site of several major research institutes, notably the International Institute of Tropical Agriculture (IITA), Cocoa Research Institute (CRIN), Forestry Research Institute, National Horticultural Research Institute, and Nigerian Institute of Social and Economic Research (NISER). Most of Nigeria's leading publishing companies are also based in the city. Inhabited mainly by the Yoruba people, Ibadan grew rapidly in the mid-19th century when Yoruba civil wars shook the region. The city attracted soldiers and refugees displaced by the wars and by Fulani incursions to the northern part of Yorubaland. The population of Ibadan is about 1,731,000, (2000 estimate) and 2,550,593 (2006 estimate). Ibadan is also close to the historic towns of Oyo, Ogbomosho, Ijebu-Ode, Ife, Ilesha, and Oshogbo.

---

[104] Toyin Falola, *Ibadan, Foundation, Growth and Change*, (Ibadan: Bookcraft, 2012)

The city is made up of eleven Local Government Areas (LGAs) namely: Akinyele, Egbeda, Ibadan North, Ibadan North-east, Ibadan North-west, Ibadan South-east, Ibadan South-west, Ido, Lagelu, Oluyole and Ona ara Local Government Area.[105] Therefore, these eleven Local Government Areas (often referred to as Ibadan Metropolitan District) represent the study area for this research.

---

[105] Alex Asakitikpi, "Environmental and Behavioural Risk Factors Associated with Childhood Diarrhoea in Ibadan Metropolis, Oyo State", in *The Journal of Environment and Culture*, Vol. 2, No. 1, (2005) 1-13

Chapter Two
The Conceptual Framework and
Literature Review

Preamble

Primarily this study sees cities as drivers of sustainable development. To drive home the sustainable agenda, chapter two is divided in two parts namely, the conceptual framework and literature review. It focuses on the conceptual underpinning and intellectual postulations by some relevant scholars about the key terms of the study. Due to the fact that these terms still largely means different things to different people and because of the importance of the environment on human survival, and the resurgence of the deteriorating conditions of the Planet Earth, several literatures have been advanced on environmental studies, particularly since the 1970s when environmental issues gained more global preponderance. It is based on this assertion that some works related to this topic have been reviewed.

The Conceptual Framework

As mentioned earlier, this study sees/envisions cities as drivers of Sustainable Development. As a corollary, it combines a historical discussion on evolution and expansion of cities (from global perspective; African/Nigerian's concerns and particularly Ibadan) with a new paradigm such as the urban ecosystem approach. The approach sees cities as part of the ecosystem and not as autonomous social units distinct from nature.[1]Cities are in no way unnatural, but represent a complex mingling of nature and society. Flows of resources are drawn in, circulated,

---

[1] United Nations Conference on Environment and Development (UNCED) *Agenda 21* [On-line], 1992; G.E. Machlis et. al, "The Human Ecosystem Part I: The Human Ecosystem as an Organizing Concept in Ecosystem Management", *Society & Natural Resources*, 10, (1997). 347-367; C.H. Nilon et. al, "Editorial: Understanding Urban Ecosystems: A New Frontier for Science and Education", *Urban Ecosystems,* 3, (1999) 3-4

metabolised and ejected from the urban ecosystem. The urban ecosystem approach provides a framework for understanding the interactions between economic, social and ecological factors in the urban environment. This knowledge can then provide a basis for untangling and restructuring the city metabolism to make the flow of material more efficient, egalitarian and sustainable. Policy makers should consider the ambitious goal of integrating the socio-synthetic cycles of cities into the cycles of the ecosystems in which the city is ultimately embedded. This approach highlights the need to shift away from linear flows of material within the city toward closing loops to cycle resources and reduce the flows diverted from "natural" ecosystems and maximise the efficiency of resources use within the city.[2]

As demonstrated by Satoshi Ishii, it is traditional in the management of urban environmental crisis and social problems to always attempt to isolate individual aspects of the challenges such as air, water, poverty. And by so doing, there were little overlaps among these management strategies. In addition, these managers generally preferred to concentrate only within city and state boundary probably due to a lack of knowledge and experience.

Arguably, the introduction of strict legal controls, guidelines, economic instruments and technological innovations provided successful management of the environment to some extent. Recently, however, rising complexity of urban problems called for more powerful management tools, such as Strategic Environmental Assessment (SEA), Life Cycle Assessment (LCA), Cost-Benefit Analysis (CBA) and Environmental Management System (EMS), which accordingly broadened the scope of traditional management.[3] As a result, these tools clearly suggested that current urban environmental and social problems could not be dealt with in isolation without taking inter-linkages among other urban components into account. Coordination of environmental management agencies and strategies is of extreme importance here.

---

[2] Satoshi Ishii, "Urban Air Pollution and Urban Management: Applicability of Ecosystem Approach and the Way Forward", *UNU-IAS Working Paper*, No. 120, (March 2003) 9
[3] Satoshi Ishii···· 3- 4

Meanwhile, urban boundary was considered to have lost its meaning because overreaching problems which have continued to plague our urban centres despite these boundaries. In this context, the policy makers and practitioners in cities are now required to confront increasingly complex management tasks.

It is the above complexities that make it imperative for this study to adopt the urban ecosystem approach in conjunction with the concept of sustainable development. Though there had been different definitions and various perception of the 'urban ecosystem approach', it has currently gained global recognition and acceptability. This is evident in its advocacy through several International initiatives and research institutes such as: Convention of Biological Diversity (CBD), International Development Research Centre in Canada (IDRC), United Nations Environmental Programme (UNEP), Millennium Ecosystem Assessment (MEA), Resilience Alliance (RA), United Nations Educational Scientific and Cultural Organisations (UNESCO) and World Health Organisation (WHO).[4] The comprehensive review and comparison for these programmes was made by UNU/IAS in 2003.[5]

It is imperative that policy makers at the national, regional and international levels recognise the crucial role of cities in catalyzing shifts toward sustainable development. This is why most meetings at the UN sponsored World Urban Fora[6] usually seek to highlight the potential for a transition toward more sustainable interactions between urban areas and the rural material "resource pools" from which they draw. By so doing, they are encouraging policy makers at all levels to reflect upon the conceptual shift toward an urban ecosystems approach. We must endeavour to deepen our understanding of the urban ecosystems approach and promote it as a framework for policy development in achieving the Millennium Development Goals related to urban areas. Right from the pre-colonial era to post-colonial times, historical developments in Ibadan reflect near-absolute disconnection among the

---

[4] M. Zurek, "Millennium Ecosystem Assessment Methods", 2002. Available at: http://www.millenniumassessment.org/en/publications /methods.pdf

[5] M. Zurek 10

[6] Available at, http://www.unhabitat.org/programmes/sustainablecities/eng_home.asp

component parts in their efforts to manage the city. It is therefore not surprising that the development of the city has become haphazard and absolutely unsustainable. This Study hopes to initiate a paradigm shift away from viewing cities as parasites and toward regarding cities as key agents in achieving sustainable development.[7] Here lies the significance of urban ecosystems approach .The urban ecosystems approach provides a framework through which to gather empirical information to inform and guide decision-making. Key issues for discussion include how policy makers balance different information relevant at different spatial and temporal scales and how to balance between conflicting priorities.

Arising from the above, this study is also guided by the urban ecosystem approach which calls for concerted human efforts at tackling environmental crises. The approach is complementary to the concept of sustainable development which is advocating for environment friendly, people oriented and intergenerational development. This is because to solve the environmental challenges facing humanity successfully, it is imperative for policy makers at the local, national, regional and international levels to recognise the crucial role of cities in catalysing shift towards sustainable development. The dawn of new Millennium in the year 2000 saw the concerted global effort at achieving this for cities around the world increased tremendously. Experts across the world have continued to highlight the potential for a transition toward more sustainable interactions between urban areas and the rural material "resource pools" from which they draw.

In seeing cities as key agents of sustainable development, the study draws lessons from the traditional environmental management strategies which could still be useful in tackling the modern day environmental problems in Ibadan. Other issues discussed include: how policy makers can balance different information relevant at different spatial and temporal scale and how to balance between conflicting priorities. This study also offers practical tools that could enable both people and governments to identify

---

[7] Apart from a personal experience of the absolutely pathetic urban tragedy in Ibadan, the inspiration for this study was also drawn from information retrieved from United Nations (UN), *United Nations Millennium Declaration.* 2000. Available at, www.un.org/millennium/declaration/ares552e.pdf

those obstacles which had been preventing positive response to different types of environmental challenges in Ibadan. It equally provides examples of policies that could increase synergies between environmental, social and economic cycles and promote the creation of positive environmental feedback flows within the city. All these are mainly to generate communal commitment and general interest in recreating Ibadan as driver of sustainable development and suggest more scientific and people oriented urban environmental management strategies.

Literature Review

It is an incontrovertible fact that, there is hardly any area of study in human endeavor that has not been researched, albeit, under different conditions by different scholars and professionals coming to differing conclusions. As a result of the importance of the environment on human survival, and the resurgence of the deteriorating conditions of the Planet Earth, several literatures have been advanced on environmental studies, particularly since the 1970s when environmental issues gained more global preponderance. It is based on this assertion that some works related to this topic have been reviewed. Unlike Europe and North America where historians have provided a wealth of literature on environmental research, relatively few works have been done on environmental issues by Nigerian historians. As a corollary, this study has benefited stupendously from the compendium of compelling collections of both local and international scholars in related fields. Areas of focus are: Environment; Environment and development; Environmental management, Globalisation and Environment; Sustainable Development and urban environment.

Environment

One scholar whose views on the environment are relevant to this study is K.A. Owolabi in his book titled, *Because of our Future: The Imperative for an Environmental Ethic for Africa*.[8] He adopted empirical and theoretical

---

[8] K. A. Owolabi, *Because of Our Future: The Imperative for an Environmental Ethic for Africa*, (Ibadan: IFRA, 1996) 1- 41

approaches to analyse the issue and the nature of the existing relationship between human beings and the environment. He emphasised that the prevailing environmental crisis is a reflection of the lack of an adequate ethic of the environment in the global culture. Disregard for the environment may be largely attributed to the anthropocentric metaphysics propagated by western cultures which is seen as being responsible for the unrestrained and careless global exploitation of the natural resources. He stressed that the future of humankind and other species in the world depend on pursuing and environmental ethic that will take into consideration not only the needs, interests and happiness of the present generation of species but that of the future.

He argued further that although poverty has made people resort to unconventional and immoral means to feed themselves: communalism and supernaturalism played significant role in the use and preservation of land in the traditional African Society, while modernity and capitalism are largely responsible for the moral crisis of the modern epoch. He therefore lamented that water pollution, blocking of access roads by improper waste disposal, air pollution from insecticides, refrigerant gases and automobile emissions, and neighbourhood congestion due to the construction of illegal commercial and industrial sectors are among the major cases of environmental crises in urban centers in Africa, while deliberate poisoning of river to kill aquatic animals, over grazing, desertification and deforestation are those peculiar to rural settings in Africa. As a result, humanity is at the crossroads, and new environmental order must manifest an ethical dimension to the solution of the crisis, and there is need to revive the communal spirit of the traditional African society in order to achieve the goal of sustainable development, he concluded.[9]

In a collection of articles titled *Women and the Environment: A reader :Crisis and Development in the Third World,*[10]the writers generally examine the relationship between women and the living systems as they affect ecology in poor areas in developing countries. There was a general consensus

[9] K. A. Owolabi…..
[10] Sally Sonthamier ed., *Women and the Environment: A Reader: Crisis and Development in the Third World,* (London: Earthscan Publication, 1991)

against the popular belief that women are mostly responsible for much of environmental destruction in the rural areas. Instead, they blame the problems on a number of intertwined factors which came directly from the result of development policies geared towards the survival of the global economy. For centuries, economic survival in most developing countries had meant providing raw natural products and agric commodities which are in high demand in the developed world. This often lead to widespread depletion of resources; massive conversion of bush and wood for agric development; indiscriminate use of water resources for irrigation; felling of trees for wood products and privatization of once communally owned lands. This was in line with the World Bank estimate that over 70% deforestation in Africa is due to conversion of wood lands to agric use.[11]

Anne V Whyte[12] tries to establish the real nexus between culture and the environment. She vehemently argued against the purported neglect of the possible cultural dimensions of the environment by the social and natural scientists' construct of environmental reductionism which attempt to narrow down the complex system to a few measurable interactions. For Example the Social Scientists have emphasised the socio-economic components (income, family size, ethnicity etc) in relation to available resources (land and water) but have tended to ignore the cultural dimensions of the environment for the resource managers. According to Whyte:

> The environment is a cultural force; it underwrites group identity by providing a shared ownership of territory; it reinforces common roots and a sense of home; it locates peoples' memories and past histories in space; it provides powerful symbols for human fears and aspirations. And the Environment with its changing allocations

---

[11] Sally Sonthamier ed..... 45

[12] Anne V. Whyte, "Women Environmental Perception and Participatory Research", in Eva M. Rathgeber ed., *Women's Role in Resource Management in Africa*, (Ottawa: IDRC) 102-110

of scarcity and plenty, hazards and resources provide us with explanations and judgments of our behaviour.[13]

As demonstrated by Tuan Yi-fu, the environment is also very rich in symbols. For examples in Western culture, mountains have been symbols of the gods, of majesty and power; caves have been seen as secret places, symbols conspiracy and prophecy. Again, when wilderness was plenty in Europe and North America, it was regarded as a place of rejection and disorder; today it is a symbol of pristine nature. He concluded that, the power of these symbols for social behaviour, however, lies in the way they are organised culturally to form codes.[14]

Mary Douglas[15] on her part developed a thesis in 1973 that cultural interpretations of the natural environments are designed to maintain political influence and control coupled with social cohesion. She further explained that, this may not only apply to moral arguments on why natural disaster strike individuals but also elaborate on the codes of conduct that are embedded in cultural taboos, oracles, divinations and myths which are put in place to ensure proper environmental behavior among the people. In many cultures, the social construction on interpretations of natural event is an environmental sign of approval or disapproval of a social behavior. This is then used to make individuals conform to social norms, to undertake great tasks or to arouse communities to war or migration. With this argument, one can rightly propose that there is close link between societal social norms and the nature of environmental problems present in such societies.

Though many environmental problems are general in their effects, others have more intense local results. With a particular focus on the poor in Jamaica, a report by the World Bank in 1993 demonstrated how poor in the Third World are bearing more immediate pressures of environmental degradation due to endemic and pervasive poverty. According to the report: they [the poor] are least likely to be served with public water, sewage and garbage services; tend to live in crowded conditions where

[13] Anne V. Whyte….. 78
[14] Tuan Yi-Fu, *Landscapes of Fear*, (New York: Pantheon Books, 2000). 262
[15] Mary Douglas, *Natural Symbols*, (London: Penguin, 1973) 3

indoor air pollution is high, and may live in close proximity to disposal sites for solid and hazardous waste. Urban Slums (which usually housed the poor) are unplanned and mostly occupied by squatters; they are frequently built on unsuitable land subject to incessant flooding and landslips.[16] Regrettably, the poor in Africa, Latin America and Asia are not able to escape these areas.

In addition, poverty is a cause of environmental degradation as the poor must meet their urgent survival needs. This is particularly true in hillside agriculture, artisan fisheries and settlements of marginal lands in forests and flood plains. It is also true in the cities where informal sector enterprises operate without the benefit of health, safety or environmental regulations. The social tensions engendered by continued or worsening poverty have negative effects on the poor countries' ability to attract long term investment (domestic and foreign).[17]

W. I. Bell-Gam et-al, in a book titled, *Perspectives on the Human Environment*[18] emphasised that man harnesses several environmental resources through agriculture and mining and tries to enhance his livelihood by attempting to organise his productive activities in space. T.E. Ologunorisa, in his work "Global Warming" in the same book, established that continued emission of greenhouse gases such as Carbon dioxide (CO_2), Methane, Nitrous oxide (N_2O) and Chlorofluorocarbons (CFCs) at the present rate could lead to increased concentrations of these gases on the earth surface. According to the Inter-Governmental Panel on Climate Change (IPCC), he said that "the balance of evidence suggests there is discernible human influence on global climate", and global average temperatures have risen by 0.60^0c above the pre-industrial average. He stressed the vulnerability of most parts of Africa – Magreb, West, Horn and Southern Africa to global warming which would result in

---

[16] World Bank, *Jamaica: Economic Issues for Environmental Management*, (Washington DC: World Bank, 1993)

[17] World bank....

[18] W. I. Bell-Gam et. al eds., *Perspectives on the Human Environment*, (Port-Harcourt: Amethyst & Co Publishers, 2004). 33-53, 54-70, 93-117 and 118-144

shifts in geographical patterns of precipitation and changes in the sustainability of the environment and management of resources.[19]

P.O. Phil-Eze and J.E. Umeudji's "impact of Man on Biodiversity" in the above stated book, explained that different biodiversities occupy different environments in which they derive specialised ecological needs and perform specialised ecological functions. In addition, biodiversity represents the pathway of energy transfer of matter and recycling of nutrients through the biosphere. By importance, biodiversity positively contribute to successful agricultural productivity through pollination of crops and control of pests. It performs environmental services and has direct and indirect importance to human health. It argued that man has impacted positively and negatively on biodiversity. Positive impacts include, hybridisation, cloning, development of disease resistant species, genetically modified organisms, while negative impact stem from variable sources resulting in the loss of biodiversity, such as loss of species, decline in the population of living organisms, loss of productive capacity, reduction of rate of turnover, pollution of habitat harmful radiation and other related crises.[20] All these negative impacts of man on biodiversity have affected the environment.

A. A. Obafemi's "Environmental Pollution" in the same book, argued that as man interacts with nature and the environment in an attempt to explore it for food, shelter and clothing and other basic socio-economic needs, certain consequences result from this man-environment relationship. Some of these actions will normally have negative or positive effects on humans, the negative implications is this referred to as pollution. This pollution manifests itself in air, land, water noise and thermal pollution. Although, pollution from natural resources including sand storms, volcanic eruptions, natural forests and other natural environmental hazards were highlighted, it was these types of pollution associated with human activities that have more impact on the environment. The impact of these pollutants especially on human, plants, wildlife, buildings and other infrastructure cannot be overstated. It contends that unless there is a sustainable spatial order, efficient

---

[19] W. I. Bell-Gam et. al eds.....33-53
[20] W. I. Bell-Gam et. al eds....54-70

environmental legislation, the challenges posed by pollution may keep increasing unabated.[21]

Another Scholar, Francis E. Bisong in his work, 'Population, Research and Environment' published in the same book, contended that increase in population growth coupled with the decline in mortality rate in both the industrial and developing countries had implications for agricultural and industrial production, expansion in health care, education, urbanisation, infrastructure and governance. Further still, he stated that the natural environmental system will stand to be severely taxed, if they are to be depended upon for the basic need of the teeming global population. Quoting Deevay, he argued that "each of the periods of accelerated population growth was a response to revolution in which the earth's capacity to support population was significantly increased. The first was the Cultural Revolution, the second was agricultural, and the third was the scientific/industrial revolution. Each of these revolutions, which accounts for the historical subdivisions in the growth and development of the human population, has increased much pressure on the human environment.[22]

Another important work relevant to this study is a book by Bruce Russett et al, *World Politics: The Menu for choice,* wherein several global ecological issues facing the human race were examined. It emphasised that the ecology naturally draws attention to global interdependencies. Some of these are closely related to matters of economic and human development. Others have to do with demography, geology and climatology etc. Critical argument was centered on global interdependence and ecological problems. With the persistence of environmental threats and their appearance in newer and more dangerous forms – acid rains, ozone holes, global warming, soil erosion and degradation, and deforestation, interest in ecological problems now has a secure place on the agenda of state leaders, IGOs and civic groups. In the same vein, it examined the impact of population and demographies on the environment, stating that increase in population inevitably increase the demands for natural resources, thus generating ever greater environmental decay. He continued that, millions

---

[21] W. I. Bell-Gam et. al eds.....93-117
[22] W. I. Bell-Gam et. al eds..... 118-144

of people in the developing world suffer from chronic under nutrition, in spite of the rise in the global food production. The loss of arable land to cultivation, deforestation, introduction of new seeds, fertilizers, and pesticides and new technologies consume a great deal of energy and cause new problems. The quality of fresh water is being threatened by a wide range of human generated pollutants, and pollutions which are contributing to the global environmental disaster. This in turn has engendered development dilemmas across the world. Resource depletion and environmental degradation are externalities produced by population growth and economic development, and are clear examples of collective 'bads'. Collective action is therefore required to tackle this global catastrophe especially in this era of globalisation.[23]

Globalisation and Environment

By all standards globalisation has overbearing and overarching dominance on global economy with dire consequences including environmental degradation. According to its proponents, globalisation has aided the integration of national economy into the international economy through trade, foreign direct investment, capital flow, migration, and the spread of technology.[24] Indeed, it has created new opportunities for many, but not without its enormous costs; most especially-the ecological costs. It has placed uncontrollable pressures on the global environment and on natural resources, straining the capacity of the environment to sustain itself.[25] Globalisation and Environment in recent years are among the hottest topics and mostly explored areas of academic research. Reasons for this lies in the inherent fact that the effects, impacts and the reaction they generate transcend national borders, and

---

[23] Bruce Russett et. al, *World Politics, The Menu for Choice*, Sixth edition, (Boston: Bedford/St. Martins, 2000). 445 – 470

[24] I. Clark, *Globalization and International Relations Theory*, (Oxford: Oxford University Press, 1999); S. O. Akinboye, "Globalization and the Challenge for Nigeria's Development in the 21st Century", available at http://globalization.icaap.org/content/v7.1/Akinboye.html

[25] C. Ake, "The New World Order: The View from Africa", in Hans-Henrik and G. Sovensen eds., *Whose World Order: Uneven globalization and End of Cold War*, (London: West View Press, 1995)

the questions raised require global answers. Also embedded in this is the fact that the issues surrounding these two concepts affect all of humanity.

Expectedly, the inextricable linkages between the two concepts have been a lot interest among scholars especially on the environmental cost of globalisation. Many works on globalisation tends to focus disproportionately only on the political, economic and technological effect on the human society while the environment has been discussed in generality within the anthropogenic context. Therefore, any work on the environment must endeavor to examine carefully what other scholars and writers as well as individuals have done or written on globalisation and environment within the available framework, and the spots which we in our modest means, hope to fill.

Kelvin T. Pickering and Lewis A. Owen in a book titled, *An Introduction to Global Environmental Issues* provides us with full description of our scientific knowledge of environmental systems and processes. The authors built on this factional base to analyse the world's major environmental challenges occasioned by globalisation.[26] The human dimension- cultural economic and political factors and impacts is explained and integrated for easy digestion. Closely related to this was a very incisive work on the term globalisation provided by Ngaire Woods in a book titled, *International Political Economy in the age of Globalisation*. The author examines what drives actors in the international system and explains events in the international economy.[27] Although, this work focuses more on the economic aspect of globalisation, yet the work serves as an insightful guide to understanding the dynamics of globalisation.

Another contribution that is useful to this study was provided by Owen Greene in an article titled "Environmental Issues." This work was directed at environmental issues as a major concern in international relations and drew the correlation between global environmental change and its impact on the world population. The work subscribed to the

---

[26] K. T. Pickering and H. A. Owen, *An Introduction is Global Environmental Issues*, (London: Butler and Tanner Ltd) 7

[27] Ngaire Woods, "International Political Economy in the Age of Globalization", available on www.questia.com

approaches and concepts developed within international relations as a contribution to the understanding of global environmental concerns. Thus, the work delved into the various studies of the causes and risks of global environmental change and the international responses to them. The work also examines efforts aimed at solving these issues internationally. Such efforts include the Earth Summit Agreement, Kyoto protocol and the likes.[28]

In a report submitted by a group of "South Commission" (a group of individuals from continents of the global south), some of the problems facing the south in the "north- south" divide were highlighted.[29] With the help of this report, we are able to understand the factors responsible for global divide, globalisation and all phases of problems especially on the environment. The commission followed the precedence of the Pearson and Brandit reports, which concluded that what the countries of the South have in common are more than their differences, and urged them to come together in order to form a collective force that could call the western nations to order and check the negative implications of their industrial activities Countries of the global South. Thus, this report served as an eye-opener for the underdeveloped countries to come up with sound methods and means of challenging the negative trends of globalisation, especially the horrendous implications of industrialisation of the countries of the North on the global environment.

In his work titled, *International Relations*, Michael Nicholson devotes a chapter on issues concerning global environment and its indivisibility with international relations.[30] In this work, Nicholson examines the problems of environmental degradation in the view of a collective antithesis. The book proved invaluable to the pursuit of solving the enormous environmental problems confronting the whole world. The book, titled, *Environmental Geology* by Carle N. Montgomery, aimed at providing a useful foundation for discussing and evaluating specific

---

[28] Owen Greene, "Environmental Issues," available at www.google.com/owengreen
[29] Report of the South Commission, *The Challenge to the South*, (Oxford: Oxford University Press, 1990)
[30] Michael Nicholson, "International Relations", available at www.ebafQuestion.com

environmental issues as well as for developing ideas about how the problems should be solved.[31]

S.T Akindele in an article titled, "Globalisation, its Implications and Consequences for Africa" concretely projects globalisation as a menace behind most of the Africa's social, political and economic problems. The article regarded the phenomenon of globalisation as nothing but a new order of marginalisation of the African continent through the universalization of communication, mass production, market exchanges and redistribution. Globalisation rather than engendering new ideas and developmental orientation in Africa; subverts its autonomy and powers of self determination.[32]

According to N.J Afari Ghamzan and R.K Trinedy in an article titled "Global And Regional Availability and the Future of Renewable Fresh-water Supplies, Demands and Human Health" about one-third of the world's population live in countries suffering from moderate to high water stress, that is, where water consumption is more than 10% of renewable fresh water resources.[33] Thus, the writer concentrated on one of the environmental issues-fresh water- a seemingly irrelevant matter but not the less a serious issues especially in oil polluted areas such as the Niger Delta. This write up was designed to sensitize the readers about the horrendous impact of globalisation on Nigeria's environment.An article written by A.O. Olanipekun, titled, "Environmental Hazards in Nigeria" provides us with information on the native, incidence, effects and control strategies of environmental hazards in Nigeria. The article also highlights the major forms of environmental hazards in Nigeria.[34] This includes soil erosion, drought and desertification, flows and environmental pollution.

---

[31] Carle Montgomery, "Environmental Geology", available at www.google.com/ montgomery
[32] S.T. Akindele, and O.R. Olaopo, "Globalization: It's Implications and Consequence for Africa", *Nigeria,* (2002) 1
[33] N. Guarzan Safari, and R. K. Trinedy, "Global and Regional Availability and the Future of Renewable Freshwater Supplies, Demand and Human Health", *Journal of World Water Resources,* Vol. 3, No. 4, 23
[34] A.O. Olanipekun, "Environmental Hazards in Nigeria", in B. R. Ismaila ed., *Problems in Nigeria,* (Oyo: Odumalt Publishers August, 2005) 62

Solomon O. Akinboye in his article titled, "Globalisation and the Challenges for Nigeria's Development in the 21st Century" pointed out the indispensability of globalisation in the contemporary world system. The paper recognises the fact that existence of a global environment is deeply embedded in interdependency. The article equally provides us with useful information on the meaning of globalisation, the powerful forces that are propelling globalisation in the contemporary world environment and the subsequent challenges those forces are posing for the Nigerian Environment.[35] G. A. Bamikole in his work titled, "Human Impacts on the Environment Dimensions and Implications, Nigeria Experience" - add to our knowledge of the concept of environment, dimension of interactions between man and his environment, and its implications. The writer made some recommendations on how to mitigate and ameliorate environmental degradation and the consequent hazards on Nigeria environment.[36]

Writing with the same line of thought, Olufemi Vaughan in his work titled *Globalisation and Marginalisation*, views the images of triumph of the free market system and its innovation of the term globalisation. He, however, directs his venom at the defects of this seemingly utopian term with harsh realities most especially as a result of the disintegration of the post-colonial states and thus centering amongst others on the new pressure that globalisation has on the environment in Nigeria.[37].

Udeme Ekpo brought into lime light the issue of environmental degradation as a result of the scourge of globalisation in a book titled *The Niger Delta and Oil Politics.*[38]This book presents a historical analysis of the Niger Delta along with the economic, social and environmental struggles that the author has witnessed and recorded. The author pointed the major culprits of these struggles as the multinational corporations pointing to them as the agents of globalisation; and the Federal Government of Nigeria for the perils in the Niger Delta. According to Human Rights and

[35] S. O. Akinboye, *Globalization and The Challenge for Nigeria Development in the 21st Century*, (Lagos: University of Lagos Press, 2008) 97
[36] A.Bamikole, "Human Impact on the Environment", in B.R. Ismaila, ed. *Problems in Nigeria* Oyo: (Odumalt Publishers, 2008) 216
[37] O. Vaughan, "Globalization and Marginalization…" available at http://goggle. csnencarta.com/nigeria
[38] U. Ekpo, *The Niger Delta and Oil Politics,* (Lagos: Orit. Egwa Press, 2004) 31

Environmental Activist, Oronto Douglas, "the multinational companies are assassins in foreign lands".[39] Their mission is to maximize profit, suck and rape their host natural resources with little or no regard for the environment. Aside the relative economic prosperity that goes along with transnational trade, the cost benefit- in terms of the heavy environmental degradation, coupled with dwindling of natural resources (especially on the part of developing countries like Nigeria) is alarming!

With the aforementioned facts, it has become crystal clear that the oil industry has a significant adverse environmental impact upon the human environment of the producing areas of the country. Its activities not only exacerbate other environmental problems but also create unique problems, which are worse than they need be because the industry as a whole is corrupt and careless, and clearly does not operate to the standards; which are exacted elsewhere in the world. Asking him why in the first place, the activities of the MNOCs in Nigeria bringing so much environmental problems, Oshita Okechukwu gave the following reasons:

> Firstly, dirtying industries are choosing the countries that do not have severe legal arrangements. Secondly, the public opinion in developing countries like Nigeria is unconscious of the harms that economical activities gives to the environment....This information gives the MNOCs the assurance that they don't face with the opposing activities of the public.[40]

In a book titled, *Tropical Africa, Land and Livelihood*, Volume One, by George H.T. Kimble, we find a real time environmental analysis on the issues of Tropical Africa.[41] A geographer by profession, Dr Kimble interprets geography to involve not merely physical facts but the whole mysterious relationship between man and the environment of which he is an integral part. His work exhibits the "real time" analysis of the Nigerian mineral deposits. As a trained geographer, he rightly foresees the implication of unchecked foreign influence in the African countries and

---

[39] Available at http://www.google.com/wikipedia.org/multinationalcorporation
[40] U. Ukpo....
[41] H. T. Kimble, *Tropical Africa, Land and Livelihood*, (New York: Twentieth Century Fund Press) 366

the consequent impacts on African people and their environment. Apart from the known oil pollution, the book also sheds light on other environmental issues such as deforestation, industrial wastes, land degradation and their impacts on environment.

An inaugural lecture titled, *The Climatic Dilemma*, delivered by Professor Abraham Olukayode Ojo examines the delicate situation that man has put himself in his environment in a bid for civilisation and urbanisation which, on the long run, brought about environmental degradation.[42] Basing his discussions on Nigeria, Professor Ojo discusses extensively on one of the major consequences of environment degradation - climate change; illustrating with figures the resultant effects on the Nigeria environment.[43] The Nigerian Bureau of statistics equally enriched our study on the Nigerian environment by providing us with some statistical figures on the Nigerian environment. This would prove to be quite essential in our pursuit of certain figures needed to pinpoint for analysis on the Nigerian Environmental concerns.[44]

In a final analysis, the current debate on globalisation has become de-linked from its environmental roots. The symbiosis that had existed between economic development and natural resources has been shelved and thus, when the term globalisation emerges, much focus was put on its economic sphere of influence. The aim here is to examine different perspectives on the pursuit of a return to the healthy relationship between economic gains, which is personalised in human interest and the national environment needs to be re-examined and recognised. To ignore these links is to misunderstand the full extent of the nature of globalisation and to miss out on critical opportunities to address some of the most pressing environmental challenges faced by the 21st century humanity.

---

[42] A. O. Ojo, *The Climatic Dilemma* Lagos State University 32nd Inaugural Lecture August 19, 2007. 7
[43] A.O.Ojo.....
[44] Y. R. Adebayo, "An Analysis of Environmental Emergencies, the African Crises and Aspects of Energy Issues", *African Journal of International Affairs and Development*, (Ibadan: College Press), 4 (2), (1999) 116 – 143

Environment and Development

The close relationship between Development and environment was recognised at the UN conference on human settlements in Vancouver June 1976.The forum concluded that problems of uncontrolled population growth, rural stagnation, migration, the inability of urban centers to cope with the rate of population increase and environmental deterioration demand corrective action at both the national and international levels. It was also recognised that the ecological crisis in Africa is reflection of overall underdevelopment. The specific issues of poverty, population growth, lack of governmental resources, unemployment and underdevelopment were mentioned and discussed.[45]

The total environment of man is composed of the natural environment-soil, water and landscape etc and the man made or built environment. The environment, in short, is where we live. Development is what we do, to improve our lot within that abode (environment). According to a Report by World Commission on the Environment commonly referred to as the Brundtland Report after the chairman of the commission – Gro Harlem Brundtland (the former Prime Minister of Norway) "environment and development are inseparable, their interrelatedness has been grasped only recently."[46] Nevertheless, in the last twenty years there have been considerable advances in environmental awareness and ecology research. A major step forward was recorded in 1980 with the publication of the world conservation strategy.

The work of Thijs De La Court titled, *"Beyond Brundtland: Green Development in the 1990s,*[47] provided an insight into the central ideas of the World Commission on Environment and Development (WCED), set up by the United Nations General Assembly in 1983, under the chairmanship of Dr. Gro Harlem Brundtland. It assessed how much progress has been

---

[45] Timothy M. Shaw and Malcolm J. Grieve, "The Political Economy of Resources: Future in the Global Environment," *The Journal of Modern African Studies*, 16 (11) 1-32

[46] World Commission on Environment and Development, *Our Common Future*, (Oxford: Oxford University Press, 1987) xi

[47] Thijs De La Court, *Beyond Brundtland: Green Development in the 1990s*, (London: Zed Books Ltd., 1990) 7-140

made since the report was published in 1987, and the effects given to them at the level of national economic policies as well as new international environmental convention, particularly concerning the greenhouse effect and the ozone layer. The author argued that two aspects distinguished the *Brundtland* report, namely: its emphasis on the strong link between poverty and environmental problems, one the one hand, and its optimism for the possibility of a new era of economic growth, one that must be based on policies that sustain and expand the environmental resource base. The author contends that the key concept of the Commission's report is sustainable development, defined as development that meets the needs of the present without compromising the ability of future generations to meet their own needs.

The work also expressed the view that the problems of 'unequal' trade relationship between the North and the South, debt servicing and chronic poverty in the south are contributing to the environmental disasters. It equally drew new conventions to manage the global commons – the air the oceans and the outer space. The fact that this is linkage between peace, security development and the environment, and that poverty, injustice, environmental degradation and armed conflict were also stressed. Finally it suggested an institutional and legal mechanism at both domestic and international level with the UNEP being given a central role. And also sustainable development objective should be incorporated in the terms of reference with national economic policy while governments, NGOs the media, International Financial Institutions, the industrial sectors and relevant UN agencies should make up the balance to preserve the global environment.[48]

Furthermore, the article of Yinka Adebayo, "An analysis of Environmental emergencies, the African crisis and Aspects of Energy Issues" published in the *African Journal of International Affairs and Development*, presented the relationship between African socio-political, economic and civil crises and the environmental emergency situations. He emphasised that "post independent wars in Africa have led to untold hardships such as famine, destruction of lives and properties without any

---

[48] Thijs De La Court....

apparent solution in sight because of the perennial division in principle between nationalists on one hand and foreign backed patriots on the other".[49] The escalation of this crisis and its attending humanitarian and economic problems had resulted in severe drought, famine and the depletion of natural resources especially in North-eastern part of Africa. In his analysis, environmental factor emerging from the generation, distribution and use of energy has also been negative. He also alluded to the fact that the combination of environmental emergencies and energy disasters, resulting from oil spillage, explosion of natural gas, hydro-electric generating facility, coal mining and so on makes it difficult for Africa to formulate and implement adequate response strategies. The refugee problem – their poor settlement and struggle for survival had imposed greater threat in the human environment. He suggested that, countries and regions, policy makers and environmentalists will have to work closer to find a more effective solution to the problem of environmental emergencies in Africa.

Apart from the above, S.I. Oluya and Henry Olu Buraimoh's "Environmental Protection" in *Compendium of issues in Citizenship Education in Nigeria,* also provided a great ideal of analysis on aspects bordering on the global environmental protection with Nigeria as a focal point. It contended that the United Nations Conference on the Human Environment in Stockholm in 1972 moved the environment issue into the centre stage of political and economic discussion which was necessitated by the attending problems of industrialisation and urbanisation.[50] It premised that Nigeria's environmental problem stemmed from its underdevelopment and poverty on the one hand and the process of accelerated development on the other. It enumerated that, filth, erosion and flooding, urban slum, industrial pollution, desertification and oil pollution are the major components of environmental disasters in Nigeria, and the activities of the multinational oil companies have impoverished

---

[49] Y. R. Adebayo, "An Analysis of Environmental Emergencies, the African Crises and Aspects of Energy Issues", *African Journal of International Affairs and Development,* Ibadan: College Press, Vol. 4, No. 2, (1999) 116 – 143

[50] S. I. Oluya and·H. Olu-Buraimoh, "Environmental Protection", in S. I. Oluya et al. eds., *Compendium of Issues in Citizenship Education in Nigeria,* (Ibadan: Remco Press, 1998) 125-126

and rendered several lands infertile through mineral exploration and exploitation. Finally, it stressed the need for the conservation of the environment, by providing a recipe for the techniques in controlling environmental degradation. In Nigeria, the effectiveness of the Federal Environmental Protection Agency (FEPA) and the National Conservation Council, in prompting sustainable development was assessed while the United Nations Environment Programme (UNEP) and other United Nations summits, and protocols such as the Earth summit of 1992 were commended for their efforts towards environmental conservation.[51]

Also of importance, is the essay of R.A. Boroffice "Environment and Development" in *Issues in Management and Development* edited by S.O. Otokiti and S.G. Odewunmi. The essay begins with proposition on relevant theories about the link between environment and development. It later focuses on the quest for a sustainable development taking a clue from the United Nations Conference on Environment and Development (UNCED) held in June 1992 in Rio de Janeiro Brazil. He linked the global environment crisis to the development crisis, both of which are the result of unsustainable system of production and consumption in the North, inappropriate development models in the south and a fundamentally inequitable world order. It premised that many cultures have adopted modes of self reproduction and sustainable ways on interacting through gradual process of learning and adjustment, but the industrial revolution saw the advent of technologies and a market economy, attempting to minimise the cost of production at the expense of the environment. Even the highly centralised economic management in the East characterised by lack of control and democratic participation by the people in managing their resources has had a degrading effect on natural resources and the environment. While the imposition of colonialism disregarded the traditional knowledge and resource management system and exploited the natural resource of the colonies, the result of this is manifested in environmental degradation and poverty, while pressures from greenhouse gases, depletion of the ozone layer, pollution, soil degradation, desertification, hazardous waste and nuclear threat have all impacted negatively on the environment.

---

[51] S. I. Oluya and H. Olu-Buraimoh....125 – 126

Finally, he posited that with the creation of the commission on Sustainable Development under the ECOSOC as a follow up to the UNCED to monitor the program and problems in the implementation of Agenda 21, the Economically Developed Countries (EDCs) are already taking steps to implement sustainable development but lack of adequate information and poverty are the major impediments to sustainable the implementation of sustainable development among the Third World. He concluded that unless the developing countries are relieved of their debt burden, by the EDCs, they will not be able to make adequate budgetary commitments to environment and sustainable development.[52]

In order to ameliorate the situation, some people demanded that "conditions of inequality were tending to increase and…. must be reversed."[53] It was contended that the models of the developed countries were no longer applicable to the undeveloped ones. Therefore, it was recommended that, the rich countries should agree to give an adequate portion of their aid to Africa countries for the realisation of sustainable development.[54] Yet many African governments have yet to appreciate fully that their underdevelopment and environmental problems are related to their roles in the world economy and that fundamental change is needed in their relations with the industrialised countries.[55] Moreover, only very few African leaders pay due and sincere attention to environmental pressures. Although, most of them perpetuate the myth that Africa is an under populated continent with inexhaustible resources, however, the prospects of solutions to the problem of achieving environmentally balanced development is further retarded by the

---

[52] R. A. Boroffice, "Environment and Development", in S. Otokiti and S. G. Odewunmi, eds., *Issues in Management and Development*, (Ibadan: Rex Charles Publication, 2001) 587-596

[53] "A Review of Energy, Water and Soil in the Third World Environment'", *The Internationalist*, (Oxford, June 1976) 40

[54] Kamack Andrew, *The Tropics and Economic Development a Provocative Inquiry in Poverty of Nations*, (Baltimore: Penguin Books, 1976)

[55] Robert S. Jordan and John Renninger, "The New Environment of Nation Building", *Journal of Modern African Studies*, Vol. xii, No. 2, (1975) 187-207

attention which decision makers must pay to political, economic and other societal crises.[56]

Having considered some of these relevant materials, it is apparent that many scholars have done extensive and critical researches into the areas bordering on the environment and development. In spite of this, our world is far from realising the urgent need to combine development efforts with safe environment. Consequently, global environmental disasters are becoming more rampant as man strives harder to develop at the expense of the natural environment. The thrust of this essay is not only to evaluate the negative impact of the environmental regime on Ibadan but to as well consider the major contributions of the major stake holders - to mitigating environmental challenges in the city, and finally suggest how to institutionalised best environmental management strategies without jeopardising economic development of the city.

Environmental Management

Environmental Management offers research and opinions on use and conservation of natural resources, how to ensure sustainable cities and urban centres (through proper urban planning and excellent city management strategies), protection of habitats and control of hazards, spanning the field of applied ecology without regard to traditional disciplinary boundaries.[57] It aims to improve communication, making ideas and results from any field available to practitioners from other backgrounds. As a multidisciplinary concept, Environmental Management draws knowledge from biology, botany, climatology, ecology, ecological economics, environmental engineering, fisheries, environmental law, environmental history, forest sciences, geology, information science, public affairs, zoology and more. Environmental management is not, as the phrase ordinarily appears, the management of the environment as such, but rather the management of interaction by the modern human societies with, and impact upon the environment

---

[56] Kamack Andrew, *The Tropics and Economic Development....*23
[57] Schaltegger Stefan et. al, *An Introduction to Corporate Environmental Management: Striving for Sustainability*, (Sheffield, Greenleaf, 2003)

Femi Ayorinde in his book titled, *Solid Waste Management in Commercial Area* sees environmental degradation as one of improper waste disposal methods through indiscriminate generation and disposal of both industrial and domestic waste.[58] The implication is reflected in high mortality, outbreak of epidemic, lowlife expectancy, development of slums and its attendant problems as seen in mile 12, Lagos. Many of the waste generated are from the goods sold in the market and composes materials such as peels, shaft, suckers, leaves, stalk and the likes. To him, it is rather unfortunate that no external body takes the management of this waste in the market into cognizance despite its importance to the national economy through revenue generation. He contended that the issue needs to be addressed collectively by the traders, all tiers of government and non-governmental agencies to ensure a clean environment. Public and private agencies should also work together in order to tackle the problems more effectively. In his opinion, a comprehensive approach should be considered to include changes in population, commercial activities, types of waste generated, and infrastructural development comprehensive planning policy will equally help to reduce the extent of the problem both in Lagos and Nigeria as a whole.

In the same vein, Glencoe McGraw-Hill in his book, *Merrill Earth Science* which drew mainly from the United States experience advocates recycling as one of the best waste management options. He posited that paper makes up more that 40 percent of our trash and if recycled, it save lots of landfill space because making brand new paper from trees uses lots of water and pollutes the air. But recycled paper takes 61 percents less water and produces 70 percent fewer air pollutants. He, however, maintains that paper is not the only thing that can be recycled. There are other things like aluminum, plastic and many more can be recycled with the energy needed to produce a single brand-new can from ore. In McGraw-Hill's account, while recycling is yet to have firm root in some countries for various reasons; some have made it mandatory. For examples, in the United States, only 10 percent of garbage are now promoting and even requiring recycling while in Japan and Germany about 50 percent of the

---

[58] F. Ayorinde, *Solid Waste Management in Commercial Area,* (Lagos: Bambee Press, 2001)

garbage is recycled. But there are ongoing efforts to promote recycle culture in the United States. For instances, in Seattle, Washington, people who recycle pay lower trash collection fees. Some states even require local businesses to use recycled products for instance, the Ohio congress proposed a law to require newspaper publishers in Ohio to use some recycled paper. Federal government in Washington, DC is becoming involved in recycling. A good example is the container law, which requires at least a five-cent refundable deposit on most beverage container nationwide. This means paying five cents extra at the store for a drink, but getting your nickel back if you take the container back to the store. If everyone in the nation would participate in this programme, the nation could save enough energy to light a large city for four years.[59]

Environmental Science, earth as a living planet by Daniel B. Bolkin et-al is also germane to this work. The authors focus on the different types of waste generated by human societies. These include: nuclear wastes, medical wastes, industrial hazardous wastes, household hazardous waste, mining wastes and Municipal Solid Waste (M.S.W).Much emphasis was laid on municipal solid waste as it consist of materials people no longer want because they are broken, spoilt or have no further use. Methods of waste disposal such as landfills, incinerations, source reduction, compositing and recycling and their inadequacies are discussed in the last part of the work.[60]

Another relevant work is *Introduction to Environmental Impact Assessment* by John Glasson, et al. The authors submitted *inter-alia* that environmental impact assessment is a process with several important purposes. It is an aid to decision making for the decision makers such as a local authority on environmental management. It provides a systematic examination of the environmental implications of a proposed action, and sometimes alternatives before a decision is taken. The Environmental impact assessment can be considered by the decision maker along with other documentation related to the planned activity. Environmental impact

---

[59] G. L. Glencoe, and M.C. McGraw-Hill, *Merrill Earth Science,* (USA, Macmillan Press 1993)

[60] D. B. Bolkin, and F. A. Keller, *Environmental Science, Earth as a Living Planet*, (Boston: Von. Hoffmann Press, 1995)

assessment is normally wider in scope and less quantitative than other techniques such as Cost Benefit Analysis (CBA). It is not a substitute for decision making but it does help to clarify some of the trade-offs associated with a proposed development action, which should lead to more rational and structured decision making. The Environmental impact assessment process has the potential, not always taken up, to be a basis for negotiation between the development public interest groups and the planning authorities. This can lead to an outcome that balances the interest of the development action and the environment.[61]

In his work, *Environmental Science: A Study of Interrelationships,* Eldon O. Enger introduces the concept of integrated waste management and applies the concepts to urban waste, hazardous, chemical waste and disposal of waste in the marine environment. He also reveals that there are still inadequate environmental management in some developed countries, where open drums and illegal roadside dumping are leading to problems such as dumping spoils scenic resources, pollution of soil and water resources and potential health hazard to plants, animals and people. Emphasis was further placed on how this inappropriate dumping of waste has remained a serious challenge to environmental managers especially the resultant air and water pollution with its enduring negative effects on the health of people.[62]

In an article on Environmental management published in the *International Journal on Environmental Studies* 1976, Vol. 8, Gordon and Breach identified three systems required in Waste management, namely analysis of waste and the natural system, waste system and social system. It also mentions eight steps on how waste management can be controlled. This include: waste generation, waste treatment, waste transport, waste interaction, waste decay, waste impact, social assessment and social response. All these are interlinked. Each system requires information from other systems and in turn produces information for other systems. The components which make up the waste management cycle have been

[61] J. A. Glasson, W. B. Lawrence, and G. B. Biddinger, ed., *Introduction to Environmental Impact Assessment,* (U.S.A: SETAC Publication, 1998)
[62] E. D. Enger, B. F. Smith, and A. Y. Bockarie, *Environmental Science: A Study of Interrelationship,* (New York: McGraw Hill Companies, 2006)

described as the way in which they may be quantified. The first five steps describe the selection of the water quality level appropriate to the prevailing condition and the eight describes the selection of the most acceptable and efficient waste control strategy. It is concluded that all waste control management decisions are made on the basis of the waste management cycle.[63]

Affordable Technology and Strategies for Waste Management in Africa by A.G. Onibokun et al is also of immense value to this study. This book emphasises the issue of solid waste management as it attracted the attention of many African nations, to the extent of multi-lateral agency support for the analysis of the problem profile. The book describes waste management services in most African nations as unreliable, ineffective and inefficient. This is traceable to some inherent limitations of adopting some strategies like budgeting and resources allocation funds through annual budgets for service delivery. Also the absence of loan and leasing opportunities to the private operators involved in solid waste management have stifled their operations and reduced their scope of management.

The book suggests ways forward towards the sustainability of waste management. Some of these include: laying down some policies directed to the development of solid waste master plans; adequate funding and financing mechanisms for capital infrastructure; ensuring private sector participation through conventional commercial investment portfolio; provision of special enabling financial institutions such as Urban Development banks etc. It concluded that solid waste mismanagement in Africa is primarily caused by the inherent limitation in the part of citizens and government at all levels.[64]

In an article "Management of Environmental Pollution in Ibadan, An African City; The challenges of Health Hazard Facing Government and

---

[63] O. Gordon and C. Breach, *"The International Journal on Environmental Studies"*, Vol. 8, (1976) 227-233

[64] A. G. Onibokun, et.-al, *Affordable Technology and Strategies for Waste Management in Africa,* (Centre for African Settlement – Studies and Development (CASSAD), 2000) 120-123

The People", Omoleke Ishaq Ishola submitted that unarguably one of the main problems facing Ibadan city and which has become an intractable public nuisance is open and indiscriminate dumping of refuse, human and animal faeces. The writer buttressed his argument by giving example of piles of decaying garbage which are substantially domestic in nature and dominate strategic locations in the heart of the city including the Ibadan Lagos express-way. To him, wastes in such dump sites are a source of air and water pollution, land contamination, health hazards and environmental degradation. The ugly situation has persisted since independence and will continue to be so in Ibadan because of the following factors: (a) high rate of illiteracy (b) ignorance (c) uncivil culture of indiscrimination waste littering (i.e. throwing of waste on bare floor) (d) peoples inability to maintain sanitarily clean environment and (e) reluctance of people to cooperate with the authority by disposing solid waste in illegal dumps, rather than using means provided by the Government, other factors that militate against decent environment in Ibadan include:

1. Uncontrolled population creating slum condition;
2. Poor planning and;
3. Violation of town planning regulations.[65]

Another notable work is Adetoye Faniran's *Solid Waste Management*. Here, the author asserts that factors such as street layout, efficiency of the transport system, the nature and level of efficiency of the local administration, the sanitation habit of the inhabitants and the level of awareness by the population are the heart of environmental issues affecting the patterns of environmental management.[66] To him, this clearly shows the difference between the developed and developing economies. He said while the institutional and infrastructural requirements for the adequate management of the environment especially the efficient refuse disposal exist in the developed economies; they are nonexistent or inadequate in undeveloped economies. The manpower,

---

[65] I. I. Omoleke, "Management of Environmental Pollution in Ibadan in An African City: The Challenges of Healthy Hazard Facing Government and The People", (2004) 265-275. Available online at www.google.com/environmentalhazards/Ibadan

[66] A. Faniran, *Solid Waste Management*, (Ibadan Region: Rex Charles, 1994) 231

administrative structure and capital needed for efficient disposal of refuse are also lacking in Nigeria like most other undeveloped countries.

Finally, the situation in Ibadan and its environs where the urban fringe is rapidly engulfing the rural areas is quite pathetic. The environmental management especially urban planning and waste management leave much to be desired. The fact that the systems of solid waste disposal have implications for the surrounding semi-urban and rural areas which provide the dumping sites, makes the design of an integrated urban rural disposal system a must.[67] The absence or ineffectiveness of the existing waste disposal design in Ibadan has been responsible for incessant floods in the city. In his article titled, "flooding phenomenon in Ibadan," F. O. Akintola discussed among other things that floods can occur through various causes but only four of these are significant in the case of Ibadan. There are the characteristic of rainfall storms, heavily open surfaces resulting in low infiltration rates, deforestation of the hills in Ibadan and the clogging of the river channels with solid wastes. If care is not taken, history of floods may be repeated in other cities. The only panacea to this is the entrenchment and institutionalisation of enduring environmental management strategies to ensure sustainable development in Ibadan and other Nigerian Cities, he concluded.[68]

Sustainable Development

As mentioned earlier, sustainable development is a term widely used by people (politicians, scholars, environmentalists, activists etc) all over the world, even as the notion is still rather new and lacks a uniform interpretation. Important as it is, the concept of sustainable development is still being developed and the definition of the term is constantly being revised and extended. Although the term may still largely means different things to different people, but this study is adopting the most frequently quoted definition from the report *Our Common Future* (also known as the Brundtland Report). The report defines sustainable development as a

---

[67] W. F. Martins et. al, *Hazardous Waste Handbook for Health and Safety*, (London: Butler and Tanner Ltd., 1987) 34

[68] F. O. Akintola, *Flooding Phenomenon*, (Ibadan Region: Rex Charles Publication, 1994) 244

development that meets the needs of the present without compromising the ability of future generations to meet their own needs."[69]Sustainable development in this sense could probably be otherwise called "equitable and balanced", meaning that, in order for development to continue indefinitely; it should balance the interests of all strata of the societies in focus. Till date, the most comprehensive and ambitious step taken by the global community to achieve sustainable development is the United Nations' Sustainable Development Goals (SDGs) adopted in 2015 by member states[70].

In a 2019 report, UNCTAD contended that with global population growth in the next three decades primarily occurring in cities, achieving the Sustainable Development Goals will depend to a large extent on meeting them in urban areas.[71]According to the UN-Habitat and UNCTAD, tremendous increase in the population of urban areas or cities is expected to continue. Indeed 68% of the world's population is projected to be urban by 2050.[72]With these projections, city governments and agencies have pivotal roles in achieving the Sustainable Development Goals (SDGs).

A book edited by Sylvia Croese and Susan Parnell *Localizing the SDGs in African Cities*,[73] brought to the fore the universal applications of the

---

[69] World Commission on Environment and Development (WCED), *Our Common Future*. (Oxford: Oxford University Press, 1987) 43

[70] The 17 sustainable development goals (SDGs) to transform our world include the following: GOAL 1: No Poverty; GOAL 2: Zero Hunger; GOAL 3: Good Health and Well-being; GOAL 4: Quality Education; GOAL 5: Gender Equality; GOAL 6: Clean Water and Sanitation; GOAL 7: Affordable and Clean Energy; GOAL 8: Decent Work and Economic Growth; GOAL 9: Industry, Innovation and Infrastructure; GOAL 10: Reduced Inequality; GOAL11: Sustainable Cities and Communities; GOAL 12: Responsible Consumption and Production; GOAL 13: Climate Action; GOAL 14: Life Below Water; GOAL 15: Life on Land; GOAL 16: Peace and Justice Strong Institutions; GOAL 17: Partnerships to achieve the Goal. For details please visit https://www.undp.org/sdg-accelerator/background-goals and https://www.undp.org/content/undp/en/home/sustainable-development-goals/background/

[71] UNCTAD ,*Promoting Investment For Sustainable Development in Cities*. The IPA Observer, Issue.7, 2019

[72] UN-Habitat and UNCTAD , Urban Economy Branch Discussion Paper #7. 2017

[73] Sylvia Croese and Susan Parnell (eds.) *Localizing the SDGs in African Cities*, Springer's Sustainable Development Goals Series (electronic/e-book), 2022, available at https://doi.org/10.1007/978-3-030-95979-1 and https://link.springer.com/bookseries/15486

Sustainable Development Goals (SDGs) as a framework to address the world's biggest challenges regarding economic, social and environmental development. It also emphasised the utmost significance of cities around the world in achieving the global goals is increasingly felt. While acknowledging the existing strategies, initiatives and policies by city governments to localize SDGs, it vehemently argued that much more is still needed to be done to achieve the 2030 Agenda and avoid a reversal of the progress made since 2015 and to support an inclusive, resilient sustainable post-COVID-19 pandemic recovery. In order to do more, the book primarily focuses on the African experiences of SDG localization with three important messages.

First, it contends that the Sustainable Development Goals remain a robust framework—and perhaps the only hope for humanity to advance the transformation towards an inclusive, resilient and more equitable post-pandemic future. However, according to this book, successful realisation of the SDGs and the 2030 Agenda equally depends to a very large extent on the mainstreaming of the goals and targets within local and national visions and the degree to which the goals have been embraced by a wide range of local and national actors. Secondly, in the opinion of the writers, localisation of SDGs goes beyond the implementation of SDG 11 (Sustainable Cities and Communities), because all SDGs are relevant to cities. Therefore, recognizing the centrality of cities for achieving the SDGs is of utmost important to the city planners and executioners of urban agenda everywhere. It also means gaining better insight into the interlinkages, accelerators and barriers that operate at different scales to impact urban SDG progress.

Finally, the authors lamented the absence of data and the weakness of existing monitoring systems as the biggest challenges across African cities, not just in reporting on the SDGs, but also in their achievement. According to them, data gaps, a lack of disaggregation and spatialisation, as well as poor data quality and its inaccessibility hamper the systematic identification of priorities, undermine evidence based policymaking and limit the extent to which precious resources can be directed to the most catalytic and impactful interventions.

In another incisive article "Tackling SDG localization in sideways of Africa's cities,"[74] Sylvia Croese and Nachi Majoe, strongly acknowledged the fact that the impact of the Covid-19 pandemic has been especially felt in cities, with the urban poor in the global South having suffered most from the disproportionate impact from the crisis. With over half of the world's global population already living in urban areas, the ways in which future urban growth and development is managed (in the cities will greatly determine the extent to which sustainable development goals are met. As a way to make cities and local governments key parts of the solution to achieving sustainable development, they passionately argue in favour of a lateral approach. In their views, such an approach recognises the significance of peer-to-peer exchange and learning between cities and urban communities at large as an accelerator of progress on the localisation of global policy agendas. This is especially important in the African context where levels of decentralization and access to urban expertise or resources are uneven. They identify city networks, national governments and academia and civil society as key actors and vectors that can support such cross-city exchange and sharing, in order to stimulate horizontal and transversal local approaches to sustainable development in a post-pandemic world.

"Localising the SDGs in African Cities: A Grounded Methodology,"[75]a paper by Omar Nagati et-al is equally relevant to this study. The article starts by pointing to the urgent need for urban Africa to develop its own roadmaps for localization of the Sustainable Development Goals. In their views, this will provide platforms toward resolving myriads of economic, social, political and infrastructural challenges confronting most African cities. As a comparative research on Cairo and Dar es Salaam, and focusing on SDG 6 (water and sanitation) and SDG 11.2 (mobility), the article develops a research methodology that helps to detect fissures between the general SDG framework and microscopic realities on the

---

[74] Sylvia Croese and Susan Parnell (eds.....

[75] Omar Nagat et-al "Localising the SDGs in African Cities: A Grounded Methodology" in *Africa Development*, Volume XLVII, 4, 157-184 ,Council for the Development of Social Science Research in Africa, 2022, (https://doi.org/10.57054/ad.v47i4.2981)

ground in African cities. Although each of the two cities has a specific set of urban realities and development paradigms, the paper develops a localisation process that is applicable across both geographies (and beyond) based on the similar prevalence of urban informality in African cities, which the current SDG framework insufficiently, or at times inaccurately, factors in. The methodology comprises three key components: 1) a top-down policy analysis of SDG responses at national and city levels; 2) grounded field research of local practices at a neighbourhood level; and 3) revising the SDG targets and indicators through a proposed 'Toolkit for Localising'.

Another interesting angle to this research is the adoption of a main strategy that tends to counter the 'universal top-down development norms in the SDGs' by exploring performance-based standards and proposing local codes toward providing alternative and more responsive modes of development. They stressed further that, evidently, the SDG framework operates in the first instance at a global universal level, only credible localisation will also make the SDGs positively address and affect local change. The writers argue that for this positive change to happen, a cohesive understanding of the framework and possibilities for its improvement and localisation is possible only while looking at it from both perspectives. The principal strategy of this research has thus hinged on the dual top-down/ bottom-up approach, bridging the divide between global concepts and local practice. Most importantly, this involved undertaking bottom-up, grounded fieldwork, through which the research shed light on the diversity of existing services in the case study areas, how these services work, their use and the stakeholders involved.

As it has been mentioned earlier, more than half of the current world population is in the urban areas. Indeed 68% of the world's population is projected to be urban by 2050. It is therefore not surprising that with these projections, city governments and agencies have pivotal roles in achieving the Sustainable Development Goals (SDGs.) This is why it is equally important to look at some scholars/experts' views the urban environment as relevant to this study

Urban Environment

Throughout history, urban centres, particularly cities across human civilisations have played (still playing) significant roles in social, political and economic transformations of man.[76] As contended by Freire and Polèse, cities are necessary centres for innovation, social change, and economic growth.[77] Consequently, cities have become more prominent part of the social and economic landscape all over the world and the most conspicuous environments in which economic capabilities are expanded or impeded and social qualities of life are fulfilled or frustrated. The city in the words of Asaju is a victim of definitional pluralism because of the presence of multiplicity of definitions.[78] It could either be defined in terms of geography which deals with size, population and density or sociologically that concerns with the activities and interrelationships of the inhabitants.[79] This is why Agbaje-Williams concludes that "the origin of, and yardsticks for identifying, urbanisation are subjects of intense debate among scholars."[80] With this complexity, it is therefore not surprising that city is often used loosely as generic term for all kinds of large or dense settlements.

Agbaje-Williams described city as "the highest point of human aggregation in any spatial context."[81] Childe considered it as the hallmark of cultural sophistication and custodian of the peak of human

---

[76] "The Role of Cities across Human Civilizations was the Central Focus of the 2007 UNU-Wider Conference on Urbanization", available online at www.wider.unu.edu.

[77] Mila Freire and Mario Polèse, *Connecting Cities with Macro-economic Concerns: The Missing Link,* (Washington, DC: The World Bank, 2003) 2

[78] A. S. Asaju, "Globalization and City Development", in *The Nigerian Tribune*, Tuesday 10th Jan 2006.40

[79] A.S. Asiwaju....

[80] See Babatunde Agbaje-Williams, "Yoruba Urbanism: the Archaeology and Historical Ethnography of Ile-Ife and Old Oyo," in Akinwumi Adediran ed., *Pre-Colonial Nigeria Essays in Honour of Toyin Falola,* (New Jersey: Africa Press World Incorporated, 2005). 215; Gordon Childe, "The Urban Revolution," *Town Planning Review,* 21 (1950):3-17; Gideon Sjoberg, *The Pre-industrial City, Past and Present,* (Glencoe, IL: The Free Press, 1960); Roderick McIntosh, "Early Urban Clusters in China and Africa: The Arbitration of Social Ambiguity", *Journal of Field Archaeology,* 18 (1991)

[81] Babatunde Agbaje-Williams, "Yoruba Urbanism: the Archaeology and Historical Ethnography of Ile-Ife and Old Oyo,".....215

civilisations.[82] The World Book Encyclopedia advocated four main features that have been conducive for the formation of cities. These are: advances in technology; a favourable physical environment (location, climate, rivers and thus water supply); social organisation (authority, government); and population growth. This is also in consonance with Asaju's submission that, throughout the recorded human history, it has been urban centres that have provided leadership, the technology, the institutions and administration for development. Thus, it has become almost a universal phenomenon to equate civilisation with urban civilisation and identify the quality of life in terms of the quality of life in the cities.[83] In the same vein, while commenting on the significance of cities to humanity, Justin Labinjoh opined that:

> Cities are urban centres....The city is an autonomous phenomenon, the exploration of whose historical, cultural, economic, and political ramifications is not only intellectually exciting, but also contributes immensely to our understanding of the larger society. Just as there have been great empires in history, there have also been great cities past and present reflecting various flourishing civilisations.[84]

However, this study strongly argues that cities can never stand alone. As reflected in the opinion of Vidrovitch, a city supposes a coupling; city and countryside[85] (which could even be found in limited cases of city-states in pre-colonial Africa such as Hausaland and Yorubaland). This is also in conformity with French concepts of 'urbanisation' and of 'ruralisation of cities.'[86] Although the relationship between the component parts of urban and rural settlements may not be easily defined, Catherine Vidrovitch Conquery still contend that city can only be fully understood not only by itself but also as a network of other cities, smaller urban centres and a

---

[82] Gordon Childe, "The Urban Revolution,"....3

[83] A. S. Asaju, "Globalization and City Development...."

[84] Justin Labinjoh, *Modernity and Tradition in the Politics of Ibadan,1900-1975,*(Ibadan: Fountain Publications, 1991)

[85] Justin Labinjoh, *Modernity and Tradition in the Politics of Ibadan....*

[86] Richard Stren, *Coping with Rapid Urban Growth in Africa: An Annotated Bibliography in English and French on Policy and Management on Urban Affairs in the 1980s,* (Montreal: Centre For Developing Areas Studies, McGill University, 1986)

whole region, including the rural parts.[87] This shows that understanding of rural realities (especially in their relationships with urban ones) is a necessity before anyone could have a full grasp of what cities entail. Contrary to Wheatley's "Concept of Urbanism", Vidrovitch submitted that, a city is and its process is not a concept but a reality.[88] The latter's preponderant view has been proven by evidences across human civilisations.

As it has been established earlier, several schools of thought exist on definitions of city and on the major criteria for identifying urban centres. As demonstrated by Agbaje-Williams, there is an increasing awareness that local historical, economic, political, cultural, and ecological conditions usually determine the forms of urbanisation, and these factors must be considered in the conceptualisation or definitions of cities in different places and at different times.[89] Among the indices that have been used to define urbanism are sizes and density of settlement and permanent residential population. It is perhaps the differences in these features that differentiate a city from a town and or a village, with the former being larger in spatial dimension and population density.[90] Other important characteristics that scholars tend to agree on are the presence of economic activities and administrators.

Across most of the past and present glorious civilisations, urbanisation as reflected in the structure and building of their cities has emerged as the highest point of human aggregation and has been considered as one of the hallmarks of cultural sophistication because it entails the most complex spatial ordering of social, economic, political, and power relations and manifest the most intricate form of social organisation. From the above

---

[87] Catherine Vidrovitch Conquery, "The Process of Urbanisation in Africa (From Origins to the Beginning of Independence)", *African Studies Review*, Vol. 34, No. 1, (April 1991) 3-4

[88] Paul Wheatley, "The Significance of Traditional Yoruba Urbanism", *Comparative Studies of Society and History*, Vol. 12, No. 4, (1972), 393-423, Catherine Vidrovitch Conquery, "The Process of Urbanisation in Africa...."4

[89] Babatunde Agbaje-Williams, "Yoruba Urbanism: the Archaeology and Historical Ethnography of Ile-Ife and Old Oyo..." 217

[90] Ronald Fletcher, "African Urbanism: Scale, Mobility and Transformation," in Graham Connah ed., *Transformations in Africa: Essays on Africa's Later Past*, (London: Leicester University Press, 1998) 105

explanations, it has become obvious that, cities are not isolated phenomena, distinct from and independent of the larger societal contexts. These societal contexts determine, in large part, the nature of cities within any given society: their organisation, functions, and form. It is also true that every generation lives and works in an inherited environment shaped, in some cases, by very distant predecessors.

Presenting one of the earliest works on urban environments in Nigeria, Wirth deeply acknowledged the presence of metropolitan cities in Yorubaland. Contrasting these Yoruba Cities with the historic city of Timbuktu, Wirth concluded that nine out of ten largest cities in Nigeria in 1931 were in Yorubaland. These included Ibadan that was described as the "largest Negro city in Africa".[91] Corroborating Wirth's account, William Bascom opined that "the Yoruba have large, dense, permanent settlements; based upon farming rather than industrialisation…they are the most urbanised of all African peoples' percentage living in large communities comparable to that of European nations."[92] In an attempt to prove the existence of urban centers in Nigeria before the advent of colonialism, Bascom contended further that urbanisation was indeed traditional to the Yoruba rather being an outgrowth of the Europeans. This particular work is significant as it warns us of the danger of confusing urbanism with industrialism and capitalism when it argued that:

> Though the development of modern cities could not be divorced from the power of technology but the cities of the earlier may have been different by the virtue of their development in a pre-industrial and pre-capitalistic order, they were nevertheless cities.[93]

Therefore, the Yoruba cities were not based on machines especially in the pre-colonial epoch but the fact that the craft form of specialisation made each individual dependent on the society as a whole qualified them as

---

[91] L. Wirth, "Urbanism as A Way of Life", *American Journal of Sociology*, 44, (1938)
[92] William Bascom, "Urbanization among the Yoruba", *The American Journal of Sociology*, Vol. 60, No. 5. (1995)
[93] William Bascom, "Urbanization among the Yoruba…." 23

cities. The two works no doubt provide essential historical background for any study on urban centers in Africa such as the current effort.

Mabogunje presents one of the earliest indigenous researches on urban environmental issues detailing urban planning in Nigeria with a combination of many approaches.[94] He gives an overview of the major development in urban life of Nigerians; especially the development of Nigerian cities and towns from the pre-colonial period dating back to the first century A. D. Four phases were identified namely: The Pre 1800 period, the 19th century, the colonial period and the post independence era. While this quintessential work provides essential background information, it was couched in pure social science mode without deep historical analysis of the problems and also neglects other important issues such as the necessity for adequate urban environmental management; its implications for development; colonial and post colonial urban crisis and renewals. The same thing could also be said of most of the earlier relevant works. Connor particularly offered incisive analysis on urban crisis and its implications for Africa.[95] Though, he suggested some ways out of the problems but also failed to historicise the issues raised.

Every generation has had to live and work in an inherited environment, shaped in some cases by distant predecessors. This is particularly true about urban environments. This is why it almost impossible to solve urban crisis without understanding the historical forces behind the problems In this regard, the work of Gerard Burke is quite germane to this study. The writer reviews town formation from earliest human settlement to the modern age; highlighting physical achievements rather than the political, economic or social circumstances in which they were founded. He stated further that planning of urban environments is not a recent phenomenon as:

> Conscious planning of land use and development has been practiced in one way or the other ever since man first acquire the

---

[94] A. L. Mabojunje, *Urbanization in Nigeria*, (London: University of London Press, 1968).
[95] A. M. O. Connor, *The African City*, (London: Hutchinson University Library for Africa, 1983)

knowledge and skills that enabled him to settle in places his choice.... By slow degrees he learnt those things that helped to shape his daily life and environment: to control and use fire, to fashion and use tools, to domesticate animals, to cultivate and store food, to build a shelter for protection.[96]

Though no particular mention of any African City, the work nevertheless provides tools and indices for comparative historical analysis of African cities; especially when corroborated with Schwab who had earlier offered comparative analysis of typology, classification and urban behaviour of African cities.[97] He compares and analyses corporate groups in Osogbo and Gwello in Rhodesia and also indicates how urban milieu has affected or altered corporate groups and *vice-versa*.

In fact, several authors have written extensively on the phenomenon of urbanisation in Nigeria. Ibadan in particular has received the attention of scholars who have documented accounts on the development of the city. However, it should be noted that despite the availability of some works on Ibadan, none has paid specific and extensive attention on the place of environmental management in the development or underdevelopment of the city. This is the lacuna the present study tried to fill.

One of such works is the one edited by Toyin Falola and Steven J Salm, *Nigerian Cities*. This book which grew out of the African urban Spaces: History and Culture Conference held at the University of Texas at Houston from March 28 to April 30 2003. The study among other things examines various issues on the management of modern Nigeria cities, the movement of people and goods, improving sanitation and minimizing ethnic tension in the cities. The book addresses the concern of scholars, experts and policy makers. In chapter one, Ayodeji Olukoju presented an in-depth and tantalizing perspective on the historical background of Nigerian cities and attempted to provide an overview of the major developments in the urban life of the peoples of Nigeria. In doing this, he discussed the pre-twentieth century period which dwelled on traditional

---

[96] Gerald Burke, *Towns in the Making*, (London: Edward Arnold, 1975) 6
[97] William Schwab, "Urbanism, Corporate Groups and Culture Change in Africa below the Sahara", *Anthropological Quarterly*, Vol. 43, No. 3, (1970)

urban centres located mainly in the territories of the Yoruba and Edo of western Nigeria and the Hausa and Kanuri of Northern Nigeria.[98]

Also, in the same book Olasiji Oshin's "Railways and Urbanisation", attempted to evaluate the impact of railway on the growth of modern urban centers in Nigeria with particular reference on the emergence of such centers as socio-political melting points, drawing peoples from diverse ethnic and cultural backgrounds in the country against the backdrop of deliberate policy of segregation pursued by the British colonial authorities.[99]In all, this book covers not only the pre-colonial period of urban development but also the colonial period of urban development. However, the work focused mainly on the colonial period and the economic development of cities in Nigeria. Particular focus was given to railways and economic interactions and their effect on urban development, without much attention given to the environmental as well as the social development of the cities. Nevertheless, the book is still relevant to this work.

Similarly, A. L Mabogunje defines urbanisation as the process whereby human beings congregate in relatively large number at one particular spot of the earth's surface.[100] Beyond his definition, he entered into discussion of theories of urbanisation and concluded that as against the industrialised countries, the situation presented by cities in the underdeveloped countries has not given rise to any formal theory of city growing and structure. He however, states that the fundamental idea in the theory of urbanisation is the specialisation of functions among human communities through division of labour. He noted that for functional specialisation to give rise to urban centres, there must be surplus food available for specialist, that for the work of the specialist to be facilitated and their needs for raw materials satisfied, there must be a class of traders and merchants. The work is important to our study because it is concerned with urbanisation as a process operating through time. The work which also concentrated on two leading Nigerian cities –Ibadan and

---

[98] Toyin Falola and Steven J. Salm, eds., *Nigerian Cities*, (New Jersey: African World Press Inc. 2004)

[99] Toyin Falola and Steven J. Salm, eds.....

[100] A. L. Mabogunje, *Urbanization in Nigeria*....21

Lagos throws light on our period. However, while the work concentrates on the structural and spatial development of Ibadan as a model of pre-colonial colonial and post-colonial city, it did not emphasise the historical underpinnings of the development of Ibadan. However, some specific works on Ibadan have attempted to fill this gap.

For examples, *The City of Ibadan*,[101] which referred to Ibadan as the largest indigenous African city south of the Sahara provided some essential information on the evolution of the city. The book gives a historical account of the emergency of Ibadan as a war-camp and how it has retained to a large extent the structure of an over-grown village. The book consists of articles by versatile historians who uncovered different aspect of the development of the city. Bolanle Awe, "Ibadan Its Early Beginning"[102] in the same book, gave an account of the early Yoruba settlement during the pre-colonial era, their interactions, rivalry for power and their ethnic composition until Ibadan was founded in 1829. She described Oyo as the dominating force in Yoruba land during the period. And as a result, Oyo was able to build a large empire stretching as far as the coast, until its collapse which led to the emergence of such newly founded settlements like Ibadan and Ijaye.

Mabogunje, also writing in the same book, "The Morphology of Ibadan", attempted to describe the physical structure and morphology of Ibadan. He explained that Ibadan represent a convergence of two traditions of urbanism-a non mechanistic, pre-industrial African tradition, more akin to the medieval urbanism in Europe; and a technologically- oriented European tradition. Hence, he described Ibadan as a combination of two cities in one- the old and the new, with each section of the city depending on the other economically, and the commercial and industrial development of the new Ibadan modifying the life of the old city.[103]

---

[101] P. C. Lloyd, Bolanle Awe and Akin Mabogunje, eds., *The City of Ibadan*, (Ibadan: Institute of African Studies, University of Ibadan, 1967)

[102] Bolanle Awe, "Ibadan its Early Beginning", in P. C. Lloyd, Bolanle Awe and Akin Mabogunje, eds., *The City of Ibadan*, (Ibadan: Institute of African Studies, University of Ibadan, 1967)

[103] A. L. Mabogunje, "The Morphology of Ibadan", in P.C. Lloyd, Bolanle Awe and Akin Mabogunje, eds., *The City of Ibadan*, (Ibadan: Institute of African Studies, University of Ibadan, 1967)

Similarly, Barbara Lloyd wrote on "Indigenous Ibadan Market"[104] focused on Oje Market, and in particular fifteen family compounds bordering it, as representative of the original non-industrialised eastern portion of Ibadan. By writing on Oje, she attempted to give a representation, though limited picture of the life of the indigenous people of Ibadan. Although an attempt was made to throw light on the social development of Ibadan and also to give an in-depth historic account on the emergence and development of the city until colonial period, however, an in-depth account of the contribution of the British colonial administration to urban growth in the city was not given and an attempt was not made to look at the impact of the colonial administration on urban growth in Ibadan.

Also, Kemi Morgan in *Akinyele's Outline History of Ibadan* gave an account on the founding of Ibadan. The book sheds light on the founding and destruction of the two previous sites of Ibadan. According to the book the site on which Ibadan and its suburbs now stand was originally called the forest of Ipara.[105] The forest belonged to a section of the Egba Agura people called Ibadan Soge. After the settlement headed by Lagelu was destroyed, a second settlement was founded around a low mound called "Yangi" around which they built houses and established a market which has now become very famous in Ibadan. It is now called "Oja Ba", named after Basorun Oluyole an Ibadan ruler. The second settlement survived until the early nineteenth century. By then as a result of the aftermath of the Owu war and the political upheaval that enveloped Yoruba land, the allied armies of Oyo, Ife and Ijebu invaded Ibadan. This book is useful to our work because of the light it sheds on the founding and rise of Ibadan during the nineteenth century.

Lola Tomori, in his book, *Ibadan: Omo ajoro sun*, gave a brief account on the founding of Ibadan. According to him, both oral traditions and documented historical evidence obtained, indicated that the allied army fought their way to settle down by displacing the original settlers (Lagelu

---

[104] Barbara Lloyd , "Indigenous Ibadan Market", in P. C. Lloyd, Bolanle Awe and Akin Mabogunje, eds., *The City of Ibadan*, (Ibadan: Institute of African Studies, University of Ibadan, 1967)

[105] Kemi Morgan, *Akinyele's Outline History of Ibadan*, Part I-IV, (Ibadan: Caxton Press, West Africa Ltd)

descendants) and the Owu settlers (Owu Ogbere) from 1821-1825. According to the author, this kind of historical migration was not peculiar to the founding of Ibadan alone, but nearly all settlements in Yoruba land including Ife, Ekiti,Owu among others.[106] This book is important to this work because of the knowledge it provides on the origin of Ibadan up till 1980. However, it did not give detailed information on the impact of colonial rule on urban development during this period in Ibadan. In view of this, this work will give detailed information on the impact of colonial rule on development in Ibadan.

In another book, *Politics and Economy in Ibadan, 1893-1945,* Toyin Falola examines the major economic and political development in Ibadan from 1893 when the British imposed colonial rule to 1945 when the Second World War ended. This book detailed the process of transformation during this period. The book reviewed early Anglo-Ibadan relations which was a major Yoruba power during this period. The book also talks about the political foundation of a colonial society which was carried out by the British. According to the author the major changes in the immediate post 1893 era owed to two factors, the dissolution of the empire and the British imposition of colonial rule. The latter was the more significant and the former made any concerted resistance to the external power almost impossible.[107]

Furthermore, the book also focused on the economic foundation of colonial society. This was with the view of creating a society conducive for the establishment of colonial economy. In other words, the new political changes and structures were not the end, but a means to an end. According to the author, a political culture that was undemocratic and dependent of the use of force was in line with an exploitative economic arrangement imposed after 1893. The new political system also made it possible to impose necessary non-political measure to create the new colonial economy.[108]The author also includes the corporation of the local economy in that of Europe, as one of the goals of colonial economic policies. According to the author, this incorporation was important, at

---

[106] Lola Tomori, *Ibadan: Omo A Joro Sun I,* (Ibadan: Penthouse Publication, 2004)
[107] Toyin Falola, *Politics and Economy in Ibadan, 1893-1945...*
[108] Toyin Falola....

least three reasons which have been outlined by Walter Rodney; (a) to protect national interest against competition from other capitalists: (b) to arbitrate the conflict between their own capitalist: and (c) to guarantee optimum conditions under which private companies could exploit Africans.[109]Again just like previous works, the book concentrated on the economy and its impact on development in Ibadan because of the general notion that development is an index of urbanisation. But the book failed to look at other areas of development (other than economy) which could also aid urbanisation in Ibadan. Hence, this work has attempted to fill this gap by concentrating on the physical, environmental and social aspects of development in Ibadan.

Also, Wale Oyemakinde in his book, *Global Economy Meltdown* traced the history of global economic recession and its impact on world economies. According to him, during the global economic depression of 1930s, the impact was felt on wage earners all over the world including Nigeria. This book is relevant because it explains the intricacies of colonial economic policies and the effect of imperialism on the Nigerian Economy. The author while trying to differentiate the era of informal empire that preceded actual colonisation of African zones by European authorities with the colonial era stated that: "but with real imperialism, the commerce of a dependency and, indeed its whole economy came under the control of its political administrators who naturally safe guarded their own national interests."[110]

As much as possible, whatever raw material were needed in the metropolitan factories were to be produced by the subject people of the territories. On the other hand, they were to provide the markets for the manufactured goods from the master's factories. The colonies were treated as mere extension of the metropolitan economic political and social system. The exploitative economic policy of the colonial masters affected development particularly during the difficult years of world economic depression. The railway which was seen as an agent of development introduced by British colonial government was particularly

---

[109] Walter Rodney.....
[110] Wale Oyemakinde in his book, *Global Economy Meltdown*, (Ibadan: Sunlight Syndicate Ventures, 2009)

affected during the depression of the 1930s. The work provides relevant information on the colonial economic policy and the railway and its impact on economic development during our period.

Filani, M. O. et.al 1994, *Ibadan Region* is another book which consist of readings by distinguished geographers analyses various aspects of the physical setting, growth, development and environment of the city of Ibadan. Adetoye Faniran "Relief and Drainage in Ibadan" in the same book wrote on the scenery of Ibadan. According to the writer three major land forms units, hills, plains and river valleys dominate the scenery of Ibadan. Also the writer states that analysis of stream order shows that the higher order stream is the 5th order streams, represented by the lower reaches of the Omi and Ona rivers. River Ona at the Eleyele reservoir is a 5th order stream, while Ogunpa and Kudeti within the built up area are both 3rd order streams.[111]

Also in the same book, Bola Ayeni "Ibadan Metropolitan Area" focuses on the growth and subsequent consolidation of Ibadan as a major commercial, educational and administrative center. By 1893 Ibadan signed a treaty of peace with the British and a resident was posted to the city. The arrival of the resident marked the end of the era of Ibadan dominance over much of Yoruba land and the beginning of the emergence of the city as a major commercial and administrative center. As a major exponent of the new system of commercial organisation, the railway whose construction began in Lagos in 1805 was extended to Ibadan in 1901. This led to the influx of aliens, Nigerian and non-Nigerian alike, into the city. This led to the establishment of a modern business center and a European reservation area.[112]

In the same book, O. Areola, wrote on the "Spatial Growth of Ibadan." The author wrote on the impact of urbanisation on the rural landscape and economy. According to the writer in 1952 Ibadan became the headquarters of the Western region. This meant a rapid expansion of

[111] Adetoye Faniran, "Relief and Drainage in Ibadan", in M. O. Filani et. al, *Ibadan Region*, (Ibadan: Rex Publications and Connel Publications, 1994). 28-48
[112] Bola Ayeni "Ibadan Metropolitan Area", in M. O. Filani et. al, *Ibadan Region*, (Ibadan: Rex Publications and Connel Publications, 1994). 72-84

buildings in Ibadan to serve the city's political, as well as economic activities. The establishment of schools, hospitals and hotels followed to serve the rapidly expanding population of the city which had risen from an estimated 60,000 in 1856 to 459,000 according to the 1952 census. Hence, the colonial administration established government reservation areas (GRAs) at the outskirts of the city to the east (Agodi), and north – west (Jericho), whilst modern residential layouts were established outside the traditional core of the city to serve the flood of migrants.[113]

These readings are particularly relevant to our study because of the detailed and analytical information they provide on settlements and urban growth in Ibadan. They are in fact relevant because they touch on almost every aspect of urban growth of Ibadan in the colonial period. In spite of its extensive research on physical and infrastructural development, the book failed to pay attention to the social development of the town. Particularly educational development and the rise of the elite group and middle class, which are features of an urban center, were neglected. Hence, this work will fill the vacuum in knowledge on the social development of Ibadan.

Furthermore, G.O Ogunremi in *Ibadan A Historical and Socio-Economic Study of: An African City*, put together a compendium of well researched studies by scholars on the historical, cultural and socio-economic development of Ibadan. Justin Labinjoh writing in this book focused on Ibadan and the phenomenon of urbanisation. He looked at Ibadan as a Micro-society within a larger entity and as a satellite of a metro pole called Nigeria. According to the writer we talk of urbanisation we usually think of trade, business of all kinds, differential allocation of space, industrialisation, migration, population growth and explosion, markets, even slums and poverty among other things.[114]

---

[113] O. Areola, "Spatial Growth of Ibadan", in M. O. Filani et. al, *Ibadan Region*, (Ibadan: Rex Publications and Connel Publications, 1994). 98-106

[114] Justin Labinjoh, "Ibadan and the Phenomenon of Urbanism", in G. O. Ogunremi, ed., *A Historical and Socio-Cultural Study of an African City*, (Ibadan: Oluyole Club, 1999) 238

Also, in the same book Dele Oketoki focuses "industrialisation in Ibadan since 1930" stated that the British administration added *fillip* to the initial social economic and geographic advantages of Ibadan by providing modern transportation facilities such as railway and roads. The first railway to the north had reached Ibadan by 1901, leading to increased commercial activities. In spite of these advantages, Ibadan had no manufacturing industry before the establishment of British administration. How then did industrialisation begin in Ibadan? According to the writer the answer is that industrialisation in Ibadan was the result of the effort of local people. But this was not industrialisation that involved many tall factory buildings. Rather, it was the use of imported machinery in the processing of local raw materials and production of consumer goods required locally. According to the writer the British administration provided the environment that made this possible. However, this is highly debatable as the British attitude to industrialisation and economic policies generally in Nigeria and elsewhere in Africa has been a contentious one.[115] This book also throws light on the colonial administration and economic development of Ibadan. Particular emphasise were placed on markets, trade and industrialisation. However, little or no reference was made to the efforts geared towards maintenance of healthy environment in the development of the city.

African cities seemed to share similar horrendous environmental crises. This was the conclusion of Herbert Werlin. With a special focus on Nairobi, he contributed to the theory of urban systems or "micropolitics" and discussed in great details: the multiplicity of African urban environmental problems; the nature of settler regime; relationship between urbanisation and development; the dangers of urbanisation; the malaise of African "sick cities" which have been overwhelmed by air pollution, noise, traffic, waste, crime, racial tension, slum conditions, maladministration and many more. This work which grew out of an earlier East African Town planning conference in Kampala concluded inter-alia that:

---

[115] Dele Oketoki, "Industrialization In Ibadan Since 1930", in G. O. Ogunremi, ed., *A Historical And Socio-Cultural Study of an African City*, (Ibadan: Oluyole Club,1999)

Nowhere in Africa has the problem been solved of enabling the mass of Africans who live in urban areas to live in healthy and socially desirable conditions.... There is no organised public waste collection service, and community clean-ups are not usually carried out until conditions have become totally unbearable".[116]

More than fifty years after this submission, environmental conditions in most African cities are getting worse as most of them remain filthy. The case of Ibadan is particularly pathetic. Hence, the need for a more comprehensive appraisal of environmental situation, with a view of improving the condition.

The prevalence of poverty in Africa is a key factor in the environmental crisis in most of her urban centers. As early as 1960s, this has engaged the attention of some scholars. For instance, P.C. Lloyd observed that "most African dwellers suffer from extreme poverty"[117] which is traceable to a number of interrelated factors such absence of basic infrastructure-(road, electricity, toilets, portable water, health care facility etc), mass unemployment, under employment, under-nourishment and so on. This would be particularly true of the illiterates who constitute more than 80% of the population in most African cities. This give rise to deep sense of frustration among the millions of these poor urban dwellers who are mostly under-nourished, ill educated, poorly housed, uninformed and constantly exploited by others. Due to the fact that most of them usually rely on nature for bare existence, keeping basic environmental laws become unthinkable or impossible.

In establishing the unbreakable nexus between poverty and environment degradation, the World Bank with a 1993 report on Jamaica concluded:

Though many environmental problems are general in their effects ....the poor in many countries bear more of the immediate pressure of environmental degradation. They are the least to be served with public water, sewage and garbage services; tend to live in crowded

[116] Herbert Werlin, *Governing an African City: A Study of Nairobi,* (New York: Africana Publishing Company, 1974) 10

[117] P.C., Lloyd, *Africa in Social Change,* (Baltimore: Penguin, 1967) 3

homes where indoor pollution is high and may live in close proximity to disposal sites for solid and hazardous waste.... the poor are not able to escape these areas.[118]

Most importantly, the report concluded that poverty is a cause of environmental degradation as the "poor must meet their urgent survival needs." This is especially true of rural, semi-urban dwellers and the urban dwellers in the informal sector that operate without the benefit of health, safety or environmental regulations. The social tension engendered by the continued and worsening poverty always has negative effects on the countries' ability to attract long term investment (domestic and foreign).

Closely related to issue of poverty is the shelter crisis ravaging most African urban centers. Indeed, of all problems confronting urban center in Africa as observed by Herbert Werlin, "the lack of decent housing is usually the most apparent." He stressed further that "in many metropolitan areas, one quarter to one half of the population are found in slums and shanties". And the situation is getting worse".[119] According to a UN report in 1996, the shortage of inexpensive, well constructed houses in Africa is so great that most urban households are accommodating 3 to 6 persons per room.[120] Consequently, African urbanites are forced to move to "peri-urban" areas where the houses are poorly constructed, inadequately furnished and lacking sanitation facilities, water supplies, paved streets, electricity, transportation and other essential services. Apart from linking the lack of decent housing in the urban areas in Africa to other problems like crime, diseases, family instability and many more; the report concluded that only few African governments will be able to meet more than one-fifth of their minimum housing requirements.

The report also condemned the persistence of and widespread of urban slums and blight; the concentration of persons of low income in older areas; and the unmet needs for additional housing and community

---

[118] Jamaica, *Economic Issues for Environmental Management*, (Washington DC: World Bank, 1993)

[119] Herbert Werlin, *Governing an African City: A Study of Nairobi*, (New York: Routledge, 1974)

[120] 1996 UN Report on Urban Africa, available online at www.unhabitat.org./ 1996urbanafricareport

facilities and services arising from rapid expansion of Third world urban population. He argued that these "have resulted in a marked deterioration in the environment of large numbers of the people while the nation as a whole shows no serious concern"[121] leading to further destruction the long existing equilibrium between man and nature.

Steve Hinchliffe has also produced an exemplary introduction to cutting edge work on the geographies of nature geared towards equalising the badly battered man/nature equilibrium. Intellectually demanding, clearly written and empirically rich, this is a book that deserves a wide readership among environmental researchers and practitioners. The author introduces readers to conventional understandings of nature while examining alternative accounts from different disciplines - where nature resists easy classification. Accessibly written, organised in ten chapters in two sections, the work demonstrates how recent thinking has urgent relevance and impact on the ways in which we approach environmental problems. The text: makes concepts like 'environment', 'conservation', and 'sustainability' accessible and applicable to readers' own experience with the extensive use of case studies.

Buckingham and Turner equally provided an indispensable tool for a clear understanding of urban environmental issues with the use of learning outcomes, text boxes, tables and figures throughout to make complex ideas accessible and relevant.[122] In section one, the authors described- the philosophies, values, politics, and technologies which contribute to the production of environmental issues. They also use cases on-climate change, waste ,food, and natural disasters in section two to provide detailed illustration and exemplification of the ideas described in section one. The piece was concluded with a case study of Mexico City. Vivid, accessible and pedagogically informed, this work has been a key resource for this study.

Additionally, this study has also gained from global environmental issues raised in Urban Affairs Reviews Vol.1-30 (1967-1980) and several reports

---

[121] 1996 UN Report on Urban Africa....

[122] S. Buckingham and M. Turner, *Understanding Environmental Issues*, (London: Sage Publications, 2008)

from the United Nations Organisation (1972, 1992, 2000, and 2006). They all attempted analysis of issues like: crime, security, gender, poverty, legal transportation, violence, globalisation and other problems related to urbanism in Third World, Africa and Nigeria. However, they failed to emphasise the role of urban environmental management, impact industrialisation and the peoples' perceptions on environmental issues. But all the cited texts only offer general /universal explanations which still need to be critically analysed and corroborated by efforts of few Nigerian historians whose works on urban history of Nigeria are germane to this study before adoption to explain Nigerian general urban environmental crisis and Ibadan's peculiar mess.

Although, all the texts reviewed for this study present copious data and engaging arguments about general odds against sustainable development and urban environments in Africa and Nigeria (urban neglect, poverty, spread of slums and diseases etc); they mostly failed to give in-depth historical analysis of the crises. Most importantly, despite its size and strategic significance, the environmental problems of Ibadan have not received the deserved scholarly attention. Compared with the previous works, this study has broken new grounds by being deliberately historical. With a special focus on Ibadan, it has also endeavoured to: analyse the relationship between sustainable development and urban centres on one hand and environment and man on the other; examine the existing planning philosophies, techniques and legislation on environment and development; push new frontiers for urban planners, environmental specialists/policy makers and enrich general readers. Above all, this study has initiated resurgence in the study of sustainable development in African/Nigerian cities from a historical perspective by trying to enhance the position of environmental history as a veritable field of study in Nigeria.

Chapter Three
Environmental Challenges in Cities from Global and African Perspectives

Preamble

Regrettably, today, no place in sub Saharan Africa seemed to be immune against tremendous urban crises, though, the impact varies from place to place. Nigeria as one of the few countries in sub Saharan Africa, which had many large pre-industrial cities could not possibly be an exception to the African urban environmental decay.[1] While the world is full of both urban environmental "successes" and "failures", most African cities are examples of urban decadence; Ibadan is particularly notorious in this regard. As a corollary, the study also recognises the fact that the ecological crisis in most sub Saharan African cities is a reflection of overall underdevelopment of the region which is directly link to global unsustainable environmental practices. Therefore, any study (such as this) that is primarily concern in proffering solutions to the myriad of environmental crises confronting cities in Africa must draw inspirations from continental and global perspectives. This chapter essentially focuses on: relationship between globalization and the environment, an overview of global environmental challenges, African Environmental Crises, the Nigerian Dimension and the impact on Ibadan.

Throughout history, urban centres, particularly cities across human civilizations have played (are still playing) significant roles in social, political and economic transformations of man. Consequently, the daily reality of people on this planet is increasingly an urban reality, with over 47% of the world's population (2.9 billion people) living in urban areas. According to the United Nations Environment Program **(UNEP)**, this proportion is expected to increase to 60% or 4.9 billion people by 2030.

---

[1] Laurent Fourchard, *Urban Slums Reports: The Case of Ibadan, Nigeria*, (Ibadan: Institut Francais de Recherche en Afrique (IFRA), University of Ibadan, Ibadan, Nigeria, 2003)

The New Urban Agenda also shows that by 2050, the world's urban population is expected to nearly double.[2]

Despite this, the situation of urban environmental issues within the internationally agreed goals and targets for sustainable development has remained marginal in the urban growth dynamics. The United Nations through several international conventions and treaties has unsuccessfully tried to reverse the negative trends in urban development. For example, chapter seven of **Agenda 21** sought to galvanise local level action to prioritize and develop human settlements, chapter twenty-one focuses on solid waste management and sewage infrastructure.

It is most worrisome to note that environmental disasters have recently become a common occurrence in the world. Throughout the continents, there is no doubt that the world is under a serious threats from the environment. But this study argues that the environment was only responding to the abuses heaped on it by man's activities. Components of the environment – maintains tropical forests, oceans, soils, the atmosphere, air, and other natural resources, considered essential to the ecosystem. Instead of the proper utilization of these free gifts of nature, the earth's resources have continued to deplete, as dirty air, global warming, polluted waters and toxic wastes are just a few of the maladies of the planet earth.

While global community has considered homicide, ethnocide and genocide as crimes against humanity, **ecocide** which is the deliberate destruction of environment is yet to attract the needed attention and sanctions. Some of the problems witnessed in this regard include: Climate Change, pollution, global warming (occasioned by ozone layer depletion), deforestation, desertification, flooding, hurricane, tsunami and so on.

The good news is that, the world is standing together more than ever before, brain-storming on how to ensure that human development strategies are always environment friendly. Till date, the most comprehensive and ambitious step taken by the global community

---

[2] See   (https://au.int/en/pressreleases/20190526/making-african-cities-more-habitable-ministerial-meeting-kicks)

though the United Nations is the Sustainable Development Goals adopted in 2015 by member states.

The United Nations' conference on human settlements in Vancouver June 1976 concluded that problems of uncontrolled population growth, rural stagnation, migration, the inability of urban centers (especially cities) to cope with the rate of population increase and environmental deterioration demand corrective actions at both the national and international levels.[3] This is why R.O. Borofice once linked the current global environment crises to the development crises, both of which are the result of unsustainable system of production and consumption in the North, inappropriate development models in the south and a fundamentally inequitable world order.[4] Indeed, globalisation has opened unprecedented opportunities which are enhancing human civilization in North almost in the same manner it is aggravating developmental challenges in the South. And cities as "the agglomeration of the best of human civilization" and "the epic centres of development" are presenting gravest challenges to environmental sustainability and sustainable development in most of the undeveloped countries (especially those in sub-Saharan Africa.

Consequently, cities rather than providing the basis for sustained economic growth therefore have become serious impediments to development in sub Saharan Africa and environmental crises are often sharpened or even triggered by glaring social and economic inequalities.[5] Regrettably, no place in sub Saharan Africa seemed to be immune against

---

[3] According to Mrs. Anna Tibaijuka, (Under Secretary-General of the United Nations and Executive Director of UN-HABITAT), it was in 1976 UN Conference at Vancouver that the agency today called UN-HABITAT was born. And ever since the aim of turning new ideas into action for a sustainable urban environment in cities around the world has been central to UN agenda. The Vancouver session of the World Urban Forum brings together many strands of global cooperation and bridge-building by providing a forum for dialogue that promotes universal values, such as tolerance, freedom, justice and equal rights."

[4] R. A. Boroffice, "Environment and Development", in S. O. Otokiti and S. G. Odewunmi, eds., *Issues in Management and Development*, (Ibadan: Rex Charles Publication, 2001) 587-596

[5] Aidan Campbell, *Western Primitivism: African Ethnicity – A Study in Cultural Relations*, (London: Cassell Press, 1997) 86-88

this tremendous tragedy, though, the impact varies from place to place. Nigeria as one of the few countries in sub Saharan Africa, which had many large pre-industrial cities could not possibly be an exception to the African urban environmental decay.[6]While the world is full of both urban environmental "successes" and "failures," most African cities are examples of urban decadence. The case Ibadan is particularly worrisome in this regard. As a corollary, the study also recognises the fact that the ecological crisis in most sub Saharan African cities is a reflection of overall underdevelopment of the region which is directly link to global unsustainable environmental practices. Therefore, any study (such as this) that is primarily concern with proffering solutions to the myriad of environmental crises confronting cities in Africa must draw inspirations from continental and global perspectives.

The world today has become 'a global village'; a term commonly used but coined by Marshall McLuhan in 1962.[7] This arose from unprecedented level of interconnectedness of political, economic, social and technological forces that permeates the contemporary international system. Globalisation is often used to refer to economic globalisation, that is, integration of national economies into the international economy through trade, foreign direct investment, capital flows, Migration and the spread of technology. Through globalisation, people of the world are unified into a single society and function together. Thomas Larson, in his book *the Race to the Top: the Real story of Globalisation*, states that globalisation "is the process of world shrinkage, of distance getting shorter, things moving closer it pertains to the increasing ease with which somebody on one side of the world can interact, to mutually benefit with somebody on the other side of the world."[8] In the consequence of modernity, Anthony Giddens defines globalisation "as the intensification of worldwide social relations which link distant localities in such a way that local happenings are

---

[6] Laurent Fourchard, *Urban Slums Reports: The Case of Ibadan, Nigeria*, (Ibadan: Institut Francais de Recherche en Afrique (IFRA), 2003)

[7] Marshall McLuhan, "Globalisation Available online at http://en.wikipedia.ord /wiki/globalization

[8] Thomas Larson, *The Race to the Top: the Real Story of Globalization*, (Washington DC: Cato Institute, 2001) 9

shaped by events occurring many miles away and vice versa."[9] Giddens's explanation did not only clearly shows the overbearing influence of globalisation on environment but also makes it mandatory for any study that concerns itself with resolving environmental issues in any part of the world (such as this) to start with the evaluation of global and regional perspectives of the problems before delving into the specifics.

Implications of Globalisation for the Environment and the vice-versa

The terms globalisation and environment are seemingly distant concepts at a first glance, nevertheless both concepts overlap as a form of residual relationship.[10]While the importance of the linkage between the environment and globalisation is obvious but understanding of how these twin dynamics interact remains very weak.[11] Much of the literature linking globalisation and the environment is vague, (often discussing generalities); myopic (focused disproportionately only on trade related connection) and partial in its purpose, often highlighting the impacts of globalisation on the environment, or the *vice versa*.[12] This interesting part of the linkage highlights a basic point that not only does globalisation impact on the environment, but also that the environment impacts on the pace, direction and quality of globalisation.[13]At the very least this happens because environmental resources provide the fuel for economic globalisation and also for the fact that our social and policy responses to global environmental challenges contain and influence the content in

[9] Anthony Giddens, *The Consequences of Modernity*, (Cambridge: Polity Press, 1999) 5

[10] Grossman Gene, and Krueger Alan, "Economic Growth and the Environment", *The Quarterly Journal of Economics*, MIT Press, Vol. 110, 2, (May 1995). 353. See also Smith Perry and J. Andres Espinosa, "Environment and Trade Policies, Some Methodological Lessons", *Working Papers*, (Duke University Press, Department of Economics, 1995).

[11] A. D.Nagam and Halle, M. Runnalts, "Environment and Globalization Understanding the Linkages". Available online at www.google/environmental-2bisuses.band.com,. See also Petra Christman and Glen Taylor "Globalisation and Environment: Determinant of Firm Self Regulation in China", *Journal of International Business Studies*, Vol. 32, (2001)

[12] A. D.Nagam and Halle, M. Runnalts, "Environment and Globalization Understanding the Linkages...."

[13] Grossman Gene, and Krueger Alan, "Economic Growth and the Environment"....354

which globalisation happens. This happens through the transfer of social norms, aspirations and ideas that crisscrossed the globe to formulate instant and emergent social movements, including global environmentalism.[14]

Globalisation and the environment are interrelating and both have over the years had significant impact on each other. It is imperative to note that the main impact of globalisation on the environment remained considerably negative; this does not, however, prove that globalisation has no positive impact environment, but such has been minimal.[15]The development in industrial machinery marked the beginning of industrial revolution over two hundred and fifty years ago, which ushered in the urgent need for Nation-states to be more interrelated and interdependent.[16]The heavy activities of the industrial corporations had created immense environmental problems especially through the overuse of natural resources due to increased demand and also the removal of ecosystems due to population growth have had a large negative impact on the environment.[17] Extensive deforestation has occurred for disposable products. More than eleven million acres a year are cut for commercial and industrial use.[18] Deforestation whether it is for an increase in demand or for expansion is causing a loss of substantial biological diversity on the planet.[19] In Australia, 90% of nature heritage; about one half of the forests that covered the earth are gone.[20] Each year, another 16 million acres disappear. Deforestation is expanding and accelerating into the remaining

---

[14] Global environmentalism is an international discourse on globalization that tends to highlight the promise of economic opportunities and also draw a parallel global discourse on environmental responsibility. According to this theory more nuanced understanding needs to be developed that will seek to actualise the global opportunities offered by globalization while fulfilling global ecological responsibilities and advancing equity. Such an understanding would, in fact make sustainable development a goal of globalisation, rather than a victim

[15] Julian Hoppit, "The Nation, the State, and the First Industrial Revolution", *Journal of British Studies,* (April, 2011) 307

[16] Julian Hoppit....331

[17] Global Deforestation, Global Change Curriculum, University of Michigan, Global Change Program, January 2006

[18] Global Deforestation,....

[19] Global Deforestation....

[20] Julian Hoppit, "The Nation, the State, and the First Industrial Revolution"....307

areas of undistorted forest. As industries continue to produce and countries compete on production without proper consideration of the environmental hazards, being one of the main sources of air pollution, these industrial activities has contributed to global warming and sudden change in climate. Climate change occurs as a result or natural conditions or anthropogenic sources changing the composition of the atmosphere or the land use type.[21]

The growth of industrialisation in the developing world, the heavy reliance and dependence on fossil fuels has led to increasing global warming which is due to greenhouse gas emissions.[22] This has consequently led to increase in sea level as polar ice cap continues to melt. A recent study in Male, Maldives gave a dreadful report that, by 2050 there might be no Maldives, because of the continued increase in sea level made which has drastically made Maldives, smaller.[23] Average global temperatures have risen approximately 0.6⁰c since the late Nineteenth Century due to humanity's emission. As a result of globalisation, transportation and the gases produced have become large contributors to global warming. About 95% of the world's traded goods are moved by maritime transport, which in turn causes about 5% of the globe's sulfur oxide and 14% of the world's nitrogen oxide emissions.[24] According to the office of National statistics, Greenhouse gas emissions from transport have been raised by 47% since 1990. Without doubt, Multinational Corporations (MNCs), Transnational Companies (TNCs) or International Corporations (INCs) control the largest share of third world economies, with a huge capital base and a body of international networks.[25] These companies hold themselves accountable to no one but their shareholders,

---

[21] Soji Oyeranmi, "Climate Change and Sustainable Development in Nigeria", in Toyin Falola and Maurice Amutabi, eds., *Perspectives on African Environment, Science and Technology*, (New Jersey: African World Press, 2012) 298- 303

[22] Grossman Gene, "Economic Growth and Environment"....350

[23] See Wikipedia Article of 9th May, 2007 on the Maldives available online at http://en.wikipedia.org/wiki/Maldives

[24] Larsson Thomas, *The Race to the Top: the Real Story of Globalization*...10

[25] Soji Oyeranmi, "Globalization as a source of Environmental Tragedy in Sub-Saharan Africa: The Role Multinational Oil Corporations in Nigeria", *Global South*, SEPHIS e-Magazine, Volume 7, No. 3, (July 2011). 47. Available online at www.sephisemagazine.org

and largely dominate many areas like oil exploration and production, extracting, mining, production and merchandise factors of their host economies.[26] They are subject to no control or regulation, and their negative impact have more devastating effects on both the lives of peoples and environment, due to their orientation towards profit- making and internationalism which is often and effective guarantee of their immunity against punishment for environmental crime including the destruction of host communities in the name of globalisation.[27] The Multinational Corporations have activities in areas like gold mining, petro, gas, chemicals and food industry which have potentially large impact on the environment in developing countries like Algeria, Angola, Congo, Libya, Indonesia and Nigeria.[28]Globalisation has drastically aided their activities, which has led to deterioration of environmental conditions in their host communities with lesser or no concerns by these corporations.[29]

Globalisation has also made warfare take a different turn in the twentieth Century, and its implications on environment are inexplicable. Creation of chemical weapons had created much accidents, heightens death tolls, injuries, damages and environmental disasters. Accidents of Bhopal, India, Chernobyl in Russia (Now Ukraine), Fereso in Italy, London smog disaster, Spanish waste spill, all claim inestimable properties and human and aquatic lives.[30]War is inherently destructive to sustainable development. Therefore, in order to minimise incidence of war, States must respect international law by ensuring the protection of the environment in times of armed conflicts.[31] The application of weapons, the destruction of structures and oil fields, fires, military transport movements and chemical spraying are all examples of the destructive

---

[26] Soji Oyeranmi....

[27] Anne Marchand, "Multi-nationals Immunity and the African Environment", *African Journal of Environmental Studies*, Vol. 24, No.74, (December, 2002) 18-21

[28] Soji Oyeranmi, "Globalization as a Source of Environmental Tragedy in Sub-Saharan Africa: The Role Multinational Oil Corporations in Nigeria".... 47

[29] Soji Oyeranmi.....

[30] Colm Gray, *Another Bloody Century: Future Warfare*, (United States: Phoenix Press, 2007) 269

[31] Colm Gray, *Another Bloody Century....*

impact war may have on the environment.[32] Air, water and soil are usually polluted, man and animal are killed, and numerous health effects occurred among those still living. The environmental effects of wars and incidents leading to war that have occurred in the Twentieth and Twenty First Centuries cannot be downplayed.[33] Easy movements and shipment of arms and ammunitions have continued to provoke wars and make armed conflicts most rampant around the World especially Third World Countries. Most striking environmental effects, including biodiversity loss, famine, sanitation problems at refugee camps, and heavy dispersal of people.[34] Congo crisis, Liberian civil war, Chadian imbroglio, Cote D'Voire's civil strife to mention a few in Africa alone, which compelled governments to divert funds into purchase of weapons and other war instrumentation.[35] Globalisation has given wider room for movement of light weapons and heavy arms into restive and volatile areas around the globe, as its impact on environment worsen. The use of landmines in crisis ridden African countries such as Eritrea and Mozambique planted in farmland, when flood occur these mines are washed into cities and its risk and casualties systematically heightened.[36] Evidently, pains of wars must not only be calculated in terms of loss of human lives and properties; the attendant heavy destruction of the environment must always be accounted for.[37] During the Second World War (1939 – 1945) United States deliberately blew off Antarctica ice sheets to provide passage for submarines and warship to circumvent Nazi Warships and submarines. All these was to escape Nazi recklessness, Nazi is gone now, but the damages are still very much active.[38]

The Vietnam war, one of various proxy wars fought over ideological differences witnessed the use of extremely environmentally damaging

---

[32] Chickering Roger, *A World at Total War: Global Conflict and the Politics of Destruction*, (Cambridge: Cambridge University Press, 2006) 64

[33] Chickering Roger....64

[34] Chickering Roger.... 65

[35] Colin Gray, *Another bloody Century* 268

[36] James Olson and Randy Roberts, *Where the Domino Fell: America and Vietnam, 1945-1990*, (New York: St. Martin's Press, 1991) 67

[37] Ibid.

[38] Haslam Jonathan, *Russia's Cold War: From the October Revolution to the Fall of the Wall*, (United Sates, Yale University Press, 2011) 512

weapons. During the hostilities, this was considered as war tactics and strategies at the detriment of the environment.[39] These tactics took its toll as United State became desperate to win the war.[40] In order to blow open the forest cover sheltering Viet cong guerrillas and simultaneously deprive Vietnamese peasants who had supported Viet cong of food, a massive use of herbicide war was carried out.[41] The spraying destroyed 14% of Vietnam's forests, diminished agricultural yields, and made soil useless for planting and replanting.[42] As development in science and technological innovations increases globally so is the dangers of provoking crisis as globalisation of arms cartel maintain flow of arms traffic.[43] Cold war which originally began as an ideological warfare later turned into proxy war when places where ideas were expected to spread tactically turned into battle fields. Consequently, war instead of ideas was globalised at the detriment of the environment.[44] It was not all strange, as industrialisation and massive production also thrive while damaging the environment.

However, globalisation has played remarkable roles in some quarters, through a collective effort aimed at solving world environmental problems. The increasing awareness and Multinational Corporations' researches into eco-friendly technologies are worthy of consideration.[45] There has been improvement in the use of resources and awareness to create greener technology. Through globalisation, efforts have been made to assist in improving the use of resources and saving the environment by promoting growth, developing education and improving incomes.[46] Unfortunately, the negative impacts of globalisation on the environment far outweigh the positives. Obviously, present trend of globalisation

---

[39] James Olson and Randy Roberts, *Where the Domino Fell: America and Vietnam 1945-1990*, (New York: St. Martin's Press, 1991) 67
[40] James Olson and Randy Roberts....68
[41] James Olson and Randy Roberts....68
[42] James Olson and Randy Roberts....68
[43] Colin Gray, *Another Bloody Century* .... p. 268
[44] Haslam Jonathan, *Russia's Cold War: From the October Revolution to the Fall of the Wall*....
[45] O. Ryan, "25 Big companies that Are Going Green", available online at www.businesspundit.com/25-big-companies,
[46] Grossman Gene, "Economic Growth and Environment".... 350

instills a kind of pressure on the built environment, which finds its forces mostly observed with the Least Developed Countries (LDCs), mainly situated in the socio-political division known as the Third World.[47] Thus, these forces, aided by the Multinational Companies, tend to become a cancer on the world, influencing governmental laws and other hindrances with financial clout to exploit the natural resources of their host environments.[48]

In other words, not only are the environment and globalisation intrinsically linked, they are also so deeply wielded together that we simply cannot address the global environmental challenges facing us, unless we are able to understand and harness the dynamics of globalisation that influence them.[49] By the same token, those who wish to capitalise on the potentials of globalisation will not be able to do so unless they are able to understand and address the great environmental challenges of our time, which are part of the context within which globalisation is operating.[50] The dominant discourse on globalisation has tended to highlight the promise of economic opportunities. On the one hand, there is a parallel global discourse on the challenges faced by nations seeking ways to develop their respective economies with the so-called magic wand of globalisation.[51]Today, globalisation more than ever before has opened so many new windows of opportunities and hindrances because of unequal access to the benefits.[52]The fact that it is impossible to separate globalisation and environment from each other makes the implications of this interaction universal for humanity. Human environment is essentially borderless; and where physical, psychological or cultural barriers exist, globalization has broken such to a very large extent. Therefore, attempt to resolve or ameliorate local environmental

---

[47] Andre Gunder Frank, *Reorient: Global Economy in the Asian Age*, (Los Angelis: University of California Press, 1998)

[48] Andre Gunder Frank....

[49] Paul Harrison, *The Third World Revolution: Population, Environment and a Sustainable World*, (New York: Penguin, 2003) 441

[50] Paul Harrison....

[51] Larsson Thomas, "The Race to the Top: the Real Story of Globalization", (Washington D.C.: Cato Institute 2001) 9

[52] Soji Oyeranmi, "Globalization as a Source of Environmental Tragedy in Sub-Saharan Africa: The Role Multinational Oil Corporations in Nigeria"....

crises must fundamentally begin with deep understanding of global perspectives and resolutions on these issues and vice versa. After all, humanity is what unites all of us, so, an injury to one, should be an injury to all!

An Overview of Global Environmental Challenges

The world today is confronting myriad of environmental problems which are undermining human existence and they have continued to escalate. This was vividly captured by the current UN Secretary-General, António Guterres in his 2020 'State of the Planet' address. According to him, '...humanity is waging war on nature. This is suicidal. Nature always strikes back -- and it is already doing so with growing force and fury....'As a grave corollary, 'the state of the planet 'has become broken, leading to the most dangerous escalation of the global environmental tragedies. In his words:

> ....We are facing a devastating pandemic, new heights of global heating, new lows of ecological degradation and new setbacks in our work towards global goals for more equitable, inclusive and sustainable development. To put it simply, the state of the planet is broken....Dear friends, Biodiversity is collapsing. One million species are at risk of extinction. Ecosystems are disappearing before our eyes. Desserts are spreading. Wetlands are being lost. Every year, we lose 10 million hectares of forests. Oceans are overfished -- and choking with plastic waste. The carbon dioxide they absorb is acidifying the seas. Coral reefs are bleached and dying. Air and water pollution are killing 9 million people annually – more than six times the current toll of the pandemic. And with people and livestock encroaching further into animal habitats and disrupting wild spaces, we could see more viruses and other disease-causing agents jump from animals to humans. Let's not forget that 75 per cent of new and emerging human infectious diseases are zoonotic. Today, two new authoritative reports from the World Meteorological Organization and the United Nations Environment Programme spell out how close we are to climate catastrophe.2020 is on track to be one of the three warmest years on record globally –

even with the cooling effect of this year's La Nina. The past decade was the hottest in human history. Ocean heat is at record levels. This year, more than 80 per cent of the world's oceans experienced marine heat waves. In the Arctic, 2020 has seen exceptional warmth, with temper....[53]

Several other global reports have demonstrated how rampant and ravaging environmental disasters have become in the 21st Century especially in cities across the World.[54]

With the great magnitude of these problems, a number of scholars and researchers have engaged in studies gearing towards abating the challenges on the environment. For instance, the environment as a concept has been variously described in the works of several scholars. Adeniyi on his own part defines environment as the outer physical and biological systems in which man and other organisms live in a wholly albeit a complicated one with many interacting components.[55] Abrahams also sees environment as the sum total of all external conditions

---

[53] For details see António Guterres, UN Secretary-General's "The State of the Planet" an address at Columbia University, New York,02 December 2020,available online at https://www.un.org/sg/en/content/sg/speeches/2020-12-02/address-columbia-university-the-state-of-the-planet

[54] Prominent among these include: "Death toll climbs above 50,000 after Turkey, Syria earthquakes," https://www.aljazeera.com/news/2023/2/25/death-toll-climbs-above-50000 -after-turkey-syria-earthquakes, February 2nd, 2023; D. Dodman *et-al*, "Cities, Settlements and Key Infrastructure. In: Climate Change 2022: Impacts, Adaptation and Vulnerability," in H.-O. Pörtner (eds. *Contribution of Working Group II to the Sixth Assessment Report of the Intergovernmental Panel on Climate Change*, Cambridge, UK and New York, NY, USA Cambridge University Press, pp. 907–1040, doi:10.1017/ 9781009325844.008. Available online at https://www.ipcc.ch/report/ar6/wg2/ chapter/chapter-6/; World Meteorological Organization, "Weather-related disasters increase over past 50 years, causing more damage but fewer deaths," 31 August 2021, available online at https://public.wmo.int/en/media/press-release/weather-related-disasters-increase-over-past-50-years-causing-more-damage-fewer; According to the WMO Atlas of Mortality and Economic Losses from Weather, Climate and Water Extremes (1970 – 2019), there were more than 11 000 reported disasters attributed to these hazards globally, with just over 2 million deaths and US$ 3.64 trillion in losses; Roger Pielke, "Tracking progress on the economic costs of disasters under the indicators of the sustainable development goals," in *Environmental Hazards* 18(1):1-6,October 2018,DOI: 10.1080/17477891.2018.1540343

[55] E. O. Adeniyi, *Environmental Management and Protection in Nigeria*, (Ibadan: NISER, 1986) 1-10

influencing the growth and development of an organism. These factors could be physical, biological, social and cultural. Meanwhile, during the past few decades, the world has witnessed the beginning of one of the most profound revolutions in the history of human society, with the emergence of the environment as a major public issue. Concern for the environment which was initially perceived as an exclusive concern of the industrialised nations, though had its advantages, brought in its wake unexpected, social cost such as air and water pollution, destruction of material resources and deterioration in the quality of urban life. This was the general perception when the United Nations took the decision to hold a conference on the human environment in Stockholm in June 1972. This initiative moved the issue of environment into the centre of world's political arena and established it as an important item on the agenda of governments throughout the world.[56]

Specifically, environmental disasters have recently become a common occurrence in the world. Throughout the continents, there is no doubt that the world is under a serious threat from the environment.[57] But this study argues that the environment was only responding to the abuses heaped on it by man's activities. Components of the environment – maintains

---

[56] A. D. Nagam and Halle M. Runnalts, "Environment and Globalization Understanding the Linkages", available online at www.google/environmental-2bisuses.band.com

[57] Disasters that are related to environment, weather, climate or water hazard occurred every day on average over the past decades with tragic losses of millions of lives and billions in economic damages/costs across the World. For examples see :the World Meteorological Organization (WMO), "Weather-related disasters increase over past 50 years, causing more damage but fewer deaths" online at https://public.wmo.int/en/media/press-release/weather-related-disasters-increase-over-past-50-years-causing-more-damage-fewer ,August 31,2021; UNEP, "About Disasters and Conflicts" https://www.unep.org/explore-topics/disasters-conflicts/about-disasters-conflicts
Hannah Ritchie, Pablo Rosado and Max Roser "Natural Disasters" available online at https://ourworldindata.org/natural-disasters, 2022; Global Volcanism Program, Volcanoes of the World, v. 4.7.3. Venzke, E (ed.). Smithsonian Institution. https://doi.org/10.5479/si.GVP.VOTW4-2013 2013; National Geophysical Data Center / World Data Service (NGDC/WDS): Significant Earthquake Database, available at: https://www.ngdc.noaa.gov/nndc/struts/form?t=101650&s=1&d=1; Dean Yang "Coping with Disaster: The Impact of Hurricanes on International Financial Flows, 1970-2002," *The B.E. Journal of Economic Analysis & Policy*, 8 (1) 1935-1682, DOI: 10.2202/1935-1682, June 2008.https://www.cfr.org/timeline/ecological-disasters, https://www.thezebra.com/resources/research/natural-disaster-statistics/

tropical forests, oceans, soils, the atmosphere, air, and other natural resources, considered essential to the ecosystem. Instead of the proper utilisation of these free gifts of nature, the earth's resources have continued to deplete, as dirty air, global warming, polluted waters and toxic wastes are just a few of the maladies of the planet earth.[58]

Meanwhile, since contamination of the planet earth is now so pervasive, it's already affecting virtually all aspects of our lives. Large sections of oceans are overfished. A report by the United Nations Environment Program says that seventy percent of marine fisheries are so exploited that reproduction cannot or can just barely keep up.[59] Deforestation has many negative sides to it. Loss of many forests results in a reduction in earth's capacity to absorb carbon dioxide, and this is said to be a cause of global warming. Certain species of plants, the potential source of lifesaving medicines are also disappearing, nevertheless forest destruction continues unabatedly. And if this persists, tropical forests could disappear in about twenty years.[60] Moreover, dumping of toxic waste both on land and in the sea is a serious problem that has the potential for bringing great harm to millions. Radioactive wastes, heavy metals, and by-products of plastics are among elements that cause abnormalities, sickness or deaths in humans and animals. Also when soil is stripped of vegetation, the bare topsoil soon dries out and is blown away by water, this process is called erosion which is another problem on the environment.[61]

Furthermore, it is apparent that the outbreak of several environmental hazards and weather – related disasters occupies the news headlines and captions across the worlds and this illustrates that if care is not taken, our

---

[58] S.I. Oluya and H. Olu-Buraimah, "Environmental Protection", in S. I. Oluya et al, eds., *Compendium of Issues in Citizenship Education in Nigeria*, (Ibadan: Remco Press, 1988) 126

[59] United Nations Environmental Programme (UNEP) *Global Environment Outlook 3: Past, Present and the Future Perspectives*, 2002. Available at: http://www.unep.org/geo/geo3/

[60] *The Environment Magazine*, (Lagos: Environment Communication Limited, Vol. 8, No. 2, February 2008) 54

[61] Gordy Slack, "Africa's Environment in Crisis", *African Journal of Environmental Studies*, Vol. 24, No. 74, (December 2002)

efforts towards development can result into utter destruction of the whole planet. In the last Century, Chernobyl, Bhopal, Valdez, three miles Island are among such names that likely conjure up images of environmental disasters that have occurred in various parts of the world. Parts of Europe sweltered during the summer of 2003, the high temperatures contributed to approximately 30,000 deaths. A pre-monsoon heat wave in Bangladesh, India and Pakistan, drought and record heat in Australia triggered bush-fires that consumed over three million hectares. According to the World Meteorological Organisation, the 2003 Atlantic hurricane season saw the development of 16 named storms, which is well above the 1994 to 1996 average of 9.8, but consistent with a marked increase in the annual number of tropical systems since the mid 1990's.[62] The pattern continued in 2004, which saw devastating hurricanes sweeping into the Caribbean and the Gulf of Mexico with a trial of destruction.[63]

Apart from these, cyclones have caused severe flooding is most part of the world particularly in central and South Asia. In 2004, a record of at least 23 typhoons brewed in Western Pacific, with extensive damage to lives and properties. Flood resulting from heavy monsoon rains, and several powerful earthquakes struck during 2003, and 2005 with Algeria, Iran and China among others grossly affected, robbing the regions of an economic and important tourist centers. Exactly in December 2004, a magnitude 9.0 quakes occurred just off the western coast of Sumatra, Indonesia, spawning by far the deadliest Tsunamis in recorded history affecting the countries bordering on the Indian ocean such as Malaysia, Sri Lanka, Indonesia, Maldives, India, Brunei to as far as the Philippines and Multitudes of Islands. Even the east coast of Africa, 4,500 Kilometers or more west of the epicenter, came within the Tsunamis fatal embrace. The killer waves claimed over 200,000 lives and left many more injured homeless or both.[64]

---

[62] _The Environment Magazine_, (Lagos: Environment Communication Limited, Vol. 8, No. 2, February 2008), 54
[63] _The Environment Magazine_.....54
[64] Several Reports such as World Bank, 1995, 1996, 2004, 2006 and UN 2006, 2007 have confirmed this. See also "Can Planet Earth Be Saved?" _Awake, February_ 8, 2005 3-7

The year 2007 saw a record number of weather – related disaster for which the United Nations office for the coordination of Humanitarian Affairs issued fourteen – emergency appeals – four more than the previous records set in 2005. In Europe, West Africa, North and South America's and Asia, floods hurricanes and cyclones destroyed many homes, farms and lives.[65] The 2008 disasters – cyclones, droughts, flood and Volcanic eruptions as well as earthquakes exacerbated the global food crisis and thus left behind the ruins of lives and properties worth billions of dollars. In many parts of the world, population growth means more urban sprawl, more Shanty towns and more environmental degradations. These developments have tended to magnify the severity of natural disasters, and thus reminded us that our planet is under attack.[66] These attacks became most ferocious in the year 2011.[67]

In terms of environmental tragedies, 2011 was extremely catastrophic for most Nigerian cities due to the "killer floods."[68] The year was particularly pathetic for Ibadan as the city experienced the worst flood disaster in its

---

[65] According to the UN office for coordination of humanitarian Affair, weather related disasters are on the increase around the world. In 2007 for example about 26 countries were affected from developed ones (Britain, US) and underdeveloped, (about 14 countries were said to be affected in West Africa)

[66] "Global Warming, Is planet Earth in Peril?" *Awake*, August 2008 3 – 7

[67] For more on the global implications of 2011 natural disasters see Jim Kavanagh," 2011,Year of billion-dollar disasters", August 20, 2011, available online at http://edition.cnn.com/2011/US/08/20/weather.disasters/index.html, See also Tim Wall, "2011 Natural Disasters Worst Ever", available at http://news.discovery.com/ earth/natural-disasters-in-2011-costliest-ever-110712.html,. The 2011 tsunami easily ranked among the worst of all. A total of 12,431 people were confirmed dead by Japan's National Police Agency, while 15,153 were missing; More than 163,000 people were in shelters around the country following evacuation, the National Police Agency said. The government has set up an evacuation area around Tokyo Electric Power Co's quake-stricken nuclear plant in Fukushima 150 miles north of Tokyo, with a 12-mile radius. More than 70,000 people lived in the largely rural area within the zone. It is unclear how many of them have been evacuated, but most are believed to have left. Another 136,000 people were within a zone extending a further six miles in which residents are recommended to leave or stay indoors; At least 46,027 buildings have been destroyed, washed away or burnt down, the National Police Agency of Japan said; The government has estimated damage from the earthquake and tsunami at 16-25 trillion yen. The top estimate would make it the world's costliest natural disaster

[68] Tell Magazine Special Edition Captioned, "Killer Floods", No 29: July 2011

history.[69] This has not only demonstrated the failure of the past and the current environmental management strategies but also call for an urgent comprehensive and constructive paradigm shift in management of environment in Ibadan. Therefore, this study is set to offer new critical insights and suggest some scientific ways of tackling environmental crisis in Ibadan and other Nigerian cities.

Furthermore, the manifestations of environmental crisis, such as industrial and nuclear pollution, land degradation, desertification, depletion of the protective ozone layer, tropical deforestation and other signs that reveal the incapacity of nature to retain its stability are externalities produced by population growth and economic development. While environmental challenges actually threaten the future of humanity as a whole, the threat appears to differ from place to place. Indeed, this study argues that it is the billions of world's poor (especially in Sub-Saharan Africa) that will bear the greatest loss.[70] This is mainly because the industrialised nations have been taken steps to implement sustainable development strategies and the poor ones (especially those in sub-Saharan Africa) are yet to have effective programmes either to solve the existing environmental crisis or avert the impending ecological cataclysm in their communities.

---

[69] Ibadan recorded 102 lives loss, thousands displaced, billions in properties lost to flood on Friday, 26th August 2011. Indeed, the calamity was described as national disaster by Deputy Senate President, Ike Ekeweremadu given its social and economic tolls on the nation and the strategic place of Ibadan city to the Nigeria's national life. The number and severity of flood related tragedies across the nation in recent times are a clear indication that the nation's environment system is ailing and.... that all is not well with the way Nigeria relates and manages its environment. Most poignantly, Ekweremadu opines that the increasing incidents of flood disasters in the country bring home the harsh realities of the dangers of climate change and environmental degradation and the frequency of this magnitude of floods portends a dangerous future for Nigeria environmentally. For more on the 2011 floods in Ibadan, see Special edition of Tell, No 29: July 2011 captioned "the killer floods" and also Emmanuel Ogala and Funso Ajewole, "Ibadan flood disaster is a bad omen", Daily Times, September, 2, 2011

[70] This was in line with the opinion of Ngozi Okonjo- Iweala, Managing Director, World Bank on why taking climate change is fundamental to development and fighting Poverty. It was extracted from issue 42, 2008 of Developments Magazine of the DFID. 14. For more information you may visit www.worldbank.org or www.development.org.uk

Environmental Challenges in Africa

From the foregoing, it is obvious that environmental crisis is a global phenomenon and is mostly arising from natural or human-made occurrence, which usually degrades the natural environment. A degraded environment is one in which not only the air and water are polluted, but also denies its inhabitants the opportunity for the development and utilisation of other full potentials and subject them to stresses and strains to which they find it difficult or impossible to adjust.[71] The various environmental disasters affect human lives and wellbeing by disrupting the normal flow of resources and activities and exercising considerable effects on socio-economic, political and cultural patterns of the affected areas. Africa is not an exception.

According to the World Meteorological Organisation, in Africa from 1970 to 2019, 1,695 recorded disasters caused the loss of economic damages 731 747 lives and US$ 38.5 billion. Africa accounts for 15% of weather-, climate- and water-related disasters, 35% of associated deaths and 1% of economic losses reported globally. Although disasters associated with floods were the most prevalent (60%), droughts led to the highest number of deaths, accounting for 95% of all lives lost in the region.[72] The majority of deaths were said to occur during the severe droughts in Ethiopia in 1973 and 1983 (total 400 000), Mozambique in 1981 (100 000) and Sudan in 1983 (150 000). Africa is the oldest continent on Earth, having remained in place since the breakup of Pangaea 200 million years ago.[73] The continent remains the second-largest and second-most populous continent in the world. Earlier in 2022, the African Development Bank released its *African Economic Outlook*, which details how the continent is bearing the brunt of climate change despite contributing to just 3% of total global carbon dioxide (CO2) emissions. With a high vulnerability to climate change and poor preparation for adaptation to climatic shocks, it is the least climate-

---

[71] S. I. Oluya and H. Olu-Buraimah, "Environmental Protection", in S. I. Oluya et. al eds., *Compendium of Issues in Citizenship Education in Nigeria*, (Ibadan: Remco Press, 1988) 126

[72] For details visit https://library.wmo.int/index.php?lvl=notice_display&id=21930# .ZEE22dfMKUk

[73] Gordy Slack, "Africa's Environment in Crisis", *African Journal of Environmental Studies*, Vol. 24, No. 74, (December 2002) 1

resilient continent in the world. The publication also presented how some of the most pressing environmental issues in Africa can be addressed by local and national governments.[74]

The array of ecosystems and organisms contained within its forests rivers, deserts, wetlands, mountains and savannah are unequalled in the world. Those ecosystems and the organisms they host are under siege. The human species evolved at the margins of the equatorial forests of Africa, and people have substantially influenced African ecosystems for hundreds of thousands of years, for example, the frantic colonial exploitation of resources that characterised the slave, rubber and ivory trades and the introduction of the exotic crop species that made rain forest agriculture possible (such as bananas, maize, cassava, taro and batata) threw African environmental change into a new gear. The escalating pace of that change now threatens the ecological integrity of the entire continent.

The prevailing poverty, exploding population growth and the lack of enforced environmental regulations combine to produce ecological degradation in Africa. That, along with the cry of great human suffering, dominates Africa's landscapes today.[75]Therefore, the causes and effects of deforestation, desertification, pollution, the impacts of population and demography, the negative activities of the multinational Corporations and Climate Change shall be concisely considered among other factors in this chapter.

Deforestation and the Threat to Bio-Diversity in Africa.

Deforestation is an act of indiscriminate felling or burning of trees in the forest for commercial or domestic uses without proper replacement through replanting.[76] Meanwhile, the majority of the world's animal species are found in tropical forests, together with an immense variety of

---

[74] For details visit https://www.afdb.org/en/documents/african-economic-outlook-2022-highlights
[75] Gordy Slack, "Africa's Environment in Crisis…." 1-5
[76] Bruce Russett et al, *World Politics, The Menu for Choice,* Sixth Edition, (Boston: Bedford/St. Martins, 2000) 445-470

plants. Many people in Africa earn their living from tropical forests. Besides providing a field for scientific research and tourism, tropical forest are commercially important for such products as timber nuts, honey, palm hearts, rubber and resin, but tropical forests are disappearing at alarming rate. There has probably been more African forest destruction in the past 60 years than in the preceding10, 000 years.[77]

While timber harvesting in virgin forests has contributed to the gradual loss of forests, the logging roads, left behind by the timber industry enable other more incursions. The road provides deep forest access to hunters seeking bush meat. They are also avenues of invasions for exotic species, some of which may out to completely wipe out the already beleaguered endangered organisms. Logging towns may act as magnets to other Africans seeking protection from civil wars (as in the Northern Congo) or seeking an income base.[78]

Droughts and civil wars contribute significantly to forest degradation as they drive refugees away from their traditional lands and livelihood far into the forest where they can only earn a living from slash – and – burn agriculture and hunting for bush meat. Increases in urbanisation and industrialisation have also raised demand for wood products, especially firewood's and charcoal.[79] In addition, consumption of forest products nearly doubled in Africa during the period from 1970 to 1998. Africa lost 39 million hectares of tropical forests during the 1980s and another 10 million hectares by 1995.[80]The figures are disputed but one fact is clear, the forests are shrinking fast, and it is posing a great threat to biological diversity. Unless energy alternatives to firewood can be found, as well as alternative sources of income for people whose lives depend on forest consumption, deforestation will continue to accelerate.[81]

Sadly, today, deforestation remains one of the main factors that harm the ecological balance of the entire planet, not just Africa. In 2023, Lei Nguyen

---

[77] Bruce Russett et al.... 445-470
[78] Bruce Russett et al.... 445-470
[79] Gordy Slack, "Africa's Environment in Crisis".....1-5
[80] Gordy Slack, "Africa's Environment in Crisis".....1-5
[81] "Who Will Save the Rain Forests?" *Awake,* June 22, 200. 4

opines that soil erosion, climate change, decreased rainfall, and many other unfavourable circumstances are partly the outcome of the clearance of forest cover for timber and agricultural land. Forested areas all across Africa have been cleared for many reasons, one of them being cocoa, one of the continent's largest cash crops. Four African nations, Côte d'Ivoire, Ghana, Nigeria, and Cameroon, use enormous quantities of land to produce about three-quarters of the world's cocoa.[82]

One of the direst consequences of deforestation is the loss of biodiversity in Africa. As noted by lei Nguyen, African diversity is crucial to the world. The continent is home to 22% of the world's mammalian species, nearly a fifth of avian species, and one-sixth of all plant species. Of the 36 biodiversity hotspots worldwide, eight are found in the continent. They include numerous endemic species and are the richest and most biologically endangered regions globally. Unfortunately, a number of factors, such as: population growth, extensive farming techniques, uncoordinated/mismanaged urbanisation, infrastructural expansion, and illegal trafficking, are causing the continent to see an unprecedented decline in biodiversity.[83]

Desertification

Desertification is the persistent degradation of dry land ecosystems by variations in climate and human activities.[84]Dry lands which harbor about one third of the human population in 2000; occupy nearly half of earth's land area.[85]Across the world, desertification affects the livelihoods of millions of people who rely on the benefits that dry land ecosystem can provide. In dry lands, water scarcity limits the production of crops, forage, wood, and other services ecosystem provide to humans. A report by the International Institute for Environment and Development estimates

---

[82] See Lei Nguyen "5 Biggest Environmental Issues In Africa In 2023," Mar 20th 2023,available online at https://earth.org/environmental-issues-in-africa/

[83] For details, please visit https://www.worldometers.info/world-population/

[84] *The Environment Magazine*, (Lagos: Environment Communication Limited, Vol. 8, No. 2, February 2008) 54

[85] The Environment Magazine …. 54

that desertification threatens more than one-third of Africa's land area.[86]It is especially bad in the south of the Saharan Desert, where the desert is said to be growing in parts by many miles each year. Even in Northern Africa, more than 432 million hectares (57 percent of the total land area) are threatened by desertification. Deforestation and overgrazing are two main contributors to desertification, and increasingly, frequent droughts exacerbate the problem, but the causes and mechanisms of desertification are still hotly debated among scientists. Some evidence suggests that the ebb and flow of the deserts occur in a natural cycle, while others argue that even if there are such cycles, human activity has sped the process up significantly.

Meanwhile, as forests shrink and desert grows, wildlife populations are forced into islands of habitat. Increasingly, thousands of tons of wild animals are hunted every year, both by commercial poachers and by subsistence hunters. But the bush meat trade, hunting for food, has an even greater impact on wildlife populations, many of which are already balanced on the brink of extinction. According to the Convention on International Trade in Endangered Species (CITES) list, hunting of wildlife for food is a major problem for 30 endangered African species including forest antelopes, monkeys, elephants, chimpanzees and gorillas.[87] From all indications, the humid tropical – forest environment is fragile and extremely complex, the myriad life-forms – plants, animals, insects and micro organisms depend on one another, the threat of desertification affects their ecosystem.[88]

An expert consulted during the fieldwork emphasised that with the felling of trees and increasing desert encroachment – the widening of the Saharan Desert – Africa's vulnerability to global warming has become acute.[89] In sum, desertification has environmental impacts that go beyond the areas directly affected. For instance loss of vegetation can increase the formation of large dusts clouds that can cause health problems in more densely populated areas, thousands of kilometers away. More so, the

---

[86] Ibid. 54-55
[87] Gordy Slack "Africa's Environment in Crisis"....1-5
[88] Gordy Slack "Africa's Environment in Crisis"....1-5
[89] Interview with Mr. Kolawole Adewale, a Lecturer at the Polytechnic, on 11/ 09/14

social and political impacts of desertification also reach non-dry-land areas, for example, human migrations from hinterlands to cities from within and outside countries can cause political and economic instability.

Pollution, Population and Demographics

The interdependent relationship among population, economic growth, energy conservation, and pollution pointed out by early Malthusians produces a dilemma that is again coming to the forefront of global attention. One way to control population growth is through economic and human development; yet to maintain high standards of living in the developed world and improve standards in the developing world means higher energy usage and the creation of ever higher levels of pollution.[90] Basically, the environment itself contains man-made elements which are directly poisonous or otherwise harmful to man. These elements undergo irreversible changes which are rapidly and drastically threatening to overwhelm many of the natural systems on which we all depend for livelihood. Surely, this could destroy precious food and water resources and make many of the limited resources of the earth unusable. Classifying pollution can be as difficult and confusing as, classifying lakes or other natural phenomena. Classification according to environment (air, water, soil etc) and pollutant (lead, carbon dioxide, solid wastes etc) are of course widely used approaches. Also from the standpoint of the totality of pollution abatement – from the ecosystem viewpoint – the two types of pollution are the non-degradable pollutants (e.g. aluminum cans, mercurial salts and DDT) and the biodegradable pollutants (such as domestic sewage that can be rapidly decomposed by natural processes).[91]

Today, pollution is ubiquitous; we can drink, breath, smell and see pollution. The environment which is the life supporting system of man is battered daily. The assaults strike many areas in Africa. For instance, gases that our refrigerators and air conditioners use are said to be responsible for depleting the protective ozone layer, and also poverty has

---

[90] Bruce Rusett, et. al eds., *World Politics: The Menu For Choice*, Sixth Edition, (Boston: Bedford/St. Martins, 2000) 465-466

[91] S. I. Oluya, and H. Olu-Buraimah..... 135

made many Africans resort to unconventional and immoral means to feed themselves. Water pollution as a result of dumping of industrial and domestic waste, blocking of access road by improper waste disposal, and deliberate poisoning of rivers, to kill aquatic animals, air pollution from insecticides, refrigerating gases and automobile emission; land pollution from the use of chemicals to destroy insects and bacteria, have all degraded the African environment, thus making the continent's adaptation to environmental changes difficult.[92]

According to a 2019 report, 780,000 premature deaths in Africa each year can be attributed to air pollution.[93] High mortality rates are brought on by the expansion of the oil and gas industry in Nigeria and South Africa, while deaths in West and Central Africa are mainly associated with fire emissions.[94] A UNICEF research also submitted that outdoor air pollution deaths increased by 60% across Africa between 1990 and 2017.[95] Closely link to this is the crisis of water pollution.

Water scarcity is equally fuelling water pollution in many parts of Africa. Access to clean, safe and affordable drinking water is a fundamental right as acknowledged by the UN's Sustainable Development Goals.[96] This is often taken for granted, but as noted Fermin Koop, for hundreds of millions of people, water is a scarce treasure.[97] This is particularly true in Africa, where access to this essential resource is still a big challenge across much part in the 21st century, with over 500 million people in 19 countries

[92] K. A. Owolabi, *Because of our Future: The Imperative for an Environmental Ethic for Africa*, (Ibadan: IFRA, 1996) 31

[93] Susanne E. Bauer *et-al* "Desert Dust, Industrialization, and Agricultural Fires: Health Impacts of Outdoor Air Pollution in Africa," 17 February 2019, https://doi.org/10.1029/2018JD029336 and https://agupubs.onlinelibrary.wiley.com/doi/full/10.1029/2018JD029336

[94] Quoted in Lei Nguyen "5 Biggest Environmental Issues In Africa In 2023," Mar 20th 2023,available online at https://earth.org/environmental-issues-in-africa/

[95] UNICEF "Silent suffocation in Africa: Air pollution is a growing menace: Hitting the poorest children hardest," 2019,available online at https://www.unicef.org/reports/silent-suffocation-in-africa-air-pollution-2019

[96] For details see "What are the Sustainable Development Goals?" available online at https://www.undp.org/sustainable-development-goals

[97] Fermin Koop, "Report: Half a billion people in Africa don't have safe access to water" March 21,2022,https://www.zmescience.com/science/water-insecurity-concern-africa-21032022/

living in areas deemed water insecure, according to a 2022 UN assessment report.[98]

There are innumerable African city dwellers without access to clean, safe water for drinking, cooking, and personal hygiene. Additionally, freshwater sources are contaminated by viruses, germs, parasites, and pollutants, creating a "water scarcity." Tropical diseases including cholera, typhoid fever, dysentery, and diarrhoea ailments can spread because of a lack of water. Trachoma, an eye infection that can cause blindness, the plague, and typhus are some frequent illnesses.[99]

In order to access clean drinking water, families must walk extremely long distances. Carrying large buckets of water back to their houses is a chore that frequently falls on the family's female members and young girls. They often have to leave school early, missing out on the chance to further their education to support their family. These water-collecting expeditions are risky and can potentially harm girls and women physically or sexually.[100] The crisis of acute water scarcity is further complicated with water pollution. African communities face a serious threat from contaminated water. According to the United Nations Department of Economic and Social Affairs (UNDESA), an estimated 115

---

[98] The Report was published on the eve of World Water Day (March 22,2022), it used a set of 10 indicators to quantify water security in Africa's 54 countries. The results are damning. For starters, data on water security in Africa is itself scarce — and where the data does exist, it often points to major problems regarding sanitation water. Overall, only 25 countries on the continent have made progress over the past three to five years, while 25 have made none.... lead author Grace Oluwasanya said in a statement. "Overall levels of water security in Africa are low. Not a single country let alone a subregion have at present achieved a state that can be seen as 'model' or even 'effective' stage of water security." For details visit https://www.zmescience.com/science/water-insecurity-concern-africa-21032022/

[99] Fermin Koop, "Report: Half a billion people in Africa don't have safe access to water" March 21,2022....

[100] Vicky Hallett, "Millions Of Women Take A Long Walk With A 40-Pound Water Can" July 7, 2016, https://www.npr.org/sections/goatsandsoda/2016/07/07/484793736/millions-of-women-take-a-long-walk-with-a-40-pound-water-can

persons die in Africa every hour from illnesses brought on by poor hygiene, inadequate sanitation, and tainted water.[101]

Generally, the twin problems of resource shortages and pollution in cities across Africa are exacerbated by the tremendous population explosion. Global population reached a record high today, as it does every day. World population, 6.2 billion in 2002, is growing by 75 million each year of this, about 97 percent emanates from the global south.[102] On 15 November 2022, the world's population is projected by the United Nations to reach 8 billion people, a milestone in human development.[103] Available records show that, our World has actually surpassed the projected population threshold. As at April 23rd 2023, the global population stands at 8,029,176,980(and still counting per second).[104]

Although Africa is not the most populated region in the world, but the geometric progression in its population has been affected by its poverty and low level of literacy. With the increase in population, there is also an increase in poverty. According to Akinjide Osuntokun:

> Because of limited availability of other modern forms of relaxation and recreation in many African countries, sex which is the only or affordable form of relaxation has increased population growth in Africa... and governments inability to control birth rate coupled with the decline in mortality has led to an outburst in Africa's population.[105]

From the statistics available, Nigeria was about 32 millions in 1956, and presently she is approximately 200 million or more in population, that is,

---

[101] See "What Causes Water Pollution In Africa?", June 28, 2019, https://thelastwell.org/2019/06/what-causes-water-pollution-in-africa/

[102] Paul Harrison, *The Third World Revolution: Population, Environment and a Sustainable World*, (New York: Penguin, 2003) 441

[103] https://www.un.org/en/dayof8billion#:~:text=On%2015%20November%202022%2C %20the,a%20milestone%20in%20human%20development

[104] For per second updates about global population visit https://www.worldo meters.info/world-population/

[105] Prof. Osuntokun A., former Chairman of the Environmental Protection Society of Nigeria and Head of Department, History and International Relations, Redeemers University, Ede, Osun State

less than a hundred year, Nigeria's population has become more than doubled.[106] The cumulative view of respondents on the factor of population explosion on the African environment contended that, the large family size in Africa, and the growth a poverty stricken population has affected its economic productivity, creating more pressure on the limited resources, and increase in human activities such as rural urban migration, vehicular congestion, over cultivation of land, poor waste disposal and increasing slums, with continuous environmental degradation.

Negative Impact of the Multinational Corporations

Multinational Corporations control the largest share of the Third World's economies, with a huge capital based and a body of international networks.[107] These companies hold themselves accountable to no one but their shareholders, and dominating mostly the extracting, mining, production and merchandize sectors of the African economies.[108] Nevertheless, they remain subject to no control or regulation, and their negative impacts is having more devastating effects on both people's lives and the environment due to their profit making orientation and internationalism – which is often an effective guarantee of their immunity.[109]

Oil spillage is a major problem in oil producing areas like Algeria, Angola, Congo, Libya and Nigeria. For instance in Nigeria, oil multinationals such as Agip, Shell, Exxon Mobil, Total and Chevron among others, are guilty of this through accidental and intentional release during production, transportation and storage. The ecological impact of this problem is enormous. Soil and water areas are heavily polluted leading to reduction in soil fertility and aquatic population. Amount of

---

[106] Ibid.

[107] Paul Harrison, *The Third World Revolution: Population, Environment and a Sustainable World*....441

[108] Ibid

[109] Anne Marchand "Multinationals Immunity and the African Environment", *African Journal of Environmental Studies*, Vol. 24, No. 74, (December 2002). 18-21

fresh water (surface and underground) is also reduced. This leads to considerable economic loss, particularly in the Niger Delta region.[110]

The mining companies have also had negative impacts on the lives of their workers and the environment. in the southern part of Africa, mining powerhouse such as Gold Fields and AngloGold have environmental policies which state that the companies will attempt to improve their environmental performance and reduce or control pollution, but this tends to be the opposite. Also in countries such as South Africa, Zimbabwe, Tanzania, Namibia, to Mozambique, local communities experience similar problems varying from water pollution, noise, displacement and destruction of agricultural production and vegetation in surrounding areas. In sum gold mining companies have a reputation for exploiting resources with little benefits for workers and the surrounding communities.[111]

Furthermore, Multinational Corporations who are involved in merchandise have not really contributed to environmental degradation, but their counterparts in the production or manufacturing sector of the economy have tremendously polluted the air and water systems in Africa by releasing untreated effluents into environment. The Lagos State Environmental Protection Agency (LASEPA) reported that due to the flexibility of environmental laws in Africa, multinationals such as Coca-Cola, Guinness, Heineken and the likes, have failed to replicate what obtains in their home countries in Nigeria, thereby causing great harm to our environment.[112] In the same vein, there are many examples of how multinationals have negatively impacted on the peoples and environment in Africa. The persistence dumping of toxic waste on the coast of Africa by these companies for example the Koko toxic waste of 1988 in Nigeria, and other secret shipment of toxic waste to Africa under corrupt deals,

---

[110] E. Omoregie and C. O. Onwuliri, "Interactions between Crude Oil and Biological Productivity in Aquatic Environment: A Case Study of the Niger Delta", in A. Osuntokun ed., *Democracy and Sustainable Development in Nigeria*, (Lagos: Frankad, 2002). 286-288

[111] Anne Marchand "Multinationals Immunity and the African Environment".... 18 – 21

[112] K. A. Owolabi, *Because of our Future: The Imperative For An Environmental Ethic for Africa*.....12-13

have exposed the inhabitants to poisons and other health related diseases. In sum, despite their little positive impacts, multinationals have contributed to keeping the North "clean" and the south "dirty" by disposing hazardous wastes without caution and exploiting our resources without plans for renewal.[113]

With the aforementioned facts, it has become crystal clear that the oil industries have multidimensional significant adverse environmental impacts on the environment of their host communities. Their activities not only exacerbate other environmental problems but also create other serious social crises due to their corporate corruption, carelessness, and brazen disrespect for operational standards in Third world countries contrary to what are obtainable in their home countries. Asking him why in the first place, the activities of the MNCs in Nigeria are bringing so much environmental problems, an environmentalist gave the following reasons:

> Firstly, dirtying industries are choosing the countries that do not have severe legal arrangements. Secondly, the public opinion in developing countries like Nigeria is unconscious of the harms that economical activities gives to the environment … This information gives the MNCs the assurance that they don't face with the opposing activities of the public.[114]

The environmental problems of Nigeria are both significant and consequential. These problems which include air pollution, water pollution, noise pollution, land degradation, oil spillage, deforestation, desertification, erosion, flooding etc. have continued to multiply due to the carelessness and nonchalant attitudes of the foreign oil firms and inability of governments to promulgate effective environmental policies. Consequently, petroleum exploitation activities in the oil producing areas of the country have had unpleasant environmental effects. Streams, creeks and other water bodies in the area have become highly polluted to the

---

[113] Paul Harrison…433
[114] Interview with Mr. Peter Ajayi, an Environmental Officer at the Oyo State Ministry of Environment on 07/09/14

extent of making them really very harmful to both terrestrial and marine lives.[115] They have caused the deaths of plants, animals, fishes and crabs.

In 1970, the quantity of crude oil production recorded was 395,689,000 barrels. Then there was just one reported case of oil spills which accounted for a loss of 150 barrels of crude oil.[116] However, the 1979 and 1980 periods witnessed crude oil productions totaling 845,463,000 and 760,117,000 barrels respectively. The quantities of oil spills during these two periods were respectively 630,405 and 558,053 barrels. These were the largest spilled during the 14 year period.[117] One source calculated that the total amount of oil in barrels spilled between 1960 and 1997 is upwards of 100 million barrels (16,000,000 m³).[118]

Oil spillage has a major impact on the ecosystem into which it is released. Immense tracts of the mangrove forests, which are especially susceptible to oil (mainly because it is stored in the soil and released annually during inundations), have been destroyed.[119]Several species of fishes, crabs and other marine lives were decimated. Moreover, large areas of agricultural land were covered and rendered useless by crude oil. An estimated 5 to 10 percent of Nigeria's mangrove ecosystems have been wiped out either by urban settlement or oil. The rainforest which previously occupied some 7, 400 km² of land has disappeared as well. Spills in populated areas often spread out over a wide area, destroying crops and aquacultures through contamination of the groundwater and soils.[120]The consumption of dissolved oxygen by bacteria seeding on the spilled hydrocarbons also contributes to the death of fish. In agricultural communities, often a year's supply of food can be destroyed instantaneously. Because of the careless nature of oil operations in the Delta, the environment is growing increasingly uninhabitable. The Niger River is an important ecosystem that needs to be protected, for it is home to 36 families and nearly 250

---

[115] A. C. Emeribe, *Policy and Contending Issues in Nigeria, National Developmental Strategy*, (Enugu: John Jacob Classic Publishers Ltd) 223

[116] A. C. Emeribe … 223

[117] A. C. Emeribe … 223

[118] Victor Ukaogo, Environmental Security and the Role of Foreign Interest in the Niger Delta. Available at http://www.google.com/Nigeriapetroleumindustry

[119] Emeribe, *Policy and Contending Issues in Nigeria*….223

[120] *Africa Renewal* (formerly *Africa Recovery*), Vol. 21, No. 2, (July 2007) 14-16

species of fish, of which 20 are endemic, meaning they are found nowhere else on Earth.[121] The Niger Delta has the third largest mangrove forest in the world, and the largest in Africa. Mangrove forests are important for sustaining local communities because of the ecological functions they perform and the many essential resources they provide including soil stability, medicines, healthy fisheries, wood for fuel and shelter, tannins and dyes, and critical wildlife habitats.[122] Oil spills are contaminating, degrading, and destroying mangrove forests.

Gas flaring is another major problem associated with oil exploitation activities in Nigeria. Nigeria flares more natural gas associated with oil extraction than any other country on the planet, with estimates suggesting that of the 3.5 billion cubic feet (100,000,000m³) of associated gas (AG) produced annually, 2.5 billion cubic feet (70,000,000m³), or about 70% is wasted via flaring.[123] This equals about 25 percent of the United Kingdom's total natural gas consumption, and in the equivalent to 40% of the entire African continent's gas consumption in 2001.[124] The compositions of the flared away gas included such toxic pollutants as carbon monoxide (CO), Sulphur dioxide (SO_2) and oxides of Nitrogen (NO_x). Apart from the injurious effects of these pollutants to health, they also add appreciably to the problem of global warming and the consequent green house effect. Moreover, the excessive heating of the soil and vegetation around the flare sites in the oil and gas field results in soil degradation. Another problem gas flaring poses is the release of large amounts of methane, which has very high global warming potential. The methane is accompanied by the other major greenhouse gas, carbon dioxide, of which Nigeria was estimated to have emitted more than 3,438 metric tons of in 2002, accounting for about 50 percent of all industrial emissions in the country and 36 percent of the total CO_2 emissions.[125]

---

[121] *Africa Renewal*....14-16
[122] Soji Oyeranmi, "Globalization as a Source of Environmental Tragedy in Sub-Saharan Africa: The Role Multinational Oil Corporations in Nigeria...." 47
[123] Jaad Mouwad, "Oil Companies in the Niger Delta, Growing List of Dangers", *International Herald Tribune*, (9 June 2007)
[124] Emeribe, *Policy and Contending Issues in Nigeria*..... 223
[125] A. C. Emeribe 223

Acid rain, a direct result of gas flaring, is taking its toll on the Niger Delta. Acid rain not only deprives people of drinkable rain water and stunts crops growth (as in Eket and other communities in Akwa Ibom State), it is also affecting people's homes.[126] In Iko, Eket, and Etagberi it was reported that Zinc roofs which formerly lasted 7-10 years (and were good alternative to labour – intensive thatched roofing), are now destroyed within one or two years by acid rain.

Gas flares can have potentially harmful effects on the health and livelihood of the communities in their vicinity, as they release a variety of poisonous chemicals. Just some of combustion by-products include nitrogen dioxides, sulphur dioxide, Volatile organic compounds like benzene, toluene, xylene and hydrogen sulfide, as well as carcinogens like benzaphyrene and dioxin. Humans exposed to such substances can suffer from variety of respiratory problems which have reported amongst many children in the delta but have apparently gone uninvestigated. These chemicals can aggregate asthma, cause breathing difficulties and pain as well as chronic bronchitis.[127] Of particular note is that the chemical benzene, which is known to be emitted from gas flares in undocumented quantities, is well researched as being a causative agent for leukemia and other blood related diseases.[128]A study done by Climate Justice estimates that exposure to benzene would result in 8 new cases of cancer yearly in Bayelsa State alone.[129]

Another problem facing the people of the Niger delta is the illicit use of land by oil companies. In the community of Umuebulu, River State, hardly 50 meters away from its perimeter, there is an unlimited chemical waste out belonging to shell.[130] The company reportedly acquired this land under the pretense of building a "life camp" – shell's lingo for an

---

[126] E. O. Adeniyi, *Environmental Management and Protection in Nigeria*, (Ibadan: NISER, 1986) 1-10

[127] E. O. Adeniyi 1-10

[128] See www.en.wikipedia.org/.../Environmental_issues_in_the_Niger_Delta, and news.surfwax.com/health/archives/benzene.html

[129] See www.en.wikipedia.org/

[130] "Oil For Nothing: Multinational Corporations, Environmental Destruction, Death and Impunity in the Niger Delta", *Essential Action*, 2000, available online at www.essentialaction.org

employee housing complex.[131] The construction of infrastructure for oil facilities is done with little or no regard for environmental consideration. To facilitate road construction, waterways are frequently diverted to the detriment of fish population. Sudden and drastic changes to the local environment by oil companies are sometimes accompanied by direct loss of human life. For example, the Egi community has reportedly lost five children in the last few years who during the rainy season drowned in "burrow pits" dig by Elf to extract sand and gravel for road construction.[132]

The Niger Delta region could experience a loss of 40 percent of its inhabitable terrain in the next thirty years as a result of extensive dam construction in the region. The carelessness of the oil industry has also precipitated this situation, which can perhaps be best encapsulated by a 1983 report issued by the NNPC, long before popular unrest surfaced:

> We witnessed the slow poisoning of the waters of this country and the destruction of vegetation and agricultural land by oil operations. But since the inception of the oil industry in Nigeria, more than 25 years ago, there has been no concerned and effective effort on the part of the government let alone the oil operators, to control environmental problems associated with the industry.[133]

According to an environmentalist from Iko village in Cross River state, nobody really cares to go fishing again because the fishes smell of petroleum when eaten.[134] It is based on these dastardly effects of the oil industry in the Niger delta that oil has been described paradoxically as Frankenstein monster that is slowly killing the areas that produce it. The effects of these spillages and gas flaring often persist and last for over two decades. In reality, the full impact of the degradation of the environment is usually felt immediately, but the nuisance caused continues for a long period, often lasting for over two decades. Moreover, the pollution of the

---

[131] Oil For Nothing: Multinational Corporations, Environmental Destruction
[132] Environmental Issues the Niger Delta in Greenpeace Internationals' Shell Shocked, 11
[133] Environmental Issues the Niger Delta in Greenpeace Internationals
[134] I. Anderson, *Niger Basin: A Vision for Sustainable Development*, (The World Bank, 2005) 1-131

environment renders a large chunk of the rural dwellers of the Niger delta redundant and unemployed.[135] While fishermen are displaced from rivers and fishing ponds; farmers' farmlands are also destroyed by the activities of the crude oil explorers.

Nigeria is known to be a top oil producer in Africa's, where over 300 million litres of crude oil are produced daily and are thought to be the source of 70% of the country's earnings.[136] Since petroleum was discovered in Nigeria more than 65 years ago, uncontrolled spills have been a common occurrence in the oil industry, the country's main source of GDP. Several reports attested to this extremely destructive trend. For examples: a 2013 publication noted that an estimated 240,000 barrels of crude oil are spilled in the Niger Delta each year, poisoning agriculture, waterways, and the atmosphere with hazardous chemicals.[137] In the same token a 2018 report also lamented that between 1976 and 2014, the oil-rich region experienced more than 12,000 oil spill occurrences. More than half of them were to be as a result of tanker accidents and pipeline degradation.[138]

Oil companies have also been accused of failing to uphold their legal obligations to clean up spills within the specified period.[139] At some point, Amnesty International charged Shell and Eni – the two biggest businesses in the Niger Delta – of negligence in how they handled local oil spills. According to the campaign group, the environmental disaster in the Niger Delta had gotten worse because of the companies' "irresponsible

---

[135] I. Anderson, *Niger Basin*

[136] Oghenekevwe Uchechukwu,"OIL THEFT: Nigeria now Africa's fourth largest oil producer," September,2022, https://thenewsguru.com/news/oil-theft-nigeria-now-africas-fourth-largest-oil-producer/

[137] B. Ordinioha and S. Brisibe, "The human health implications of crude oil spills in the Niger delta, Nigeria: An interpretation of published studies," Niger Med J. 2013 Jan;54(1):10-6. doi: 10.4103/0300-1652.108887. PMID: 23661893; PMCID: PMC3644738. https://www.ncbi.nlm.nih.gov/pmc/articles/PMC3644738/

[138] E. Chinedu and C.K. Chukwuemeka, "Oil Spillage and Heavy Metals Toxicity Risk in the Niger Delta, Nigeria," J Health Pollut. 2018 Aug 21;8(19):180905. doi: 10.5696/2156-9614-8.19.180905. PMID: 30524864; PMCID: PMC6257162. https://www.ncbi.nlm.nih.gov/pmc/articles/PMC6257162/pdf/i2156-9614-8-19-180905.pdf

[139] Chris Stein, "Shell Accused of Failing to Clean Up Nigeria Oil Spills," November 03, 2015 https://www.voanews.com/a/amnesty-says-shell-did-not-clean-up-spills-in-niger-delta/3034447.html

response" to oil spills.[140] Beyond the unending accusations, these Multinational Oil Corporations (MNOCs) have demonstrated their absolute disdain for environmental justice in the Niger Delta by their failure to abide by pronouncements/judgements of courts or international Organisation of competent jurisdictions regarding environmental restoration via clean-up oil spillages.

For instance, in 2011, the United Nations Environmental Programme published a report where the cleanup of Ogoniland in the Niger Delta was recommended with Shell as the chief protagonist.[141] Sadly but not unexpected, in 2023, it has been discovered that the oil company is still foot dragging on the fulfilment of this extremely important obligation. Indeed, several recent reports quoting UNEP, have demonstrated that 'mismanagement, waste and lack of transparency are making the cleanup in the Niger Delta's Ogoniland anything but exemplary,' indicate.

....when the $1 billion Ogoniland cleanup began in 2019, backed by Shell's funding pledge and support from the United Nations, it was heralded as the most ambitious initiative of its kind anywhere in the world. But now, UN Environmental Programme documents seen by Bloomberg and reported for the first time indicate that the project — far from being exemplary — is making one of the earth's most polluted regions even dirtier.[142]

---

[140] Amnesty International, "Nigeria: Amnesty activists uncover serious negligence by oil giants Shell and Eni," March 16,2018, https://www.amnesty.org/en/latest/news/2018/03/nigeria-amnesty-activists-uncover-serious-negligence-by-oil-giants-shell-and-eni/

[141] United Nations Environmental Programme (UNEP), *Environmental Assessment of Ogoniland*, 2011, Web: http://www.unep.org

[142] For details ,visit : Neil Munshi and William Clowes, "One of World's Most Polluted Spots Gets Worse as $1 Billion Cleanup Drags On," August 31, 2022 https://www.bloomberg.com/news/features/2022-08-31/shell-s-1b-oil-cleanup-left-one-of-world-s-most-polluted-spots-dirtier-for-now#xj4y7vzkg;Chris Stein, "Shell Accused of Failing to Clean Up Nigeria Oil spills, "November 03, 2015 https://www.voanews.com/a/amnesty-says-shell-did-not-clean-up-spills-in-niger-delta/3034447.html;Amnesty International, "Nigeria: Amnesty activists uncover serious negligence by oil giants Shell and Eni," March 16,2018, https://www.amnesty.org/en/latest/news/2018/03/nigeria-amnesty-activists-uncover-serious-negligence-by-oil-giants-shell-and-eni/ and Philip Andrew Churm ,"Shell's bid to clean-up polluted

The result of the identified crises is the evolution of what Blackwell in 1987 called 'Parasitic City' in most parts of sub-Saharan Africa. Rather than providing the basis for sustained economic growth, cities have become serious impediments to sustainable development in sub Saharan Africa. Some of the problems include: unsustainable city development plans, inadequate urban environmental strategies poor urban governance, wide spread extreme poverty, ubiquitous waste, pollution, traffic jam, slum conditions, urban decay. Of all the above mentioned environmental problems none is as devastating as the global warming induced Climate Change which is regarded as the most catastrophic environmental crisis confronting the 21st Century humanity.[143]

Climate Change

The dangers posed by climate change are nearly as dire as those posed by nuclear weapons.[144]This seems like an alarmist projection or speculation but with the damming evidence around us, all well meaning people around the World will be worried. Cyclone Nargis in Burma, April 2008 (over 200,000 deaths and 1.5 million displaced); Earthquake in China, May 12, 2008 (50,000 deaths, almost 30,000 missing and nearly 300,000 injured), Tsunami in 2004 claimed almost a million lives in more than five nations; floods and drought had continued to ravage many African countries (with attendant loss of lives and properties); the global food crisis and the 2011 Japan tsunami.[145] Many earlier cited reports have attributed this

---

Ogoniland labeled "incompetent"'"03/09/2022, https://www.africanews.com/2022/09/03/shells-bid-to-clean-up-polluted-ogoniland-labelled-incompetent//

[143] Soji Oyeranmi, "How Fourth Industrial Revolution Can Aid Sustainable Development in African Cities: Lessons from China," available online at https://ascir.org/2023/01/21/how-fourth-industrial-revolution-can-aid-sustainable-development-in-african-cities-lessons-from-china/ , January 21, 2023

[144] Quoted in Soji Oyeranmi, "Climate Change and Sustainable Development in Nigeria", in Toyin Falola and Maurice Amutabi eds., *Perspectives on African Environment, Science and Technology*, (New Jersey: Africa World Press, 2012). 309

[145] According to the UN office for coordination of humanitarian Affair, weather related disasters are on the increase around the world. In 2007 for example about 26 countries were affected from developed ones (Britain, US) and underdeveloped, (about 14 countries were said to be affected in West Africa. For the details on the impact of 2011 Japan Tsunami see Tim Wall,"2011 natural disasters worst ever" available at http://news.discovery.com/earth/natural-disasters-in-2011-costliest-ever-110712.html,.

seemingly irreversible human tragedy to climate change occasioned by the catastrophic impacts of human activities on nature and the environment. And evidently the worst is yet to happen. Therefore, humanity cannot afford to waste any further time in saving "our collective future."[146]

While climate change actually threatens all our futures,[147] the threat appears to differ from place to place. In reality, it is the billions of world's poor (especially in Sub-Saharan Africa) that have been bearing the greatest loss. Several reports have attested to the extreme vulnerability of Africa to the vicissitudes of climate change. For examples, IPCC 2007, 2011 and 2012 reports have shown that climate change is real. Indeed in its 2022 edition, the body of Climate Experts lamented that Africa is the most vulnerable continent to climate variability and change, a situation that is aggravated by the interaction of 'multiple stresses', including high dependence on rain-fed agriculture, widespread poverty and weak adaptive capacity.[148] In their own views Olasunkanmi Habeeb Okunola

---

According to this report, the 2011 tsunami easily ranked the worst of all. A total of 12,431 people were confirmed dead by Japan's National Police Agency, while 15,153 went missing; More than 163,000 people were in shelters around the country following evacuation, the National Police Agency said. The government has set up an evacuation area around Tokyo Electric Power Co's quake-stricken nuclear plant in Fukushima 150 miles north of Tokyo, with a 12-mile radius. More than 70,000 people lived in the largely rural area within the zone. It is unclear how many of them have been evacuated, but most are believed to have left. Another 136,000 people were within a zone extending a further six miles in which residents are recommended to leave or stay indoors; At least 46,027 buildings have been destroyed, washed away or burnt down, the National Police Agency of Japan said; The government has estimated damage from the earthquake and tsunami at 16-25 trillion yen. The top estimate would make it the world's costliest natural disaster

[146] Thijs Dela Court, *Beyond Brundtland, Green Development in the 1990s*, (London: Zed Books Ltd.,1988). 7-11; see also see António Guterres, UN Secretary-General's "The State of the Planet" an address at Columbia University, New York,02 December 2020,available online at https://www.un.org/sg/en/content/sg/speeches/2020-12-02/address-columbia-university-the-state-of-the-planet

[147] This was the opinion of Ngozi Okonjo- Iweala, Former Managing Director, World Bank on why taking climate change is fundamental to development and fighting poverty. It was extracted from issue 42, 2008 of Develop*ments Magazine* of the DFID. P.14. For more information you may visit www.worldbank.org or www.development.org.uk

[148] For details of the 2022 IPCC Report please visit https://www.ipcc.ch/report/ar6/wg2/chapter/chapter-9/

and Prof. Mulala Danny Simatele argued that "the impact of climate change is growing on the African continent and It's hitting the most vulnerable hardest and contributing to food security , stress on water resources."[149]

Earlier in 2022, the African Development Bank also released its African Economic Outlook 2022, which details how the continent is bearing the brunt of climate change despite contributing to just 3% of total global carbon dioxide (CO2) emissions. With a high vulnerability to climate change and poor preparation for adaptation to climatic shocks, it is the least climate-resilient continent in the world. Here are some of the most pressing environmental issues in Africa and how local and national governments are addressing them.[150]

Even if there are differences on their views and recommendations, one common denominator of all the reports is that Climate change will lead to changes in extreme weather and climate events such as drought, floods, sea level rise, storm surges, among many other hazards. Future risks associated with demographic trends, environment degradation, among others further increase climate change related risks. Africa is already experiencing these: In 2022, we saw devastating floods in Nigeria, South Africa, Benin, Madagascar and Central African Republic. The floods damaged livelihoods, critical infrastructure, education and economies.[151]

Nigeria's population is said to be on the increase and by 2030, more than 60% of the country's population will be living in cities. As the country's population continues to grow, so does its environmental crisis across multiple ecological zones.[152]Expectedly, cities across the country have

---

[149] Olasunkanmi Habeeb Okunola and Prof. Mulala Danny Simatele "Climate change in urban Nigeria - 4 factors that affect how residents adapt" February 26, 2023, https://theconversation.com/climate-change-in-urban-nigeria-4-factors-that-affect-how-residents-adapt-198802

[150] https://www.afdb.org/en/documents/african-economic-outlook-2022-highlights

[151] Olasunkanmi Habeeb Okunola and Prof. Mulala Danny Simatele "Climate change in urban Nigeria - 4 factors that affect how residents adapt" February 26, 2023, https://theconversation.com/climate-change-in-urban-nigeria-4-factors-that-affect-how-residents-adapt-198802

[152] Nigeria: Urbanization from 2011 to 2021,https://www.statista.com/statistics/455904/urbanization-in-nigeria/

been witnessing the manifestation of this prediction, partly due absence of sustainable urban management strategies and majorly because of the resurgence of climate change.

In the Southern Nigeria, flash and seasonal flooding, erosion, landslide etc which are been aggravated by climate change are ruining cities there. 2022 presented the most recent examples of the crises, in that year, according to the National Emergency Agency (NEMA),flooding displaced more than half a million people. In another report, between September and October 2022, floods caused the deaths of over 600 people and displaced 1.3 million from their homes across Nigeria in 2022.[153] On July 1, 2022, landslide also hit indigenes of Ogwuma Edda in Afikpo South Local Government, Ebonyi State. It led to a mass of earth collapsed in the community, burying some houses and destroying 15 others. It was also reported that 38 families were rendered homeless as a result of the disaster. In a place like Lagos, Nigeria's most populous city and economic nerve centre, frequent rising sea levels has almost made flooding an annual ritual in the Lekki, Ikoyi, Epe, and some parts of the Mainland areas in the state. Its impacts on the city are usually far-reaching, as business owners and residents count their losses (in human and economic senses).[154]

In the Northern part of the country, the climate crisis is also unleashing over-grazing, increasing desertification, deforestation and drought. Most times the people also experience constant wind erosion that has been sweeping away houses and farms especially in many of the towns along the Sahelian zone. Floods also destroyed about 200,000 houses, while 266,000 acres of farmland were completely or partially washed away. Though there are other extreme weather events that impacted the urban environments, the flood disasters overwhelmed the others with the negative effects spread to the nook and crannies of Nigeria.

---

[153] Terkula Igidi, "Major events that shaped environment sector in 2022," 29 Dec 2022 https://dailytrust.com/major-events-that-shaped-environment-sector-in-2022/

[154] Abiola Durodola et-al "Nigeria's cities are at severe risk from climate change. Time to build resilience, and fast," November 10, 2022,https://climatechampions.unfccc .int/nigerias-cities-are-at-severe-risk-from-climate-change-time-to-build-resilience-and-fast/

Climatic emergencies are increasing in the Third World countries like Nigeria due to the fact that the industrialised nations have been implementing sustainable adaptation and mitigation strategies and the poor ones are yet to have specific program to mitigate the climatic cataclysm. Due to this terrific fact, the international community need to fashion more adequate ways of responding to the complex global challenge of climate change (especially in saving the poorest arms of humanity from this climatic doom).With the current speed of unstable climatic condition, then, poor countries in Africa (Nigeria inclusive) are suffering more due to: geographical locations, fragile economies, low income, heavy reliance on climate sensitive sectors such as extractive industries and agriculture and lack of adequate modern conservation skills and facilities. And this definitely are already having catastrophic consequences on their already battered economies.

Obviously, climate change has been reversing any progress made by these countries towards achieving the Millennium Development Goals, especially the goals of eradicating extreme poverty, combating communicable diseases and securing environmental sustainability.[155] As noted in several international conferences, including Conferences Of the Parties (COPs) especially the latest COP27 (which took place in Egypt from 6-18 November 2022); if the global North fail to support the global South financially and technically; most of them won't be able achieve Sustainable Development Goals (SDGs). Consequently, it will become impossible for these poor countries to fight the resurgent climate change which will then expose the whole world to the risk of being overwhelmed by climatic cataclysms.[156]

---

[155] Developments Magazine of the DFID... P.14

[156] For examples see : United Nations Climate Change, "Five Key Takeaways from COP27,"November, 2022 https://unfccc.int/process-and-meetings/conferences/sharm-el-sheikh-climate-change-conference-november-2022/five-key-takeaways-from-cop27; United Nations Climate Change , November 2022,"Mobilizing more financial support for developing countries," https://unfccc.int/process-and-meetings/conferences/sharm-el-sheikh-climate-change-conference-november-2022/five-key-takeaways-from-cop27/mobilizing-more-financial-support-for-developing-countries; Sustainable Development Goals Center for Africa, "Africa 2030: Sustainable Development Goals Three-Year Reality Check,", June 2019, https://sdgcafrica.org/wp-content/uploads/2019/06/AFRICA-2030-SDGs-THREE-YEAR-REALITY-CHECK-REPORT.pdf ;Belay Begashaw,

The security threat resulting from climate change is a global phenomenon, and of course the greatest environmental problem in the world.[157] The overwhelming scientific evidence attributed climate change to anthropogenic (human) activities. Increase in the carbon dioxide content of the atmosphere resulting from the burning of fossil fuels like coal, oil and natural gas; create the greenhouse effect which traps heat into the atmosphere. The result is global warming, a gradual increase in world temperature. Estimates vary, but the UN's Intergovernmental Panel on Climate Change (IPCC) projects that global temperatures will rise by 1 to 3.5 degrees Celsius over the next 100 years. Climate will change in erratic and unpredictable ways: forests will move pole ward, arid zones in the middle of continents will grow and make grain cultivation uneconomical; the ice caps will melt and the sea level will rise, drowning low-lying coastal areas. The increased frequency and duration of heat waves will claim more lives, as will malaria, dengue and other diseases carried by mosquitoes and rodents that will proliferate along with warmer and damper climates. Harmful effects will be different at different latitudes and in different regions.[158]

Meanwhile, as noted by the then IPCC chairman Rajendra Pachauri, countries can reduce harmful emissions and halt warming, by investing in cleaner "green" technologies and changing consumer habits.[159]The IPCC also lamented that Africa is faced with increasing aridity and this is complicated by of its poverty, which has also drastically lessens the continent's capacity to adapt to climate change.[160] Although sub Saharan Africa produces less than 4 percent of the world's greenhouse gases, scientists have predicted with "very high confidence" that the region's

---

"Africa and the Sustainable Development Goals: A long way to go," Monday, July 29, 2019, https://www.brookings.edu/blog/africa-in-focus/2019/07/29/africa-and-the-sustainable-development-goals-a-long-way-to-go/ and SDGs Implementation in Africa, https://www.youtube.com/watch?v=96Obg0Quk2Y

[157] See the *Bulletin of the Atomic Scientists* as quoted in August 2008 *Awake* Magazine of the Jehovah witness titled, "Global Warming is Planet Earth in Peril?" 3. and IPCC, Climate *Change,*(Cambridge: Cambridge University Press, 1996)

[158] Bruce Russett et-al, eds. ….. 466 – 467

[159] *Climate Change 2007:The Physical Science Basis*, Contribution of Working Group 1 to the fourth Assessment Report of the Intergovernmental Panel on Climate Change(IPCC), (Cambridge: Cambridge University Press, 2007).

[160] Climate Change 2007: The Physical Science Basis

diverse climates and ecological systems have already been altered by global warming and will undergo further damage in the years ahead. Among the most worrying effect of global warming is the impact on water supply. Africa is fortunate to have large reserves of untapped water, and some dry areas are likely to benefit from increased rain, but the Sahel and other arid and semi arid regions are expected to become even drier. A third of Africa's people already live in drought – prone regions and climate changes could put the lives and livelihood of an additional 75-250 million people at risk by the end of next decade. Flood prone areas in southern Africa on the other hand, are likely to become wetter as rainfall patterns shift, causing floods to become more frequent and severe diverting resources from development to relief.[161]

Moreover, Africa's agricultural sector is already hampered by its reliance on rain-fed irrigation, poor soils and antiquated technology and farming methods. It is likely to be hit hard as droughts and flooding worsen, temperatures and growing season's changes, and farmers and herders are forced off their lands. This could spell humanitarian and economic disasters on a continent where farming accounts for 70 percent of employment and is often the engine of national economies. The fishing industry is also likely to suffer as lakes and rivers dry up and rising water temperatures destroy commercial species of fish – a situation currently affecting the Lake Chad. Coastlines and islands around the world are threatened by rising ocean levels as warming temperatures melts the earth's polar ice-caps. But East Africa's low-lying islands and coastal regions are at particular risks of frequent flooding or becoming submerged. Higher water temperatures are expected to increase the power and frequency of hurricanes and other violent ocean storms. Africa's coastal fisheries and the fragile ecosystems that support them could also be damaged if higher sea levels push salt water inland and destroy freshwater estuaries and coastal farmland.[162]

Other effects include the spread of parasitic diseases such as malaria. Malaria is the single greatest killer of African children and imposes a $12

[161] Africa Renewal (Formerly Africa Recovery), United Nations Department of Public Information, Vol. 21, No. 2, (July 2007) 14 – 16
[162] *Africa Renewal*.....14-16

billion annual drain on African economies through death, medical cost and lost productivity. Nor will Africa's diverse plants and animals be spared. A study of nearly 5,200 plant species throughout the continent found that about 5,000 of them would lose much of their natural habitat to climate change. Of these some 2,100 could lose their entire native habitat by 2085. Wildlife will fare no better in South Africa's famous Kruger game reserve, for example, studies show that two-thirds of animal species could disappear.[163]

Earth temperature is rising and the climate is changing, no matter the cause or causes. Though the world is already experiencing some the dire consequences but the worst is yet to happen in most African countries. And if we are going to ameliorate the already worsening situations; and prevent the impending catastrophe-then Nigeria and Africa must act very fast. One of the best ways is to replenish the existing but neglected indigenous conservation techniques and improve the modern ways of conserving the environment and natural resources. Indeed, it is very significant to note that in Africa, various communities have various forms of traditional mechanisms usually adopted to conserve their environment. Popular among these methods include: Shifting cultivation, which very common among the People of Niger delta and other forest zones in Nigeria. Though many reasons have been given for these methods but the main one remains allowing land to have rest and regain its productive powers. Other local methods include: seasonal fishing; erosion control mechanism; the use of folklores and myth.[164]

Over the time, African governments have adopted several methods of conserving the environment especially by making laws to protect the environment. For example in Nigeria: there are Federal Environmental Protection Act of 1988, The National Park Decree of 1991 and many more. Only recently, the Federal government inaugurated anti climate change

---

[163] UNEP's Report on Africa's Vulnerability to Climate Change, UNEP Climate Change Resources available online at http://www.unep.org/themes/climatechange/
[164] Ukoha Ukiwo, "Indigenous and Received knowledge in the conservation of the Niger-Delta Environment", in Kayode Fayemi et. al eds., *Toward an Integrated Development of the Niger Delta*, (Ikeja: Centre for Democracy and Development, 2005). 224

agency-Building Nigeria's Response to Climate Change (BNRCC).[165]At the launching ceremony, the then Minister of Environment and Urban Development, Fatima Alao admitted that "Nigeria is more vulnerable to impacts of Climate change because of its size and population, amongst others."[166] It has led to greater soil erosion and flooding in areas of higher rainfall, heightened drought and desertification in Northern parts, accelerated sea level rise and water salt erosion in coastal areas, she concluded.[167]This clearly demonstrates the fact that no part of the country is save from the impending catastrophe. Therefore, no effort should be spared by individuals, groups and Government at all levels to save Africa's collective future.

In terms of policies for remedying the situation, the key is the diversification of the economy away from weather dependent activities such as mining and oil production earlier mentioned. This will inter alia ensure the global switch away from fossil fuels and the consequent reduction in fossil fuel demand will have little impact on the African economy and foreign exchange earnings. Also research on climate change and the socio-economic implications for Africa is necessary for developing adequate response strategies. Further, study of the science of climate change and its potential impacts on Africa is very important for creating awareness and providing the background information for targeting policies adequately. This point is made in view of the recognition that the major constraint to adequate forecasting and formulation of adaptation policies is the paucity of climate data in Africa. Long term studies on national and regional climate change in Africa should be embarked upon and vigorously pursued. The findings of such studies will be crucial for the formulation of adequate response and adaptation policies such as adequate resettlement programs for those that may be displaced by climate change.[168]

---

[165] Soji Oyeranmi, "Climate Change and Sustainable Development in Nigeria"....307
[166] *Tell Magazine,* August 11, 2008 Special Edition Captioned, "Environment: Nigeria's Nightmare" 20-31
[167] *Tell Magazine,* August 11, 2008
[168] Soji Oyeranmi, "Climate Change and Sustainable Development in Nigeria".... 307

Furthermore, increased government participation in the global climate change deliberation in order to negotiate a better deal for Africa is necessary. The suggestion that oil-producing countries should be compensated for their projected income losses in the event of the implementation of the Kyoto protocol and assisted in their economy diversification attempt should be vigorously argued and canvassed. Africa can only be sure that its interest is protected in the emergent global abatement strategy if it increases its level of participation. Its participatory capacity in turn will be enhanced by findings from studies and research into various ramifications and dimensions of the climate change issue as suggested above. Most times Nigerian government does complain of shortage of fund but there is also the problem of the deepening corruption that has attended the management of ecological funds across the country. But this should not deter more commitment from international community in resolving fundamental ecological problems in Africa. In fact, they should take clue from the Food and Agriculture Organisation (FAO) has recognised since 2000 that the key to arresting deforestation and had been involving in the efforts to implement sustainable forest development.[169]

The most important step to be taken is to raise environmental consciousness of Africans. Every citizen, no matter the status must know that the earth is fast losing its ability to nurture the natural resources we all dependent on. This is no doubt jeopardizing the national economy, so, the people should get more serious about saving the environment. This is just because it is the right thing to do but because nature does not need us to survive but our own existence depends on nature. Therefore by saving the environment, we are only saving ourselves. Environmentalism should therefore move from being a philosophy only championed and promoted by a passionate minority to a way that governs mainstream behavior of the people and public policy. Collaboration among individuals, NGOs, other groups and government agencies is the key achieve this. This is why groups like the Nigerian Environmental Study/Action Team (NEST) must

---

[169] Africa Renewal (Formerly Africa Recovery) United Nations Department of Public Information

be replicated to take care of every nook and crannies of the African continent.

As Prof. David Okali, chairman of NEST once suggested in combating the menace of climate change successfully; Africa's poverty reduction strategies must be made sustainable. To attain this level, African countries must strive to meet international climatic standard and adapt to climate change through "improved governance, promoting gender equality and more sustainable natural resource management."[170] All these become necessary because any further efforts at reducing poverty by African governments without paying due attention to issues and impacts of climate change will ever remain futile. For example, there is no way any government in Africa can effectively tackle poverty and climate change, if the current acute poverty is not seriously resolved. The all pervasive level of poverty in the continent has not only crippled national economies but has further engendered hopelessness especially among the poorest of the poor in Africa.

Consequently, most Africans are culpable of mounting too much pressure on the environment and the natural resources because they are forced to rely on alternative energy supply (electricity generating sets) and cheaper means of energy such as coals and fuel wood which are said to be some of the greatest sources of Global Warming with the attendant Climate Change. With this seemingly unabating ugly trend, achieving sustainable development in Africa will remain a mirage. It has been demonstrated that Africa, cannot afford to continue ignoring the potential impacts of the global climate change response measures on her economies and peoples. It was also made clear that though Africa should capitalize on the emission headroom afforded it for its low historical contribution to the climate change problem, it is in her interest to begin to introduce measures to reduce its greenhouse gas emissions, due to the negative impacts of climate change on its economic, social and environmental resources. It is imperative that full attention is paid to ways through which the African economies can be diversified and steered away from fossil fuels both in terms of production and consumption. Only such a

---

[170] *Tell Magazine,* August 11, 2008, Special Edition Captioned, "Environment: Nigeria's Nightmare", 20-31

strategy will save the economies from certain collapse in the event of implementation of climate change abatement measures.

From the foregoing, sub Saharan African countries, (especially Nigeria) are bound to the trend of globalisation (with its grievous environmental implications), wittingly or unwittingly. It is equally obvious that the dimensions of environmental challenges are too great to be catalogued; the intensity and magnitude of these problems – deforestation, desertification, pollution and climate change etc–have been compounded on Africa due to its poverty and technological underdevelopment.

As noted by noted by Okunola and Werners, poverty drives vulnerability to climate risk in different ways, from increased risk of exposure to extreme events such as floods and storms to lower access to coping mechanisms that can support resilience. Consequently, government and civil society organisations should find ways to help residents learn about and use climate change adaptation techniques. This includes building construction in line with the national building codes, protection and restoration of wetlands and tree planting. State governments and the federal government, development partners, and private sector actors all have a part to play. Their roles include investing in infrastructure, planning effective land use, creating wealth from waste, campaigning to change behaviour and reclaiming green spaces. Citizens can act too. For example they can build climate-resilient buildings, take up home insurance, warn or help neighbours at risk, and use plants to improve drainage and create a more comfortable environment.[171]

As we have advocated earlier, if Nigeria and other sub-Saharan countries are actually thinking of achieving sustainable development, serious attention must be paid to the dangers posed by Climate Change. As lamented earlier, Nigeria (like most of other undeveloped economies) are bearing the greatest burden of Climate Change related catastrophe due mainly to the prevailing poverty, weather depended economic activities

---

[171] Olasunkanmi Habeeb Okunolaa and Saskia Werners "Nigerian cities are under-prepared for flooding caused by climate change," March 16th, 2023, https://blogs.lse.ac.uk/africaatlse/2023/03/16/nigerian-cities-are-not-prepared-for-flooding-caused-by-climate-change/

and lack of adequate mitigation and adaptation strategies. We are joining other credible global voices to call on developed countries to urgently come to the rescue of the underdeveloped parts in terms climate finances and the underdeveloped countries should also get serious about establishing national action plans towards combating Climate Change.

Before we blame all the environmental crises in cities in sub- Saharan Africa on climate change; we must remind the current African city dwellers that most of their cities are still like what late Tai Solarin called 'civilizations without toilets.'[172] This is because many of them like Ibadan (the primary focus of this study) still battles open defecation and annual flooding without a single sea or ocean in the 21st Century. General absence of right environmental attitude among the people and unsustainable environmental management strategies/policies by city managers (especially in urban planning and waste management) could be also blamed for these ugly trends. This has made application of education/environmental ethics equally important as technologies towards achieving sustainable development in African cities.

From all indications these array of problems are interlocked with domino effects on economic, political, physical and socio-cultural conditions of Africa making collaboration among African nations an urgent necessity and in-depth intervention of international community inevitable. And Nigeria with her gigantic human/material potentials and strategic position must muster courage to lead sub-Saharan Africa out of the current and impending environmental catastrophe. This inevitably must start in Nigerian cities and Ibadan must play a significant role due to its historical relevance and current strategic national importance.

The Impact on Ibadan

As discussed earlier in this chapter, cities across the continents of the World as the drivers of globalisation and sustainable development are bearing the most of both positive and negative consequences especially

---

[172] Soji Oyeranmi, "A Civilization without Toilets? Ibadan and her Environment in the Postcolonial Era" *Sociology and Anthropology,* 6(2): 187-202. 2018 available online at https://www.hrpub.org/journals/article_info.php?aid=6771

the environmental costs. It is therefore not surprising that African Cities have fair share of these virulent consequences. Ibadan with its strategic location and importance in the Nigerian national life could not have possibly escaped series of monumental environmental tragedies in this era of overarching globalisation.

Ibadan, which is the primary focus of this study, has a robust metropolitan tendency. As reflected in the thought of Toyin Falola, writing on Ibadan requires no justification because the city "has always been great with an outstanding history." [173] The city is reputed to be one of the largest African cities. A glimpse at its history reveals that Ibadan is one of the few large pre-colonial cities in Africa. Indeed, according to Mabogunje "in a very real sense, Ibadan is regarded as the pinnacle of pre-European urbanism in Nigeria, the largest purely African City...."[174] Due to its significance, the city has also attracted various epithets as 'Black Metropolis', 'the largest city in black Africa'....[175] Unfortunately, Ibadan has a long history of urban arbitrariness, which explains clearly the reasons for the present urban crisis in the so called "ancient city". For example a 1945 report revealed that though Ibadan was founded in 1829, it took successive administrators a century to commence a comprehensive planning for the city.[176]

Consequently, development had proceeded in Ibadan without the needed urban management philosophy or sound environmental policies that could result in standard street systems; parks; well structured buildings and so on. Therefore, Ibadan has become what Labinjoh calls "epitome of planlessness...."[177] Laurent Fourchard further corroborated this fact in a special report on Ibadan in 2003 that "after taking a deep historical search,

---

[173] Toyin Falola, *Ibadan, Foundation, Growth and Change, 1830-1960*, (Ibadan: Bookcraft, 2012)

[174] Akin Mabojunje, *Urbanization in Nigeria*, (London: University of London Press, 1968) 186

[175] Akin Mabogunje, *Cities and African Development*, (Ibadan: Oxford University Press, 1976) 35

[176] For more on this see Maxwell-Fry-Farm Report on Ibadan Town Planning in File 1400, National Archives, Ibadan

[177] Justin Labinjoh, "Ibadan and the Phenomenon of Urbanism", in G. O. Ogunremi ed., *A Historical and Socio-Cultural Study of an African City*, (Ibadan: Oluyole Club, 1999) 238

what Ibadan reflects is a near total absence of urban management and urban planning."[178] While one may not totally agree with this assertion, it will be foolhardy to deny the enormous challenges the present generation is facing with the appalling environmental situation in Ibadan, which is a reflection of national urban decay in Nigeria.

With the above mentioned environmental crises in Ibadan, it is not surprising that storms and floods have become recurring decimal in Ibadan with the attendant devastating effects. For example, floods claimed over 200 lives and displaced about 50,000 people on 31 August 1980 alone.[179] Contrary to great hope for positive change that heralded the year 2000; environmental tragedies increased with the dawn of the new millennium in Ibadan and other major cities in Nigeria. The year 2000 was significant in the environmental history of Ibadan, with the creation of Oyo State Ministry of Environment. Apart from coinciding with the end of military rule, and marked the beginning of a new democratic rule, it equally heralded the 'magical' new millennium with great expectations of unprecedented development in every aspect of life. This included a hope for urban environmental rejuvenation. Ibadan with its national significance was expected to benefit from the so-called 'monumental millennium miracles' which has largely turned out as a mirage.

Although, Nigerian cities have witnessed varying degrees of environmental disasters; the year 2011 was also extremely catastrophic for most of these cities due to the "killer floods."[180] The year was particularly pathetic for Ibadan as the city experienced the worst flood disaster in its history. The flood that day was caused by an all-time high of 187.5mm rainfall and indiscriminate dumping of solid wastes on water channels. This eventually left over one thousand people dead and destroyed properties worth millions of naira.[181] Before this historic flooding, the city

---

[178] Fourchard Laurent, *Urban Slums Reports: The Case of Ibadan, Nigeria,* (Ibadan: Institut Francais de Recherche en Afrique (IFRA), 2003) 237

[179] A. B. Oguntala and J. S. Oguntoyinbo, "Ibadan Urban Nature", *Urban Ecology,* Vol. 7, No. 1, (September 1982) 39-46

[180] Special Edition of *Tell,* No. 29, July 2011 Captioned "The Killer Floods."

[181] Abiola Durodola et-al "Nigeria's cities are at severe risk from climate change. Time to build resilience, and fast," November 10, 2022, https://climatechampions.unf

had already witnessed varying degrees of flooding in areas along the Ogunpa and Kudeti streams in the city in 1955, 1960, 1961, 1963, 1978 1980, with the most recent in 2011.[182]Since then, Ibadan has continued to witness flash floods caused by environmental degradation and heavy rain.

This has not only demonstrated the failure of the past and the current environmental management strategies but also call for an urgent comprehensive and constructive paradigm shift in management of environment in Ibadan. Therefore, in subsequent chapters, this study has discussed the historical background to these environmental problems; offered new critical insights; and suggest some new strategic ways of tackling environmental crises in Ibadan which could also be useful in other Nigerian cities.

---

ccc.int/nigerias-cities-are-at-severe-risk-from-climate-change-time-to-build-resilience-and-fast/

[182] During this disaster Ibadan recorded loss of lives in their hundreds, thousands displaced, billions in properties lost to flood on Friday, 26th August 2011. Indeed, the calamity was described as national disaster by Deputy Senate President, Ike Ekeweremadu given its social and economic tolls on the nation and the strategic place of Ibadan city to the Nigeria's national life. The number and severity of flood related tragedies across the nation in recent times are a clear indication that the nation's environment system is ailing and ….. that all is not well with the way Nigeria relates and manages its environment. Most poignantly, Ekweremadu opines that the increasing incidents of flood disasters in the country bring home the harsh realities of the dangers of climate change and environmental degradation and the frequency of this magnitude of floods portends a dangerous future for Nigeria environmentally. For more on the 2011 floods in Ibadan, see Special Edition of *Tell*, No 29: July 2011 Captioned "The Killer Floods" and also, Emmanuel Ogala and Funso Ajewole, "Ibadan Flood Disaster is a Bad Omen", *Daily Times*, September 2, 2011

Chapter Four
Ibadan and its Environment in the Pre-Colonial Era

Preamble

Today, Ibadan appears like "a crippled city" in terms of environmental management and sustainable development. A glimpse to its history reveals that Ibadan (founded around 1829) is one of the few large pre-colonial cities in Africa. Indeed, Ibadan according to Mabogunje "in a very real sense, is regarded as the pinnacle of pre-European urbanism in Nigeria, the largest purely African City...."[1] Due to its significance, the city has also attracted various epithets as 'Black Metropolis', 'the largest city in black Africa'....[2]

But this supposedly rich history is not reflecting in the nature of Ibadan environment as the city is reputed as one of the dirtiest African cities. This ugly reputation is believed by many as a reflection of 'culture of filth' rooted among the natives and residents of Ibadan. Therefore, this chapter critically examines and evaluates the indigenous environmental practices in Ibadan which covered: the relationship between culture and nature (how they perceive land and taking care of it); indigenous environmental laws; sanitation and personal hygiene before the incursion of the British colonialists.

So far, it has been demonstrated that there are various successive approaches to the definition and explanation of the city by historians, anthropologists, architects, city planners among others, globally and particularly in Africa. This in the opinion of Vidrovitch has resulted in the multiplicity of opposing views: politics and religion versus economics; parasitic city versus generative one; pre-industrial versus industrial; pre-

---

[1] A. L. Mabojunje, *Urbanization in Nigeria*, (London: University of London Press, 1968) 186

[2] A. L. Mabojunje, *Urbanization in Nigeria*..... 186.

capitalist versus capitalist.[3] These arguments and contentions notwithstanding, one fact has been established by the available archaeological records that the first cities emerged from the Near East, specifically in southern Mesopotamia during the fifth Millennium BC.[4] For this reason, the parameters of the early Near Eastern cities have been used to judge and measure the manifestation of urbanism in Africa and some authors have even attempted to prove that urbanisation cannot be found in Africa (even in Yorubaland) before the advent of colonial rule. Indeed, the status of pre-colonial Africa has arguably been perceived as excessively rural in previous scholarship.[5]

Consequently, in order to have a full grasp of the environmental challenges that present generation of Ibadan residents are confronting: there is the need to critically evaluate the history of the growth and development of Ibadan into a pre-eminent urban centre. However, the discussion on pre-industrial Ibadan shall be preceded with a concise evaluation of African/Nigerian and Yoruba urbanism in pre-colonial epoch.

African/Nigerian Pre-colonial Cities: A Concise Assessment

As contended by Justin Labinjoh, writing about cities in sub Saharan Africa has been in the context of the discussion of the process of urbanisation. And often reading such literature one gets impression that the only motive for writing such pieces is to tell us how poor the region has been and how much it lags behind Asia, Latin America, Europe and North America.[6] There are two main reasons for such a bias: one is ideological, the other is intellectual. The first has to do with the cultural chauvinism of the European and American writers on Africa. The other, is a revelation of the confusion of urbanism with industrialism and modern

---

[3] Catherine Vidrovitch Conquery, "The Process of Urbanisation in Africa…." 4

[4] Robert M. Adams, *Heartland of Cities*, (Chicago: University of Chicago Press, 1981).

[5] Elizabeth Kitto, "Before European Colonialism was Africa Essentially Rural?" *Identity Academic Winter 2012* (Online Edition) http://retrospectjournal.co.uk/portfolio/before-european-colonialism-was-africa-essentially-rural/)

[6] Justin Labinjoh, *Modernity and Tradition in The Politics of Ibadan, 1900-1975*, (Ibadan: Fountain Publications, 1991) 1

capitalism.[7] On that confusion, Louis Wirth has rightly stated in 1938, that, different as the cities of the earlier epochs may have been by virtue of their development in pre-industrial and pre-capitalistic order from the great cities of today, they were, nevertheless, cities.[8]

Interestingly, based on archaeological evidence, other scholars have largely demonstrated not only the existence of cities in pre-colonial Africa but also that cities and towns developed in different parts of Africa as a result of indigenous initiatives and local conditions long before the colonial incursion.[9]

Based essentially on individual experiences and encounters in fieldworks, some early scholars have tended to conceive history of cities outside of Africa. They generally perceived Africa as an entirely rural world (a dark continent). It was therefore not surprising that city has been conceptualised as non African phenomenon.[10] With probably, Max Weber[11] as the only exception, to most historians of Western cities, Neolithic revolution marked the beginning of urban revolution. To them, the emergence of agriculture allowed for the accumulation of surpluses in the primary cities (Western cities) where civilisation blossomed, as opposed to the barbarism of savage life based on wild food. From his archaeological background and knowledge of Mesopotamia, Gordon Childe based his definition of urbanisation on ten criteria which began with the adoption of writing as a necessary base for administration and the use of arithmetic.[12] Definitely, the pre-colonial Africa without standardized art of writing was excluded here.

---

[7] Justin Labinjoh....1

[8] Louis Wirth, "Urbanism as a Way of Life", *American Journal of Sociology*, 44, (1938).

[9] See Graham Connah, *African Civilization, Pre-colonial Cities and States in Tropical Africa: An Archaeological Perspective*, (Cambridge: Cambridge University Press, 1995); R. W. Hull, *African Cities and Towns before European Conquest*, (New York: W. W. Norton and Co., 1976); S. K. MacIntosh, and R. J. MacIntosh, "The Early City in West Africa: Towards an Understanding", *The African Archaeological Review*, 2, (1984). 73-98.

[10] Catherine Vidrovitch Conquery, "The Process of Urbanisation in Africa...." 4

[11] Max Weber quoted in Catherine Vidrovitch Conquery, "The Process of Urbanisation in Africa...." 4

[12] Gordon Childe, "The Urban Revolution", *Town Planning Review*, 21, (1950) 3-17

Although Fernand Braudel saw universality in the nature and character of city, he contended that wherever it is, "only primitive or underdeveloped societies have not experienced urban phenomenon...."[13] According to him, even as black Africa had its towns, "city quintessentially exist in modern mercantilist societies due to the expanding economies in these European towns."[14] So, 'a true city' in pre-colonial Africa was non-existent as cities could only be European modernity. Gideon Sjoberg in a similar thesis to Gordon Childe's proposal, differentiated pre-modern towns of all pre-industrial societies from industrial metropolis which emerged from the 19th Century technological revolution. To him, if urbanisation is a proof of civilisation, only literate pre-industrial societies can claim to be civilised. In essence, writing is a sine qua non for civilisation.[15] Though Sjoberg also made reference to traditional Dahomey, Ashanti and Yoruba in Africa; he categorized them as having 'quasi-urban nature.'[16]Gladly, some Western Scholars and their African counterparts have successfully debunked these Eurocentric views of cities and Afro-pessimist concoctions of cities.

In the opinion of Hakeem Tijani, historians, anthropologists, archaeologists and ethno-linguists adequately documented the transformation of Africa from Paleolithic Age to the Neolithic Age to the Bronze Age, when more cities and urban centers emerged all over Africa.[17]According to Cox,[18]the theory of folk-urban continuum which viewed the concept of 'pre-industrial city as a negative construct was the major antagonism to Euro-centric perception of cities. This theory was derivable from the works of anthropologist Robert Redfield[19] which focused on rural Mexico and sociologist Louis Wirth's works on American

---

[13] Fernand Braudel, "Pre-Modern Towns", in Peter Clark ed., *The Early Modern Town, a Reader*, (London: Longman, 1976) 53-90
[14] Fernand Braudel....53
[15] Gideon Sjoberg, *The Preindustrial City, Past and Present*, (Glencoe, IL: The Free Press, 1960)
[16] Gideon Sjoberg....33
[17] Hakeem I. Tijani, "Reflection on Nigeria's Urban History", in Hakeem Tijani ed. *Nigeria's Urban History: Past and Present*, (Oxford: University Press of America, 2006).3.
[18] Oliver C. Cox, "The Preindustrial City Reconsidered", *The Sociological Quarterly*, 5 (1964):133-147
[19] Robert Redfield, *A Mexican Village*, (Chicago: Chicago University Press, 1930)

cities.[20] The theory assumed a dualism or polarity between the so-called traditional societies (rural ones: the folk society) and modern societies – urban ones.[21] According to Vidrovitch, "in a way, the continuum theory did represent a transition in European scholarship from the belief that urban Africa was inconceivable to the supposition that it was both possible and desirable.[22] The theory went hand in hand with the position of modernisation theorists in the 1950s, who believed that modernity was possible in Africa. Even if not totally afro centric; the theory offered more positive signs as against the colonial concept of true African society as "the rural society-naturally tranquil and immutable." The second generated from an 'orthogenic transformation' that is primary urbanisation. The third was the modern one, when 'heterogenetic transformation' gave rise to secondary urbanisation.

Years later, Redfield and Sigler offered a more nuanced approach to explain the nature of African city in what was generally referred to as 'trichotomy' of urbanisation.[23] The first part was the folk society-untouched, isolated, self-contained, homogeneous, and non-literate Vidrovitch contends that a first generation of cities appeared everywhere in Africa with variety of forms according to time and space.[24] While Davidson[25] and Hull[26] offered comprehensive historical accounts of initial major urban waves in Africa with some degree of exaggeration. While Winters[27] presented a sort of correction to the exaggerations of the earlier two narratives, Wirth[28] and Bascom[29] acknowledged the presence of metropolitan cities in Yorubaland.

---

[20] Louis Wirth, 'Urbanism as a Way of Life'....

[21] Catherine Vidrovitch Conquery, "The Process of Urbanisation in Africa...." 6

[22] Catherine Vidrovitch Conquery....7

[23] Robert Redfield and Sigler Milton, "The Cultural Role of Cities", *Economic Development and Cultural Change*, 13, (1955), 53-73.

[24] Vidrovitch, ...8-10

[25] Davidson Basil, *The Lost Cities of Africa*, (Boston: Little Brown and Co.,1970).

[26] R. W. Hull, *African Cities and Towns before European Conquest*, (New York: W.W. Norton and Co., 1976)

[27] Christopher Winters, "Traditional Urbanism in the North Central Sudan", *Annals of The American Geographers Association*, Vol. 67, No. 4, (1983). 500-520.

[28] Louis Wirth, "Urbanism as a Way of Life...."

[29] William Bascom, "Urbanization among the Yoruba", in Gerald Burke, *Towns in the*

With the above mentioned facts, it has become obvious that Africa (especially Western part, where Ibadan belongs) had a rich history of urbanisation which pre-dated colonialism by centuries. For examples, many of the famous pre-colonial African urban civilisations are found in this group of states: Kumasi in Ghana; Bamako, Timbuktu in Mali; Ibadan, Ife, and Benin in the South of Nigeria; and Kano, Zaria, and Sokoto, among others, in the emirate North of Nigeria. It is therefore not surprising that today West Africa remains a center of great urban civilisations— as most of the cities have continued to play critical roles in more recent history of the people much more prominent than their roles in the past. This is due largely to the unprecedented rapidity of global phenomenon of urbanisation with regional and geographical disparities. Expectedly, cities in Nigeria (like most of their other African counterparts) are characterised by a "multitude of faces: those of tradition, of modernity, of globalisation, of order, of chaos, and of many more, both pretty and ugly."[30]

Two propositions are generally espoused by scholars in explaining the process of urbanisation in Nigeria. In the first place, urbanisation patterns in Nigeria is said to fit into the popular model called 'central place theory.'[31] This implies the dominance and growth of a single primate city, which became the central place (the political and commercial center of the nation). Lagos (and many other administrative centers) seemed to be the perfect central place in the Nigerian context. Its emergence was largely a result of geographical location and early contact with Europeans.[32] Secondly, Tijani further submitted that urban development in Nigeria is multidimensional just as in other parts of the world. It is largely a function of politics, trade and commerce, population movement, government investment in new areas, and growth. For instance, in the northern part of the country, the transformation of the rural communities to the great centers of Kano, Katsina, Zaria, Sokoto, the early Borno capitals (Gazargamo and Kuka),and other cities served as entry ports to

---

Making, (London, Edward Arnold, 1955)
[30] Hakeem I. Tijani, "Reflection on Nigeria's Urban History….." xi
[31] Hakeem I. Tijani…. 4 and 5
[32] Hakeem I. Tijani…. 4 and 5

the Saharan and trans-Saharan trade, and as central citadels of political capitals for the expanding states of the savannah. According to Metz:

> The northern savannah cities grew within city walls; at the center of each were the main markets, government buildings, and central mosques. Around them clustered the houses of the rich and powerful. Smaller markets and denser housing were found away from the core, along with the little markets at the gates. There was also cleared lands within the gates that were used for subsistence agriculture. Groups of specialised craft manufacturers (cloth dyers, weavers, potters, and the like) were organised into special quarters, with the enterprises often being family based and inherited....[33]

According to Eyoh and Stren because urbanisation is one of the most powerful, and insistent, emerging realities of the early twenty-first century; developing regions are desperately striving to catch up with the developed countries that have been largely urbanised for many years. Africa, one of the least urbanised continents, is also striving to urbanise rapidly. In 2003, Africa's total population (estimated at 851 million) was 38.7 percent urbanised—the lowest of any major continental region. Unlike the pattern in most developed countries, urban growth in individual African countries has not always been accompanied by parallel economic growth for the country as a whole. This has been generating a lot of intractable crises in most African cities due to inadequate urban environmental management strategies and lack of effective urban planning.[34]

As Kayode Oyesiku and Wale Alade have submitted Spatial planning in a general sense was part of local indigenous administration in Nigeria, long before the colonial administration.[35] By the middle of 1800s, many indigenous cities though not urbanised in the real sense of 20,000 people

---

[33] H. C. Metz ed., *Nigeria: Country Studies*, (USA: Library of Congress, 1991)

[34] Dickson Eyoh and Richard Stren eds., *Decentralization and the Politics of Urban Development in West Africa*, (Washington D.C.: Woodrow Wilson International Centre for Scholars) 1

[35] Kayode Oyesiku and Wale Alade, "Historical Development of Urban and Regional Planning in Nigeria", in *State of Planning Report*, The Nigerian Institute of Town Planners, available online at www.nitpng.com

had a form of arrangement of land uses in their domain. For instance, the Sokoto Caliphate and much part of Oyo Kingdom seats of governments had one form of deliberate spatial arrangement of land uses around the palaces. Therefore, the Nigeria landscape to some extent had some rudimentary of planning in different parts of Nigeria before the colonial period which may not be as sophisticated as the modern ones. [36] A typical earlier example of this is traceable to Eko' (Lagos).[37]

Elsewhere outside Lagos area, there were strong local planning of settlements in line with traditional land tenure system that varied from one locality to another, the existing agrarian nature of the economy and foot-path nature of mobility and circulation. Land was mostly vested in the traditional rulers and families and therefore most Nigerian settlements were established around palaces of traditional rulers, which were then the focus of community activities.[38] Some of the settlements predating colonial period had the pattern of layout that were as a result of the need for defence or in line with religion as the case of many Northern and South-Western cities. For example, some settlements in the Northern and Western parts are located because of the factors of defense, religion or trade. The Yoruba settlements were noted for their general pattern of having a central area accommodating the king's palace together with the King's market and the most important place of worship.[39] The homes of High Chiefs are also located close to the town centre. From the centre, arterial roads that divided the town into wards radiated to the outskirts of the town.[40]

[36] O. K. Oyesiku, R. A. Asiyanbola and J. A. Sokefun, "Review of the Nigerian Urban and Regional Planning Law". *Journal of Public Law and Practice*, Vol. 1, No. 1, (June, 1999) 180 - 191

[37] A. Aduwo, "Historical Preview of Town Planning in Lagos before 1929", in A. M. Olaseni ed., *Urban and Regional Planning in Nigeria*, (Lagos: Nigerian Institute of Town Planners, Lagos State Chapter, 1999) 8-17

[38] NITP, 1991 *Twenty-Five Years of Physical Panning in Nigeria*, (Lagos: Nigeria Institute of Town Planners

[39] Lekan Sanni, "Forty Years of Urban and Regional Planning Profession and National Development in Nigeria", *Journal of the Nigerian Institute of Town Planners,* Vol. XIX, No. 1, (November, 2006) 1-6

[40] A. L. Mabogunje, *Urbanization in Nigeria*, (London: University of London Press, 1968)

Similarly, northern settlements like Zaria and Kano have walls around them for the purpose of defense and religion with gates provided in strategic locations to facilitate trade and communication. The city of Kano was a centre of trade and Islamic scholarship and there is a magnificent palace for the Emir who was the religious and political leader of the town. Indeed, as customary laws vary from one locality to another, land use patterns respond accordingly. Sanni noted that though there were no professional planners as we do at present, physical development and growth even in villages were coordinated and regulated by considering the relationship of any proposed development to the existing structures, and making adequate provision for circulation and other conveniences.[41] In the opinion of Obialo, planning and control of development in the pre-colonial period in Nigeria was effectively done.[42] Interestingly some of the cities such as: Kano, Zaria, Koton-Karfi, Toro in the North and Abeokuta and Ondo in the South-West; due to their peculiar environments still maintain their pre-colonial inner-city settlement structure.[43]

Environmental factor in the opinion of Rueben K. Udo was not only significant in relation to its effects on the character of the vegetation, but also because climate has played a dominant role in the ways of life, including architectural designs and the pattern of the economic activities of various peoples of pre-colonial Nigeria.[44] This is in consonance with the submission of Obateru that:

> ....that the founding and growth of cities rest on physical resources is axiomatic. The rise and growth of cities depend not only on the natural resources of its site and proximate socio-economic region but also on its location and external relations. So, study of any city must commence with an outline of the environmental endowments:

---

[41] Lekan Sanni, "Forty Years of Urban and Regional Planning Profession and National Development in Nigeria..." 5

[42] D .C. Obialo, *Town and Country Planning in Nigeria,* (Owerri: Asumpta Printing and Publication, 1999) 168-182

[43] D .C. Obialo, 168

[44] Rueben K. Udo "Environments and Peoples of Nigeria, A Geographical Introduction to the History of Nigeria", in Obaro Ikime ed., *Groundwork of Nigerian History,* (Ibadan: Heinmann, 1980) 8 & 10

its location, size, relief, geology, water resources, climate, vegetation and soils.[45]

Although at this period, agriculture was generally the main occupation of the people, pattern of rainfall was the most important element of climate distinguishing the regions from one another in terms crop production and period of farming activities. In essence as Udo earlier submitted, the rhythm of economic activity which is revealed in the farming calendar of the various parts of the country was controlled by the incidence and distribution of rainfall as well as by the length of the rainy season. This season usually decreases from south to north and became a critical factor in pre-colonial Nigerian agriculture because most farmers do not practise irrigation.[46] As a corollary, the crops produced in various parts of the country differ considerably, mainly as a result of the difference in the length of the growing season within one region and across the regions.

For examples, within Southern Nigeria (under heavy influence of rain), internal climatic cleavages still exist which differentiated their economic activities. In the vast low-lying of Niger Delta swamps, in the South-south with innumerable water ways and creeks, the traditional economy has been largely restricted to fishing and salt making from sea water. In Southwest with (abundant forest and arable land) and double maxima rainfall regime (not less than seven months with July and September as the rainiest months) with a short dry season; it was possible to grow crops of maize and vegetables per annum. In the Southeast due to the wetter nature of the land, it was only possible to raise one crop per year. In sharp contrast, the far North which is under the influence of the north-east trade winds from Sahara Desert is hot and dry for most of the year, and unlike the deeply forested South, it supports open savanna vegetation. It is as a result of this environmental diversity that the peoples of North cultivate crops like groundnut, cotton, millet, guinea corn, acha and engage in cattle rearing/animal husbandry, while the forest peoples of the South cultivate cocoa, palm oil, rubber, yams, cassava, cocoyams and many others. The Middle Belt which has a transitional climate between the

---

[45] O. I. Obateru, *The Yoruba City in History, 11th Century to the Present*, (Ibadan: Penthouse Publications (NIG), 2003) 32

[46] Rueben K. Udo "Environments and Peoples of Nigeria..." 10

North and South stands as a zone of a mixed culture in which crops found in South are grown side by side with those in the far North.[47]

Apart from the influence of the environment on the economy of pre-colonial Nigeria, it also impacted greatly on housing pattern of the peoples. As with most peoples of Nigeria at this period, the extended family is the basic unit of the Yoruba. And since the family lives together, each extended family has a territorial existence. That is each family is housed in the large traditional compounds within the towns as well as a district area of farmland in the rural areas.[48] Like the Yoruba and the Edo of the west, the Ibo, and the Ibiobio are settled agricultural peoples. But unlike the Westerners, the Ibo and Ibiobio live in small villages and not in towns due to the fact that urbanisation was a largely colonial invention in most eastern states with the exception of Onitsha Ibo and riverine Ibiobio (EFik) of Calabar. The complete dispersal of family compounds over the village territory is one the most prominent features of cultural landscape of parts of Awka, Owerri, Nsukka, Abak, and Ikot Ekpene. However, compact village settlements have survived in the sparsely settled forest areas of Ikom, Calabar, Obubra, Bende areas and Niger Delta.[49]

As demonstrated earlier, the grassland or savanna peoples of Nigeria fall into two distinct geographical groups, namely, the Middle belt peoples and the peoples of the far North. The Middle Belt is noted for large number of very small ethnic groups, including the hill dwelling peoples of Jos Plateau region and eastern highlands of Adamawa. The two largest and most prominent of the Middle belt peoples are the Tiv of the Benue valley in the east and the Nupe of the middle Niger valley in west. Tiv political structure is highly fragmented (stateless society) with compound as the basic social unit which rather serve domestic purpose than a political one.[50] The Nupe unlike the Tiv, have a rather integrated political organisation similar in some ways to the Yoruba system.[51]

---

[47] Rueben K. Udo "Environments and Peoples of Nigeria....11

[48] Rueben K. Udo "Environments and Peoples of Nigeria....17

[49] Rueben K. Udo "Environments and Peoples of Nigeria....18

[50] C. P. Bohannan and L. Bohannan, *The Tiv of Central Nigeria*, (London, 1962).

[51] Rueben K. Udo, "Environments and Peoples of Nigeria, A Geographical Introduction to the History of Nigeria..." 19

In the open grassland areas of Nigeria, the most numerous and politically dominant groups are the Hausa, the Fulani and the Kanuri. The Hausa traditional architecture displays a remarkable adaptation to the environment. In the drier areas of Kano, Katsina and Sokoto with plenty of good building earth, the houses consist of square or round mud-walled structures with flat roofs of mud supported by framework of timber. In the wetter parts of Hausaland, the house may also be square or round. Although the walls are also made of mud, the roofs consist of grass thatch similar to those in the rainier central districts of Nigeria. Hausa houses are usually arranged in closed family compounds which are enclosed by walls of mud or grassmat, to ensure adequate privacy for the family. The Fulani of Nigeria (who could be describe as the only ethnic group in Africa with no definite or distinct territory as they can be found anywhere and everywhere in the Sudan zone) also mainly inhabit Hausaland. They migrated into Nigeria from the west settle amongst the Hausa and subsequently conquered them during the Fulani jihad led by Uthman dan Fodio in 1804. Two main types of Fulani are usually recognised. They are the cattle Fulani and the settled of town Fulani. The cattle Fulani is essentially a nomadic cattle rearer who migrates regularly with his cattle in search of water and good grazing especially during the dry season. He leads a simple life and lives in a camp or tents or grass shelters which is often deserted as soon as a death occurs. The Town Fulani or Settled Fulani live in the conquered towns of Hausaland, where he is essentially an administrator or a hoe cultivator. Many of them own large herds of cattle which they entrust into the care of their nomadic brethren.[52]

As in the north the earlier towns in southern Nigeria often centered on the palace of the rulers. For instance, some pre-colonial Yoruba towns usually centered on the Aafin (the kings' palace) surrounded by large open space and a market. This arrangement as observed by Tijani is still evident in older cities such as Ile-Ife.[53] However, many of the important contemporary Yoruba cities, including the largest, Ibadan, were founded during the period of the Yoruba civil wars in the first half of the 19th Century.[54] Based on their origins as war camps, they usually contained

---

[52] Rueben K. Udo …..20
[53] Hakeem I.Tijani, "Reflection on Nigeria's Urban History…." 5
[54] H. C. Metz ed., *Nigeria: Country Studies*

multiple centers of power without a single central palace. Instead, the main market often assumed central position in the original town, and there were several separate areas of important compounds established by the major original factions.[55]

An Overview of Yoruba Urbanism before Colonialism

As demonstrated by Agbaje-Williams, ethnography and oral historical information are invaluable in reconstructing the structural profile of Yoruba towns. Though, the information was scanty, the European eye witness accounts also provide useful glimpses into the nature of Yoruba urbanism.[56] For example, Hugh Clapperton and Lander Brothers, in the accounts of their travels across the Yoruba country between 1829 and 1832, provided information on sizes and layout of Yoruba Towns and cities, the types of residential structures, commercial activities and aspects of the political culture of urban centres.[57]

Presenting one of the earliest works on urban environments in Nigeria, Wirth acknowledges the presence of metropolitan cities in Yoruba land. Contrasting these Yoruba Cities with the historic city of Timbuktu, Wirth concluded that nine out of ten largest cities in Nigeria in 1931 were in Yoruba land. These included Ibadan, described as the "largest Negro city in Africa."[58] Corroborating Wirth's account, William Bascom opined that:

The Yoruba have large, dense, permanent settlements; based upon farming rather than industrialisation...they are the most urbanised

---

[55] Hakeem I.Tijani, "Reflection on Nigeria's Urban History...." 5

[56] Babatunde Agbaje-Williams, "Yoruba Urbanism: the Archaeology and Historical Ethnography of Ile-Ife and Old Oyo..." 216

[57] See Hugh Clapperton, *Journal of a second Expedition into the Interior of Africa*, (London: John Murray, 1829); Richard Lander, *Records of Captain Clapperton's Last Expedition to Africa*, (London: Henry Colburn and Richard Bentley, 1830); Richard and John Lander, *Journal of an Expedition to Explore the Course and Termination of the Niger*, (London: Thomas Tegg and Son, 1832)

[58] L. Wirth, "Urbanism as a Way of Life..."

of all African people's percentage living in large communities comparable to that of European nations.[59]

In an attempt to prove the existence of urban centres in Nigeria before the advent of colonialism, Bascom contended that urbanisation was indeed traditional to the Yoruba rather than being an outgrowth of the Europeans. This particular work is significant as it warns us of the danger of "confusing urbanism with industrialism and capitalism". Although the development of modern cities could not be divorced from the power of technology, the cities of the earlier epoch may have been different by the virtue of their development in a pre-industrial and pre-capitalistic order; "they were nevertheless cities."[60] Therefore, the Yoruba cities especially in the pre-colonial epoch were not based on machines. However, the fact that form of specialisation in different crafts within these urban centres at this period made individuals depended on the society as a whole qualified them as cities.

As reflected in the thought of Atanda, during the so called 'Golden age' (which ended around 1800 A.D.), Yorubaland contained a number of major kingdoms such as Ife, Oyo, Ijebu-jesa, Ketu Popo Sabe Dassa, Egbado, Igbomina, Ondo and the sixteen Ekiti principal towns.[61] Prior to about 1550 A.D. the kingdoms were apparently inhabited by "homogeneous" ethnic groups which had a paramount ruler Oba (king). The seat of the potentate was the capital city which was the religious, political, administrative an economic centre of the kingdom of the ethic group. The ruling dynasty of most, if not all of these kingdoms traced their origin to Ile-Ife and their descent directly or indirectly to Oduduwa.

The account of the foundation of these kingdoms revealed that their founders left Ile-Ife at different times and for different reasons rather than by common decision taken in normal circumstances according to tradition. Therefore, cities established in those days were crucial to the development of the early urban centres in Yorubaland as they were in

---

[59] William Bascom, "Urbanization among the Yoruba…"
[60] J. Labinjoh, "Modernity and Tradition in the Politics of Ibadan…." 2
[61] J. A. Atanda, "Government of Yorubaland in the Pre-Colonial Period", in *Tarikh*, Vol. 2, No. 1, (1974)

Mesopotamia, India, Egypt, China, Central Andes and Mesoamerica. The urban centres became knowledge based societies where information and the conscious regular and systematic collection of data became an integrated part of maintaining controls. Yoruba city at this period was usually the royal capital established by the kings. As there was only one oba (king) in the kingdom, his seat was the only city in the kingdom. Only the seat of the king was designated as city and was the largest. The towns were the seats of Baales who by tradition did not wear beaded crowns. However, the civil wars of the 19th century had changed the concept with the emergence of other large settlements made up of refugees whose previous settlements have been destroyed during the wars. Examples of these include: Ibadan, Ede, Osogbo, Ikirun, Saki, Okeho and Ogbomosho while new towns emerged such as Abeokuta, Modakeke and Ilero.

Although the Yoruba are predominantly and agricultural people, they have a unique and enduring tradition of living in large towns, the largest of which include Ibadan (627,380 in 1963) Ogbomoso (319,880), Osogbo (201,380), Ilorin (218,550, Abeokuta (187,290), Ilesha (165,880) and Ede (134,550).[62] The Growth of these large urban centres derived in part from the need for defence, but also largely as a result of the highly centralised political organisation of the people. Urbanisation among other things, facilitated the growth of trade among the Yoruba as well as between them and other peoples of Nigeria, particularly the Nupe and Hausa.

As demonstrated earlier, the traditional urban settlements which existed in Nigeria before the coming of the British colonial administration differed tremendously in structure and conditions from those which emerged following the introduction of the colonial rule. The traditional towns were agrarian and were neither over-populated nor restricted in terms of spatial expansion.[63] Yorubaland crystally exemplified the precolonial Nigerian urbanism. The early Yoruba towns were among Africa's ethnographic anomalies. Other ethnic groups have had higher densities of population and yet lived dispersed settlements; others again

---

[62] Rueben K. Udo "Environments and Peoples of Nigeria, A Geographical Introduction to the History of Nigeria..." 14

[63] Andrew G. Onokerhoraye, *Urbanisation and Environment in Nigeria: The Implications for Sustainable Development*, (Benin: The Benin Social Science Series for Africa, 1995) 6

have had more highly developed political systems yet the capitals of their kingdoms have been small. In the pre-colonial era, Yoruba country comprised a large number of kingdoms (Oyo, Ife, Ijebu, Ijesha, Owu), the capitals of which were often towns of substantial size. [64]According to Wheatley, the tendency of the Yoruba to live in large, permanent, compact aggregations were probably the most distinguished feature in the spatial expression of their culture.[65] The People's concept of urban residence also has a great consequence on the evolution of Yoruba cities. This concept in the opinion of Oluremi Obateru certainly has a cosmic or religious bearing if not foundation.[66] This was evident in the Yoruba legend of creation of earth and the foundation of Ile-Ife. He argues further that Yoruba concept of city tallies graphically with the city concept of H. Pirenne which Lewis Mumford confirms as follows:

> The city, as it took form around the royal citadel, was a man-made replica of the universe. This opened an attractive vista: indeed a glimpse of heaven itself. To be resident of the city was to have a place in man's true home, a great cosmos itself, and this very choice was itself a witness of the general enlargement of powers and potentialities that took place in every direction. At the same time, living in the city, within sight of the gods and their king, was to fulfill the utmost potentialities of life.[67]

The permanent home of the Yoruba man is the city: there he builds his permanent residential dwelling. Yoruba man rarely admits that he comes from a village. His pride is deeply hurt if you characterise him as a villager or countrysider. Almost every Yoruba man (especially in modern times) is an urbanite, even if he resides in a distant village of 15-20 kilometres from the city for the greater part of the year for agricultural production. His village life is regarded as merely temporary. His real

---

[64] Bolanle Awe, "Ibadan, Its Early Beginnings", in P. C. Lloyd, Bolanle Awe and Akin Mabogunje eds., *The City of Ibadan*, (Ibadan: Institute of African Studies, University of Ibadan, 1967) 11

[65] Paul Wheatley, "The Significance of Traditional Urbanism, Comparative Studies", in *Society and History*, 12, (1970) 396

[66] Oluremi I. Obateru, *The Yoruba City in History, 11th Century to the Present*, (Ibadan: Penthouse Publications (Nig), 2006) 102

[67] Lewis Mumford, *The City in History*, (New York: Penguin Book, 1961). 104-105

home is the city to which he has deep religious, social and cultural commitment and attachment. The most permanent physical features of cities are the plans of their roads and streets. Once established, the street plan of a city is extremely difficult to alter. The inflexibility of street systems is explained by a number of factors such as: the heavy capital invested in their establishment, the buildings facing them and problem of traditionalism and persistence of culture. Consequently, the original street plan of a city usually survives largely unaltered even when circumstances that engendered them have changed or disappeared. Today, the street plans of some eastern Yoruba cities like Ife, Ilesa, Akure, Ondo and Owo established during the so called 'Golden Age' (1086-1793) largely remain intact. No Yoruba city was without a broad integrated layout plan. The street plan of the city was a combination of the radial and grid plans.[68] The primary road system is radial in form. It usually run from the countryside and cut through the defence wall of the city and main radial roads converge on the royal palace at the city centre. The city form was roughly wheel-shaped with royal palace as the hub and the defence wall as its perimeter. This pattern apparently gives the impression that the city was circular. The residential quarters of the classic Yoruba cities were laid out on the grid plan. The scheme was clearly apparent in the layouts of eastern Yoruba cities of Ile-Ife, Ilesa, Ondo, Akure and Owo which depicted the residential developments during the height of pre-colonial urbanism.

As observed by Obateru, the grid plan was neither casually used nor precisely fortuitous.[69] Whatever was achieved was done as far as the relief and people's survey technology permitted at that period. The scheme was and reflected the physical expression of the ideal environment of a well organised society and space organisation also was the physical embodiment of the people's religious orientation and social-cultural values. The conclusion that classic Yoruba cities were planned and laid out on the radial and grid plans are based on the following facts: firstly, the cities of during the peak of Yoruba indigenous urbanism exhibited similar street plans. Secondly, these cities in some instances emulated

---

[68] Oluremi I. Obateru, *The Yoruba City in History, 11th Century to the Present*....162
[69] Oluremi I. Obateru.....170

each other. For example, Samuel Johnson informed us that Ilesa was laid out after Old Oyo: "It is said that when the town of Ilesa was to be laid out, a special messenger was sent to the Alaafin to ask for the help of one of the princes to lay out the town on same plan as the ancient city of Oyo."[70]

However, such new cities such as New Oyo, Ibadan, Abeokuta and Sagamu established during the 19th Century were not planned. This was due to the civil wars which did not give room for a conscious integrated plan. At this period, the people were preoccupied with civil wars, so they could not spare time, resources and efforts for urban planning. In general, the transformation and revolutions of the 19th Century intensified the expansion of urban centres in Yorubaland, and with, the creation of new opportunities and new social problems. New cities developed on sites of small villages, and the refugees coming from the northwestern area of Yorubaland swelled the old cities in significant numbers. Warrior-leaders sustained these 19th century cities in the sense that they held reins of power.[71] Despite the changes in the political order and preoccupation of many of these urban centers with warfare, local economic production and long distance trade were also important features of new towns and cities. Cities continued to be centres of cultural and political innovations, and many thronged to Ibadan and Abeokuta, among others, not only for safety but also for better opportunites and social advancement. Many of these 19th century towns and cities became the nuclei of the British colonial administration at the beginning of the 20th century, and they continue to be relevant today in postcolonial economy and society.[72]

Traditional Environmental Management Strategies in Ibadan before Colonial Incursion

The city of Ibadan is located approximately on Longitude 3^0 5′ East of the Greenwich Meridian, and latitude 7^0 23′ North of the Equator at a distance

[70] Samuel Johnson, *The History of the Yorubas,* (Lagos: CSS Press, 1969) 22
[71] Toyin Falola, *The Political Economy of a Pre-colonial African State, Ibadan, 1830-1900,* (Ile-Ife: University of Ife, 1984); Toyin Falola, and G. O. Oguntomisin, *Yoruba Warlords of the 19th Century,* (Trenton, NJ: Africa World Press, 2000) 1
[72] Agbaje-Williams ….236

some 145 kilometres northeast of Lagos. Ibadan is directly connected to many towns in Nigeria, as its rural hinterland, by systems of roads, railways, and air routes. The physical setting of the city consists of ridges of lateritised quartzitic hills that run approximately in northwest-southeast direction. The largest of these ridges lies in the central part of the city and contains such peaks as Mapo, Mokola and Aremo.[73] These hills range in elevation from 160 to 275 metres above sea level, and thus afford the visitor a panoramic view of city. The area occupied by the metropolitan area of Ibadan is drained by two important rivers the Ogunpa and the Ona Rivers. The former drains the eastern while the latter drains the western parts. Thus, rising in the northeastern section of the area, the Ogunpa flows southeast-wards breaking through the central ridge before turning south along a course that is parallel to the ridge. Its major tributary is the Kudeti which drains eastern part of the ridge. On the other hand, the western parts of the city which consists of more recent residential and other developments is drained by River Ona and its numerous tributaries that include Alalubosa, Osun and Yemoja streams.[74]

Ibadan city is fascinating to know and study.[75] It was for a long time the largest city in tropical Africa. Although, it has also been surpassed by Lagos, it remains a truly African city characterised by a small non-African population. Furthermore, the city preserves in an unadulterated form all those characteristics of the typical Yoruba city- a central market, a remarkable social structure and the rather unusual (by European Standards) pattern of urban-rural migration whereby farmers spend on the average, three days in the city and four days on the farms.[76]The locations of Ibadan in the words of Ayeni definitely add to its fascination. This is because the present site of the city is known to have had some factors which have made it a suitable and favourable location for the

---

[73] Bola Ayeni, "The Metropolitan Area of Ibadan: Its Growth and Structure", in M.O. Filani et. al eds., *Ibadan Region*, (Ibadan, Rex Charles Publication, 1994) 72

[74] Bola Ayeni....72

[75] Bola Ayeni....72

[76] A. L. Mabogunje, *Urbanization in Nigeria*, (London: University Press, 1968); N.C. Mitchell, "Some Comments on the Growth and Character of Ibadan Population", *Research Notes*, (Ibadan: Department of Geography, University College, 1953). 4, & 9-10; G.J.A. Ojo, "The Journey to Agricultural Work in Yorubaland," *Annals of the Association of American Geographers*, 197.

concentration of human groups. Thus, three times, the site served as a point for the foundation of cities, the present city being the third.[77] It would seem that the presence of ranges of hills makes the site of the city easily defensible while its location close to the boundary between forest and grassland makes it a melting pot for people and products of the forests as well as those of the grassland areas. Of course, the location close to the forest also provided some protection against aggressive neighbours.[78]

According to Akinyele, a local historian of Ibadan, the first group to inhabit this area was comprised of fugitives from justice, wild and wicked men expelled from adjacent towns, rebels and robbers. However, this motley group did not found a town. That was left for Lagelu (later regarded as the founder of Ibadan) a leader of the colonizing group from Ife who, in the general exodus from the hearth of the Yoruba people came in this direction. Akinyele stated further that Ibadan began to derive great advantage from its frontier location. It became a market for the Ijebu and Egba whose territories lie partly or wholly in the forest belt, on one hand, and the Oyo who mostly inhabit the grassland region on the other.[79]

However, whatever the importance of Ibadan at this time, it did not prevent it from attacks by stronger neighbours. Oral tradition has it that because of failure to maintain certain strict propriety in connection with the Egungun cult, Ibadan was placed under a siege and was completely destroyed. The survivors, which included Lagelu and his children, together with whom he had escaped earlier and took refuge on one of the hills. After the horrible episode, Lagelu descended the hill to found a second Ibadan. The site of this second town was a few miles farther south. It was never a very big town and its area was probably less than three square miles in extent.[80]

The second Ibadan survived until the early nineteenth century. By then Yoruba land was enveloped in an unprecedented struggle in which many

[77] A.L. Mabogunje, *Urbanization in Nigeria*....187
[78] Bola Ayeni "The Metropolitan Area of Ibadan: Its Growth and Structure....." 73
[79] A.L. Mabogunje, *Urbanization in Nigeria*....186
[80] A.L. Mabogunje, *Urbanization in Nigeria*....186

towns and kingdoms were destroyed. The Egba who inhabited Ibadan and an extensive region all around were a target of the allied armies of Oyo, Ife and Ijebu after the destruction of Owu which was thought to be an ally of Egba. The only exception to the destruction of Egba settlement that followed was Ibadan which, nonetheless, the inhabitants deserted. Hence, the armies quartered their forces within it. As Bolanle Awe shows, from this war camp developed the town we now know as Ibadan. What began as a war camp, however, grew so rapidly that it no longer was reasonable to think of breaking it up. In this way the present Ibadan was founded around 1829.[81]

Furthermore, it must be noted that Ibadan was distinguished by the heterogeneity of its population. This comprised the armies of Ife, Ijebu, Oyo and some friendly Egba. The Ife and the Oyo settled around the present Oja Iba and Mapo hall, while the Ijebu settled at Isale Ijebu towards the south. A few of the Egba inhabitants who returned with the allied armies took up residences to the west in a quarter now known as Yiosa. This new arrangement eventually came to be dominated largely by the Oyo elements. However, this did not necessarily make it easy for Ibadan to adopt the customs and traditions of Oyo.[82]In fact, as a primarily military settlement, the city attracted the adventurous young men and numerous others anxious to escape the stifling traditionalism of other Yoruba cities. This influx of people definitely improved the metropolitan nature of Ibadan as virtually all parts of Yoruba became represented in the city. According to Bolanle Awe:

> As the settlement grew, there came not only from among the Oyo-Yoruba, but from all parts of the Yoruba country: indeed a survey of the various compounds in Ibadan indicates that every Yoruba town had a son in Ibadan...[83]

---

[81] P. C. Lloyd, Bolanle Awe and Akin Mabogunje (eds.) *The City of Ibadan*, (Ibadan: Institute of African Studies, University of Ibadan, 1967) 3

[82] A. L. Mabogunje, *Urbanization in Nigeria*....186-189

[83] P. C. Lloyd, Bolanle Awe and Akin Mabogunje (eds.) *The City of Ibadan*, (Ibadan: Institute of African Studies, University of Ibadan, 1967)

This population included: ambitious young men eager to get on; craft men looking for better opportunities for their trade; rich men bored with life in their own towns and a handful of Sierra-Leonean and Brazilian immigrants, descendants of ex-slaves who found their way back to the Yoruba country and later thronged into Ibadan. It is therefore not surprising that the city soon expanded and by then covered an area of nearly 16 square miles. With such a population, new and novel forms of organisation prevailed. This was particularly evident in the form of government that the city evolved. With no historical traditions, ascriptive claims based on heredity paved way for competition; the conferment of titles and recognition was based purely on a man's wealth. By 1851, the result was that government in Ibadan was divided into two sphere, the military headed by the Balogun, and the civil headed by the Baale.

In his own account, Olajire Olaniran postulated that the environmental\physical setting of Ibadan comprises of its topography, drainage, rock types, soils and the climate. The hills are of two types: the quartzite and gneissic inserbergs. The ridges are quite impressive and widespread (about seven in all). Prominent examples are Agodi Hill, the Mapo Hill, the Premier Hill, the Oke Aremo and Mokola-Oremeji Ridges.[84] The sustained growth of Ibadan over the years, from its sudden emergence about 1829 to become the largest urban centre in Nigeria as from about 1860 owes a lot to its location in the heart of Yoruba ethnic territory as well as its location in relation to the other prominent Yoruba city states of Abeokuta, Oyo, Ijebu Ode and Ife. Its central location and accessibility from the colonial capital city of Lagos were major considerations in the choice of Ibadan as the headquarters of the western provinces which later became the Western Region of Nigeria.[85]

As far as Nigerian cities are concerned, Ibadan is a city of relatively recent origin. The evolution of Ibadan as an urban centre and later a regional power is traceable to the reign of Bashorun Oluyole from about 1836.[86]

---

[84] Olaniran Olajire, "The Geographical Setting of Ibadan", in G.O. Ogunremi, *Ibadan: A Historical, Cultural and Socio-Economic Study of An African City,* (Ibadan: Oluyole Club, 1999) 3

[85] R. K. Udo, "Ibadan in its Regional Setting", in M.O. Filani....8

[86] Justin Labinjoh, "Modernity and Tradition in Politics of Ibadan......" 78-101

According to Femi Osofisan, although he was regarded by some historians as a most ruthless ruler but he also paid serious attention to agricultural innovation and municipal administration just as he did to war.[87] To demonstrate this, he founded the famous Oja'ba market directly in front of his compound. Apart from the fact that it was then the main market; it still flourishes till date in Ibadan and it is now known as Oja Oba market. Under Iba Oluyole's leadership, Ibadan was transformed from a mere war camp to a significant city which further attracted settlers from other parts of Yorubaland at this period. Based on Oluyole's successes and the republican nature of Ibadan, many people were encouraged to migrate to the city which led to a tremendous increase of its population. By 1851, the city's was put between 60,000 and 100,000 and its surrounding walls stretched out for up to ten miles.[88] This is no doubt an evidence of pre-colonial urbanisation built upon by colonial urbanism. With this unprecedented feat, Oluyole could be credited for the establishment of Ibadan as one of the important commercial and urban centres that existed in Nigeria before colonial rule.

As reflected in the opinion of Mabogunje, Ibadan claim to city status carries none of the customary sanctions of a crowned head, a palace or a hereditary line of chiefs and yet in a very real sense, it is the pinnacle of pre-European urbanism in Nigeria, the largest purely African city and the emporium for the commerce of an extensive region.[89] This seeming contradiction of the evolution and urban development in Ibadan was also observed by P.C. Llyod when he submitted that:

> Several existing Yoruba towns were in existence before the first Portuguese visits to West Africa-Ile Ife and Ijebu Ode are proven examples. But Ibadan does not belong to this group. Though, as the discovery of stone axes indicates, men have been settled in the area for centuries. Ibadan is a city-village. It is a city of a million inhabitants; the capital of a Region of 8 million and a larger and

---

[87] Femi Osofisan, "Ibadan and the Two Hundred Snails", in Dapo Adelugba et. al eds., *Ibadan Mesiogo: A Celebration of a City, its History and People*, (Ibadan: Bookcraft Ltd., 2001) 2-3

[88] Femi Osofisan, "Ibadan and the Two Hundred Snails".....6

[89] A.L. Mabogunje, *Urbanization in Nigeria*....186

more wealthy territory than many African states. Yet the core of Ibadan, settled in the 19th century, is peopled by farmers, traders and craftsmen living in large compounds organised on principles of common descent-a society more resembling the villagers of Africa than the urban areas of modern world.[90]

In traditional Ibadan, people live in huge compounds, often containing several hundreds of inhabitants. In the past, these were structures of series of enclosed rectangular courtyards. With few exceptions, the descendants in the male line of one of the more powerful immigrants of the early or the mid-19th century together with their wives lived in those compounds.[91] Ibadan evolved as a small village on the less elevated southern extremity of a ridge of hills running from the northwest to the southeast, around the present Mapo hall. Its position on the edge of the grassland, and the protection which it was afforded by the large expanse of lateritic outcrop in the area, made an ideal place of refuge from the marauding Fulani cavalry attacking from the north, and the hostile Egba in the neighbourhood.[92]

However, if the environment had offered protection against external aggression; the aggressive nature and heterogeneity of the early inhabitants, who were mainly soldiers of fortune, rendered internal peace a herculean task. As demonstrated by Bolanle Awe, the republican nature of the early Ibadan exposed it to frequent civil wars due to the incessant struggle for leadership (A common Yoruba saying about Ibadan up to the present day is *Aki i wa saye ki a ma ni arun kan lara, ija igboro ni arun Ibadan,* meaning: We all have defects, Ibadan's shortcoming is constant internal strifes).[93] This even inspired a book by Ruth Watson, *'Civil Disorder is the Disease of Ibadan'-Ija Igboro Larun Ibadan: Chieftaincy and Civic Culture in a*

---

[90] P.C. Lloyd, Bolanle Awe and Akin Mabogunje eds., *The City of Ibadan*, (Ibadan: Institute of African Studies, University of Ibadan, 1967) 3
[91] P.C. Lloyd, Bolanle Awe and Akin Mabogunje....5
[92] P.C. Lloyd, Bolanle Awe and Akin Mabogunje....5
[93] Bolanle Awe, "Ibadan, Its Early Beginnings", in P.C. Lloyd, Bolanle Awe and Akin Mabogunje eds., *The City of Ibadan*, (Ibadan: Institute of African Studies, University of Ibadan, 1967) 14

Yoruba City, which was a historical account of the warlike origin of the city and define it as both an urban center and military polity.[94]

As observed by Watson, the historical genesis of a civic Ibadan is in the city's military origins. The third verse of the Ibadan City Anthem also acknowledged this:

Ibadan, Ilu jagunjagun	Ibadan, city of warriors
Awon to soo d'ilu nla	They who transform it into a great city
Awon omo re ko nije	We its children will not allow
Ko'ola ati ogo won run	That their honour and glory perish.[95]

History has is that from 1820s onwards, the area now known as Southwestern Nigeria was engulfed in a bitter political conflagration. This was occasioned by series of events, which led to near-continuous violent upheaval and wars throughout most of the 19th century among various rival polities and refugees that flooded the area. In the aftermath of one of the wars Ibadan was founded in 1829 initially as war camp, which grew to a large city within twenty years with remarkable population. In 1851, the Anglican missionary David Hinderer suggested a figure between 60,000 and 100,000; six years later, his Baptist counterpart proposed 70,000.[96] Assistant Colonial Secretary Alvan Millson, who visited Ibadan on a peace mission to Ibadan in 1890, viewed Ibadan as 'a bellicose society, where control over people was militarily and politically vital' but nevertheless 'an impressive urban centre.' At a meeting of Royal Geographical Society in 1891, he described Ibadan as:

The London of Negroland.... Surrounded by its farming villages, 163 in number, Ibadan counts over 200,000 souls, while within the walls of the city itself at least 120,000 people are gathered. Its sea of brown roofs covers an area of nearly 16 square miles, and the

---

[94] Ruth Watson, *Civil Disorder is the Disease of Ibadan - Ija Igboro Larun Ibadan: Chieftancy and Civic Culture in a Yoruba City*, (Ibadan: Heinemann Educational Books Nigeria Plc, 2005) 5

[95] Ruth Watson, *Civil Disorder is the Disease of Ibadan* 5

[96] CMSB, CA2\049\104: Hinderer Journal, 23 October, 1851 and Awe 15

ditches and walls of hardened clay, which surround it, are more than 18 miles in circumference.[97]

As the city grew in size, its commercial significance also increased. Environment played a major role in this as its geographical location between the coast and the interior made it a convenient melting pot for traders from all parts of the Yoruba land.

As early as 11th century, most of Yoruba towns and cities were generally bounded and fortified by walls and Ibadan was not an exception.[98] In 1857, Ibadan was surrounded by walls estimated to be about 10 miles in circumference. There were four main gates in these walls, leading to Abeokuta, Ijebu, Oyo and Iwo.[99] Some small farms were within the walls, the main farms were outside them; these stretched as far as Lalupon and beyond to the north-east, and as far as Apomu in the south-east. Indeed, they were sometimes as far as 30 miles away and embraced more deserted Egba towns such as Ojo, Ika, Iroko and Ikeye which became Ibadan farm villages in 1858.[100]

Mabogunje argued that the city of Ibadan represents a complex phenomenon, baffling in its size, unusual in the nature of its economic base and bewildering in its physical lay-out and morphology.[101] He stated further that, until the late 1950s was easily the largest city in Nigeria and attracted various epithets such as 'a Black Metropolis', 'the largest city in Black Africa' and not uncommonly 'a city-village.' Although today, Lagos is regarded as the most populous city in Africa but Ibadan remains the more intriguing of the two cities. It is still more a traditional African creation in a sense that Lagos is fast ceasing to be and its morphology continues to exert considerable influence in preserving traditional ideas of urban living and interpersonal relations. Indeed, Ibadan represents a

---

[97] A. W. Millson, "The Yoruba Country, West Africa", *Proceedings of the Royal Geographical Society*, 13, (1891) 583

[98] Ruth Watson, *"Civil Disorder is the Disease of Ibadan" - Ija Igboro Larun Ibadan*....1-8

[99] Bolanle Awe, "Ibadan: Its Early Beginnings"....15

[100] Bolanle Awe, "Ibadan: Its Early Beginnings"....15

[101] A.L. Mabogunje, "Morphology of Ibadan", in P.C. Lloyd, Bolanle Awe and Akin Mabogunje eds., *The City of Ibadan*, (Ibadan: Institute of African Studies, University of Ibadan, 1967) 35

convergence of two traditions of urbanism namely, a non-mechanistic, preindustrial African tradition more akin to the medieval urbanism in Europe and a technologically orientated European tradition.[102] Needless to affirm here again that, Ibadan in the pre-colonial period was predominantly pre-industrial as it reflected Gideon Sjoberg comprehensive description of pre-industrial urbanism. According to him in many urbanised pre-industrial societies, the city is often:

> A mechanism by which society's rulers can consolidate and maintain their power, and more important, the essentiality of a well developed power structure for the formation and perpetuation of urban centres.... Often these are fortified places to protect the upper class against the local marauders or invading armies. But invariably they are the focal points of transport and communication, enabling the ruling element not only to maintain surveillance over the countryside but also to interact more readily with members of their own group in other cities as well as within a city. The congestion that defines the city increases personal, face-to-face communication therein, essential if the heads of the various bureaucratic structures- governmental, religious, and educational- are to sustain one another. So too, craftsmen and merchants prosper in the urban milieu, whose density and occupation heterogeneity foster economic activity. [103]

Ibadan is indeed is a city of pre-colonial epoch that seems to have refused to change. Just as empires rose and fell in history, some cities have developed tremendously while others simply decayed. Ibadan is a curious mixture of the two experiences: it has not developed economically and physically, but it has also not decayed.[104] Ibadan grew as a garrison and fortress; it soon became a centre of administration and (like all cities) a market. It was never dominated by a real bourgeoisie interested in production; rather it was dominated, first by an indigenous aristocracy who were mainly consumers, then by middlemen merchants and much later by a stratum whose common link was literacy (education but not

[102] A.L. Mabogunje, "Morphology of Ibadan"....35
[103] Gideon Sjoberg, *The Pre-industrial City*, (Glencoe, Illinois, 1960) 67-68
[104] Labinjoh, "Modernity......"2

necessarily erudition) and whose concern was the modernisation of the city. Consequently, industrial capitalism never developed there. Therefore, Ibadan has remained underdeveloped till today.[105]

Ibadan residents like most other Yoruba people during the pre-colonial period relied heavily on indigenous or traditional knowledge and informal ideas to protect and manage the environment. These traditional ideas are generally referred to as 'taboos.' Taboos were introduced to regulate the moral order of the society. They took their origin from the fact that people discerned that there were certain things which were morally approved or disapproved by the deity.[106] These are not contained in any written law but are preserved in the tradition. Mabawonku and Agboola sometimes ago submitted that taboos and superstitions were often regarded as integral part of traditional education.[107] Every society in the world cares for its tradition because in a society without schools, a type of education known as traditional/informal education goes on. In Yoruba societies, traditional education is supported and encouraged because of its contribution to the growth, renewal and development of the society.[108] In traditional Ibadan (like in other pre-colonial Yoruba cities), the environment is people, and people are the environment.[109] The environment and the people are in close relationship in Yoruba religion. The Yoruba concept of the environment is all embracing; humans, animals, plants, and "non-living beings", form the entire human society or

---

[105] Labinjoh, "Modernity 3

[106] Cecilia Omobola Odejobi, "An Overview of Taboo and Superstition among the Yoruba of Southwest of Nigeria", *Mediterranean Journal of Social Sciences*, Vol. 4, No. 2, (May 2013). 221-223

[107] Tunde Agbola, and A. O. Mabawonku, "Indigenous Knowledge, Environmental Education and Sanitation; Application to an Indigenous African City", in D.M. Warren, Layi Egunjobi and Bolanle Wahab eds., *Indigenous Knowledge in Education*, (Ibadan: Indigenous Knowledge Study Group, University of Ibadan, 1996) 78-94

[108] E. B. Idowu, *Olodumare: God in Yoruba Belief*, (Ibadan: Longman, 1962); O. Oduyale, "Traditional Education in Nigeria", in O. Y Oyeneye and O. M. Shoremi, eds. *Nigerian Life and Culture*, (Ogun state University, 1985) 230-244

[109] Raymond Ogunade, Environmental Issues in Yoruba Religion: Implications for Leadership and Society in Nigeria, a paper prepared for "Science and Religion: Global Perspectives", June 4-8, 2005, in Philadelphia, PA, USA, a program of the Metanexus Institute. Available online at www.metanexus.net

community.[110] Therefore, for a peaceful co-existence of all of these beings, humans who consider themselves to be in charge must be careful not to provoke or destabilize their environment and their "co-tenants." For instance, the tiniest of insects is regarded as having rights to life.

Mabogunje further contended that traditional or pre-industrial cities such as pre-colonial Ibadan broadly consisted of four categories or classes which include: the farmers, the traders, the artisans, and the rulers. The farmers often have largest population and mostly an undifferentiated mass and because of this, they closely occupied large areas of the city with little effect on the structure of the city. The traders were also a fairly large group. Since Ibadan traders (as in other Yoruba cities) at this period were women, their effect on the morphology of the city was limited to a number of open spaces, usually in front of the compounds of quarter chiefs which served as markets.[111] The craftsmen seemed to be fewer but as Callaway, once noted, their absolute number was larger than often realised. If one were to consider the number of people required to make and repair the hoes and other implements needed by farmers; to weave the cloth for the older members of the community; to construct the houses and fashion the wood; to make leather, and bead articles to the rich; to carve door posts, decorative calabashes, and various objects of religious worship; to smelt and make brass and bronze objects: it would be found that artisans formed a considerable proportion of the population of most Yoruba cities, including Ibadan.[112] As applicable to other traditional cities, Ibadan rulers exercised enormous control and most significant influence on the morphology of the city. This could be seen in the unusual size of their compounds, often covering more than an acre of land; in the existence of a wide verandah measuring the length of the house frontage and used frequently for public meetings; in the special gabled gateway to the compound; in the line of well-carved wooden pillars and in the generally more substantial nature of the buildings.[113]

---

[110] Raymond Ogunade, Environmental Issues in Yoruba Religion....
[111] A. L. Mabogunje, "Morphology of Ibadan"....40
[112] A. Callaway, "Nigeria's Indigenous Education: The Apprentice System", *Odu*, 1, (1964) 62-79
[113] A.L. Mabogunje, Morphology....40

Generally, as observed by Onokerhoraye, the streets in pre-colonial Ibadan (as in most notable pre-colonial urban centres in Nigeria) were well laid out and urban dwellers were organised into various quarters.[114] Within each quarter effort was made to ensure that all the residents had a right to residential accommodation provided by their family. The head of each household which was then very large took responsibility for keeping the compound in good shape and general sanitary conditions. On the other hand, the head of each quarter was responsible for ensuring that dwellers within the quarter have access to daily needs for various goods and services. This was generally done by providing the neighbouring markets.[115]

One major type of environmental problem associated with land use development in most Nigerian cities (especially Ibadan) is the uncontrolled intensification of land use within the built up areas of pre-colonial cities. Generally, the land use development in these areas took place without any attempt at regulating it. Thus, all the available spaces which existed in the pre-colonial cities before colonial rule were covered with residential buildings without any effect to ensure the allocation of space for other essential activities such as community services, transportation, recreation and commercial activities. Land use development in the central part of Ibadan illustrates the environmental problems created by uncontrolled land use development in pre-colonial Nigerian cities.[116] When Ibadan was created in the 19th Century, little land use planning took place as blocks of land were arbitrarily demarcated by families who moved into the city and many houses or compounds were built without reference to any type of systematic lay-out.[117]

The crisis generated in Ibadan as a result of this lopsided arrangement did not even subside after the establishment of the colonial rule. Rather, the social changes which accompanied this development were felt on the

---

[114] A.G. Onokerhoraye, *Urbanisation and Environment in Nigeria: Implications for Sustainable Development*, Benin, (Benin: The Benin Social Sciences Series for Africa, 1995) 39

[115] Toyin Falola, *Politics and Economy of Ibadan, 1893-1945*, (Lagos: Modelor Design Aids Limited, 1989)

[116] A.G. Onokerhoraye, *Urbanisation and Environment in Nigeria......*

[117] Toyin Falola, *Politics and Economy of Ibadan, 1893-1945...*

traditional extended family system which had been the basic residential unit in the city. The traditional compound houses were broken up and replaced in the city by multiple housing units; and more buildings were infused into the already densely crowded areas. Thus, a complex, chaotic and highly dense array of compound houses became the dominant feature of the physical landscape of the central area of Ibadan. This area has been referred to as the traditional 'core sector' of Ibadan.[118]

Ibadan residents like most other Yoruba people during the pre-colonial period relied heavily on indigenous or traditional knowledge and informal ideas to protect and manage the environment. These traditional ideas are generally referred to as 'Taboos.' Taboos were introduced to regulate the moral order of the society. They took their origin from the fact that people discerned that there were certain things which were morally approved or disapproved by the deity.[119] These are not contained in any written law but are preserved in the tradition. Mabawonku and Agboola sometimes ago submitted that taboos and superstitions were often regarded as integral part of traditional education.[120] Every society in the world cares for its tradition because in a society without schools, a type of education known as traditional/informal education goes on. In Yoruba societies, traditional education is supported and encouraged because of its contribution to the growth, renewal and development of the society.[121] In traditional Ibadan (like in other pre-colonial Yoruba cities), the environment is people, and people are the environment.[122] The

---

[118] Mabogunje, 1968, Onibokun, 1973 quoted in Benin 40

[119] Cecilia Omobola Odejobi, "An Overview of Taboo and Superstition among the Yoruba of Southwest of Nigeria", *Mediterranean Journal of Social Sciences*, Vol. 4, No. 2, (May 2013) 221-223

[120] Tunde Agboola, and A. O. Mabawonku, "Indigenous Knowledge, Environmental Education and Sanitation; Application to an Indigenous African City", in D.M. Warren, Layi Egunjobi and Bolanle Wahab eds., *Indigenous Knowledge in Education*, (Ibadan: Indigenous Knowledge Study Group, University of Ibadan, 1996) 78-94

[121] E. B. Idowu, *Olodumare: God in Yoruba Belief*, (Ibadan: Longman, 1962); O. Oduyale, "Traditional Education in Nigeria", in O. Y Oyeneye and O. M. Shoremi, eds. *Nigerian Life and Culture*, (Ogun state University, 1985) 230-244

[122] Raymond Ogunade, Environmental Issues in Yoruba Religion: Implications for Leadership and Society in Nigeria, a paper prepared for "Science and Religion: Global Perspectives", June 4-8, 2005, in Philadelphia, PA, USA, a program of the Metanexus Institute. Available online at www.metanexus.net

environment and the people are in close relationship in Yoruba religion. The Yoruba concept of the environment is all embracing; humans, animals, plants, and "non-living beings", form the entire human society or community.[123] Therefore, for a peaceful co-existence of all of these beings, humans who consider themselves to be in charge must be careful not to provoke or destabilize their environment and their "co-tenants." For instance, the tiniest of insects is regarded as having rights to life. This is the reason why the ants are considered "aafa inu igbo" (the alfa of the forest).[124]

In Ibadan (like in other Yoruba cities), the environment is people, and people are the environment.[125] The environment and the people are in close relationship in Yoruba religion. The Yoruba concept of the environment is all embracing; humans, animals, plants, and "non-living beings", form the entire human society or community.[126] Therefore, for a peaceful co-existence of all of these beings, humans who consider themselves to be in charge must be careful not to provoke or destabilize their environment and their "co-tenants." For instance, the tiniest of insects is regarded as having rights to life. This is the reason why the ants are considered "aafa inu igbo" (the alfa of the forest).[127]

In order to proper enunciate the efficacy of traditional environmental ethics (taboos) in solving modern environmental among Yoruba in Ibadan and elsewhere, a number scholarly works have been done to identify and propagate such taboos at different times and circumstances.[128] Some of the

---

[123] Raymond Ogunade, Environmental Issues in Yoruba Religion....

[124] Raymond Ogunade, Environmental Issues in Yoruba Religion....

[125] Raymond Ogunade, Environmental Issues in Yoruba Religion: Implications for Leadership and Society in Nigeria, a paper prepared for "Science and Religion: Global Perspectives", June 4-8, 2005, in Philadelphia, PA, USA, a program of the Metanexus Institute. Available online at www.metanexus.net

[126] Raymond Ogunade, Environmental Issues in Yoruba Religion....

[127] Raymond Ogunade, Environmental Issues in Yoruba Religion....

[128] See Tunde Agbola and A. O. Mabawonku, "Indigenous Knowledge, Environmental Education and Sanitation; Application to an Indigenous African City...."; G. Parrinder, *West African Religion*, (London: Epworth Press,1969); J. Mbiti, *African Religions and Philosophy*, (London: Heinemann, 1969); O. Oduyale, "Traditional Education in Nigeria....." 230-244; L. M. Foster, M. Osunwole, A. Samuel and W. Wahab, "Imototo: Indigenous Yoruba Sanitation Knowledge Systems and their Implications for Nigerian

identified Yoruba traditional ethics /taboos that could be utilised in environmental management towards sustainable development in Ibadan are listed below: *A ko gbodo we owo sinu awo ti a fi jeun, ki aya ma rin eni ti yoo tun fi awo naa jeun* (we should not wash our hand in the plate we used to eat so that those who will use the plate after may not feel nauseated). This taboo is to keep ethics of cleanliness. It is a dirty habit to wash hand in the plate after eating in it because of the filthiness already in the hand after the meal. Anyone who saw this dirty habit may feel nauseated to use the plate again. *A ko gbodo subu ni baluwe, ki iru eni bee ma ba ku* (one must not fall down in the bathroom, so that the person will not die). Everybody in the house usually take their bath and urinate in the bathroom. This frequent use may make the room to be slippery. If the bathroom is not washed properly and become slippery it may cause accident which may actually lead to death. So this taboo was put in place to keep the rules of cleanliness by washing the bathroom regularly.[129]

Others include: "No drawing of water from the well at night" as one could easily slip and fall into the well; "No sitting on the edge or top of a well", as one could release foul-air and pollute the well water; "No drawing of water from a well while using a chewing stick". The chewing stick could drop into the well and contaminate the water; "No standing at a cross-junction or road, as doing so will disturb "esu", Cross – junctions are traffic nodes with too many vehicles and one may be knocked down; "A woman when observing her monthly menstrual period must not prepare; meals". This prevents food contamination with menstrual blood/discharge; "No plaiting or shaving of hair along the passage and around cooking area-This will anger "eleda" – "the creator". This is to prevent hair from getting into foods, drinking water or edible things and cause food poisoning. "No leaving of melon peels unclear over-night. Leaving them out will cause the housewife to have bad dreams in the night". Unclear melon peels attract many flies, which carry diseases; "No

---

Health Policy", in Frank Fairfax III, Bolanle Wahab, Layi Egunjobi and Michael Warren D. eds., *Alafia and Well-Being in Nigeria Studies*, (IOWA: Technology and Social Change, No. 25, Ames Centre for Indigenous Knowledge for Agriculture and Rural Development, IOWA University, 1996)

[129] Cecilia Omobola Odejobi, "An Overview of Taboo and Superstition among the Yoruba of Southwest of Nigeria", ...223

walking or kneeling on refuse heaps". The germ infection on the knee called "kulatan" could develop from kneeling on refuse; "No carrying of refuse with the bare hand. Doing so will cause the shaking or trembling of the hand". This prevent touching of contaminated/poisonous materials which may cause diseases; "No eating of a food item that dropped on the floor. One will be afflicted with "segede" – the abnormal swelling of the check/neck". This is to prevent eating of contaminated food items; cleanliness ("imototo") is the foundation for hygiene and having a bath per day is a compulsory routine. Not having a bath is believed to result in illness.[130]

However, environmental Management in the traditional (classic) Yoruba city such as Ibadan, collective or communal sanitation arrangement appeared to be lacking.[131] According to Johnson, each family was concerned with the cleanliness of only its compound and its surrounding.[132] Similarly, each seller in the market-place swept only its stall and its surroundings. Thus, there was no comprehensive system of keeping the entire city clean notwithstanding the existence of quarter and city councils. The depositories of refuse and garbage included:

a) The borders of principal roads;
b) The periphery of market-market places;
c) The open pits from which building mud was dug;
d) Several open spaces; and
e) The city outskirts.

Fire was intermittently set to the dumps.[133]

As regards sewage, the intra-urban bush lands (open spaces over grown with bush) were notable disposal areas. In addition to the city bush lands, "important chiefs have large areas of land enclosed within their

---

[130] B.O. Olabode, and S.O. Siyanbola, "Proverbs and Taboos as Panacea to Environmental Problems in Nigeria, a Case of Selected Yoruba Proverbs", *Journal of Arts and Contemporary Society,* Vol. 5, No. 2 (2013) 56 – 66
[131] O.I. Obateru, *The Yoruba City in History, 11th Century to the Present,* (Ibadan: Penthouse Publications (NIG), 2003)
[132] Samuel Johnson 92
[133] Samuel Johnson 92

compounds within which spots were selected for sanitary purposes."[134] The spots were possibly enclosed pit latrines. We are not informed of other sewage disposal sites. But one could speculate that intra-city streams and rivers were employed as sewers. The city outskirts, the inner greenbelt in particular, must also be key zones of sewage dump. In addition to the intermittent use of fire to destroy refuse garbage and sewage, nature provided its own sanitary system: the sun was an effective and efficient antiseptic, while the pig, the dog, and the vulture were ready and eager scavengers. These natural systems rendered the city cleaner than what would have been the case.[135]

Johnson described the sanitation system of the Yoruba city as "the most primitive imaginable."[136] It is noteworthy that the Yoruba traditional city was not unique in the primitive sanitary system. It was also reported sanitation problem beset Athens and plagued Rome during the pre-industrial era. In essence, it could be argued that poor sanitary systems usually bedeviled pre-industrial cities. On the sanitation dilemma of Athens, Lewis Mumford once observed as follows:

> Thus the highest culture of the ancient world, that of Athens, reached its apex in what was, from the standpoint of town planning and hygiene, a deplorably backward municipality. The varied sanitary facilities that Ur and Harappa had boasted two thousand years before hardly existed even in vestigial form in fifth century Athens....Refuse and ordure accumulated at the city's outskirts, inviting diseases and multiplying the victims of plague....it was at such municipal dumps that unwarranted babies in Athens were exposed and left to die.[137]

The insanitary condition of imperial Rome was even more grandiose, a horrible condition that provoked Lewis Mumford to offer the following lamentation:

---

[134] Samuel Johnson 92
[135] Samuel Johnson....294
[136] Samuel Johnson 292
[137] L. Mumford.... 154.

In sum, in the great feats of engineering where Rome stood supreme, in the aqueducts, the underground sewers, and the paved ways, their total application was absurdly spotty and inefficient. One cannot leave the subject of sewage disposal without noting another feature that casts serious doubt on the intelligence and competence of the municipal officials of Rome, for it records a low point in sanitation and hygiene that more primitive communities never descended to. The most elementary precautions against disease were lacking in the disposal of the great mass of refuse and garbage that accumulates in a big city....If the disposal of faecal matter in carts and in open trenches was a hygienic misdemeanor, what shall one say of the disposal of other forms of offal and ordure in open pits? Not least, the indiscriminate dumping of human corpses into such noisome holes scattered on the outskirts of the city, forming as it were a *cordon malsanitaire*.[138]

In the pre-industrial society, the larger the city, the more difficult its sanitation problem, the critical factor being the inaccessibility of rural land at the urban fringe to every urbanite for the disposal of refuse, garbage and sewage. While the hygienic condition of larger Yoruba urban settlements was poor, the smaller centres were relatively clean. For instance, pre-colonial Ilaro was described as clean by Clapperton.[139] The sanitation problems of smaller centres did not reach intractable dimensions as they did not surpass the human scale.

Occurring here and there in the traditional Yoruba city were unorganised open spaces particularly between courtyard compounds and quarters. The open spaces comprised: vacant land tracts too small for compound construction; pits and depressions from which the mud used for building was excavated; areas unsuitable for building developments: steep slopes, irregular sites, rocky areas, wetlands and water bodies. The open spaces were not improved for outdoor recreational activities. Some of them were used for gardening; while the rest were left vacant generally outgrown

---

[138] L. Mumford.... 252
[139] H. Clapperton, *Journal of Second Expedition into the Interior of Africa*, (Philadelphia: Carey, 1829) 8

with bush. It was these open spaces that were first exploited for residential developments by refugees of the 19th century Yoruba wars.[140]

Ibadan no doubt was a creation of series of wars but all these conflagrations were not offensive as Ibadan itself was not attacked especially at the initial stages of the crises. The city thus became safe haven for inhabitants and a favoured place for refugees. As Falola once observed, indeed, Ibadan maintained an open door policy; several migrants came as adventurers and wanderers.[141] Migrants were not restricted to northern Yorubaland, others were from other Yoruba groups who wanted to live in a new town and practice their occupations. Initially, each of the major Yoruba sub-groups in the city at this period had its separate quarter and autonomous administration. For example: the Ife and Oyo, who were the majority occupied Oja'ba and Mapo areas; Ijebu lived in Isale-Ijebu and Egba inhabited Yeosa.[142] It was therefore not surprising that by 1851, Ibadan had grown into a huge urban centre with an estimated population figure of between 60,000 and 100,000.[143]

Ibadan at the Eve of Colonial Rule

From the forgoing, it has become crystal clear that, right from the pre-colonial period, Ibadan had been exhibiting characteristics akin to modern urbanisation and similar to European oriented urbanisation. Significantly, the arrival of the British in Ibadan in1893 opened a new phase in the history of urbanization and expansion in Ibadan. With the colonial incursion, alien ideas of urban existence as well as alien institutions were introduced to the scene. Since most institutions tended to be space-oriented, it was inevitable that they could not be integrated into, or contained within, the city. They had therefore to find new location beyond the limits of the existing built up areas. First to come were the administrators, with their ideas of specialised roles in a hierarchy of authority. At the bottom were the messengers, the clerks, the chief-clerks,

---

[140] Samuel Johnson, *The History of the Yorubas*, (Lagos: CSS Ltd., 1997) 296

[141] Falola Toyin, *Politics and Economy of Ibadan, 1893-1945*....16

[142] Falola Toyin, *Politics and Economy of Ibadan, 1893-1945*....16

[143] David Hinderer provided these figures in Anna Hinderer, *Seventeen Years in Yoruba Country*, (London, 1873, Contained in Falola, 4)

and so on; at the top were the Assistant District Officers, the District Officers, The Provincial Officer, The Resident and the Governor. With each of these ranks went not only to a residence but an office, often separate from the residence. The latter represented a clear departure from the traditional pattern. It meant increased in space consumption within the city and it began the process of a major separation of places of work from places of residence. Eventually, it was to impose on Ibadan the familiar problem for workers who must embark on daily long distant journey from their residence to work. The administration located itself at the north-eastern end of the city and this today has grown to become the centre of Government for the whole of Western Region.

The Railway came after the administration in 1901, with its need for hundreds of acres of land. This, in turn, was located at the western extreme of the city, just outside the city wall. With the railway came the business community - the shops, the banks, the insurance companies, the motor sales and repair-establishments, produce stores, wholesalers and light industries such as bakeries. These institutions constitute the important new elements introduced into the traditional urbanism of Ibadan. But each in its own way was to effect significant transformation in the life of the city and help in creating the complexity which is now Ibadan.

It has been established in this chapter that, African/Nigerian societies including in the pre-colonial period, had by and large succeeded in evolving urbanisation systems that suited their meaningful existence as humans. In the case of Ibadan, before colonial incursion, it has begun to exhibit latent characteristic of an urban centre which the British-oriented eventually built upon.[144] Unfortunately,, however, these systems were disrupted during the colonial period in some parts of Africa by the expropriation of land for white settlers and for plantations, commercialisation of agriculture, inappropriate macro-economic policies and ill-conceived infrastructural projects; and in other areas through enactment of urban and environmental laws and policies segregating the Europeans from the Natives. Nowhere in colonial Nigeria bore the

---

[144] Isiaka Raifu, "Urbanisation in Ibadan, 1951-2000", M.A. Dissertation, Department of History, University of Ibadan, 2003 16

burden of 'environmental racism' exemplified by policy urban segregation more than Ibadan.

Despite the pre-colonial pre-eminence of Ibadan, the arrival of the British in 1893 opened a completely new phase in the history of urban development of the city. Although, Ibadan witnessed a number of unprecedented urban expansion so much that it could be argued that modern Ibadan was a colonial creation but the haphazard nature of the expansion by the British imperialists out-rightly negated sustainable urban development in Ibadan. The most debilitating effect was the evolution of a segregated society resulting into polarisation of Ibadan into two unequalled worlds (just as in other colonies). Many of these policies were continued in the post-independence period. The effects of these colonial environmental policies/laws on the development of 'modern' Ibadan shall be the focus of the next chapter.

Chapter Five
The Impact of Colonial Environmental Policies on Sustainable Development of Ibadan

Preamble

On the surface, a research on urban environmental situations (urban planning, waste management and sanitation) in colonial Africa/Nigeria at first glance seems to have little to do with the contemporary environmental challenges (e.g. unsustainability and climate change). However, as rightly noted by Jennifer Hart:

> the histories of infrastructural and sanitary failure in {African} cities like Accra help us better understand the contemporary proliferation of what Brenda Chalfin terms "Wastelandia"—urban space that is "a site of political evacuation, devoid of order or formal government...a dumping ground for domestic, industrial and human waste. This historical perspective provides important context for contemporary debates about sanitation, citizenship, and civic responsibility in the city, and raises a number of powerful questions about the colonial roots of development and public health...."[1]

More importantly as she contended further, "if sustainable development and just, resilient cities are the goals of the future, we must first reckon with these structural inequalities of the past."[2]

---

[1] Jennifer Hart "Fruity" Smells, City Streets, and the Politics of Sanitation in Colonial Accra," *Urban Forum* (2022) 33:107–127, 111-112 https://doi.org/10.1007/s12132-021-09446-4

[2] Jennifer Hart "Fruity" Smells, City Streets, and the Politics of Sanitation in Colonial Accra...."112

As demonstrated by Mamadou Diouf, the most commonly used notion to characterise the African city is that of crisis.[3] Whatever the preferred answer might be, the crisis of the African urban model poses the problem of the state. In the opinion of Pourtier R. "nowhere else is the genetic link between the city and the state seen more clearly than in sub-Saharan Africa: the great majority of cities are indeed daughters of the state."[4] Although the reproduction of the colonial city model seems to have reached the point of no return but the attempts to 'Africanise' most African cities during the colonial era which became accelerated with independence almost ended the role of the state as planner due to the economic and financial crises of the 1970s.[5] The problem it poses is that of the relevance of both concept and reality covered by the term *African city*. African cities in the reckoning of Herbert Werlin seemed to share similar horrendous environmental crises.[6] With a special focus on Nairobi, he contributed to the theory of urban systems or "micro-politics". The author discussed in great details the multiplicity of African urban environmental problems; the nature of settler regime; relationship between urbanisation and development; the dangers of urbanisation; and the malaise of African "sick cities". He submitted that most of these cities have been overwhelmed by air pollution, noise, traffic, waste, crime, racial tension, slum conditions, maladministration and many more. This work, which grew out of the 1956 East African Town Planning Conference in Kampala, concluded:

---

[3] Mamadou Diouf, "Social Crises and Political Restructuring in West African Cities", in Dickson Eyoh and Richard Stren eds., *Decentralization and the Politics of Urban Development in West Africa*, (Washington DC: Woodrow Wilson International Center for Scholars, 2007). 95. Available online at www.wilsoncenter.org/sites/default/files/Stren.pdf. See also Mamadou Diouf and Rosalind Fredricks eds., *The Art of Citizenship in African Cities: Infrastructures and Spaces of Belonging*, (New York: Palgrave Macmillan, 2014); I. L. Lindell, *Walking The Tight Rope: Informal livelihood and Social Networks in a West African City*, (Stockholm University Press, 2003); Laurent Fourchard, *Urban Slums Reports: the Case of Ibadan, Nigeria*, (Ibadan: Institut Francais de Recherché en Afrique (IFRA), Institute of African Studies, University of Ibadan, 2003).

[4] R. Pourtier, "Migrations et dynamiques de l'environnement", in G. Pontier and M. Gaud eds., *Afrique contemporaine*. (L'environnement en Afrique, Paris, La Documentation française, 1992)

[5] Coquery-Vidrovitch, C *Histoire des villes d'Afrique noire. Des origines à la colonisation*, (Paris: Albin Michel, 1993)

[6] This was the conclusion of Herbert Werlin, *Governing an African City: A Study of Nairobi*, (New York, Routledge, 1974)

Nowhere in Africa has the problem been solved of enabling the mass of Africans who live in urban areas to live in healthy and socially desirable conditions…. There is no organised public waste collection service, and community clean-ups are not usually carried out until conditions have become totally unbearable.[7]

Several decades after this submission, environmental conditions in most African cities are not getting better (the case of Ibadan which is the primary concern of this study is particularly pathetic). This requires a more comprehensive appraisal of the environmental situation in those cities, with a view of improving the condition.

A cursory evaluation of the conditions of these cities reveals that the greatest cause of the enduring crisis lies in the fact that, while industrialised countries are already taking critical steps to implement sustainable development policies, most underdeveloped countries, especially in sub-Saharan Africa, are yet to have effective programmes or sustainable national action plans on how to ensure provision of the basic necessities of life and maintaining decent and healthy environment. For this reason, most of sub Saharan African cities that ought to be "epic centres of sustainable development" are bogged down with severe environmental problems. This has led many to tag most urban centres in Africa as "sick cities" overwhelmed by many urban malaises and concluded that "today, the hearts of many cities in Africa are like islands of poverty in the seas of relative affluence."[8] The general consensus is that most people living in African cities today are facing great challenges such as: the deterioration of basic services, housing and environment, mass unemployment and underemployment, the virtual absence of State

---

[7] Herbert Werlin, *Governing an African City*….5

[8] A.G. Onibokun, *Public Utilities and Social Services in Nigerian Urban Centres: Problems and Guides for Action,* (Canada: IDRC and Ibadan: NISER, 1987). Others who shared similar view about urban centres in Africa include: Anne V Whyte, "Women Environmental Perception and Participatory Research", in Eva M. Rathgeber ed., *Women's Role in Natural Resource Management in Africa,* (Ottawa: The International Development Research Centre, IDRC, 2004); Hakeem Ibikunle Tijani ed., *Nigerians Urban History Past and Present,* (Maryland: University Press of America, 2006); I.L. Lindell, *Walking The Tight Rope: Informal Livelihood and Social Networks in a West African Cities*; Laurent Fourchard, *Urban Slums Reports: The Case of Ibadan, Nigeria,* (Ibadan: Institut Francais de Recherche en Afrique (IFRA), 2003)

welfare and many more. All of this culminated in what Ilda Lourenco Lindell called "urban crisis."[9] Although the impact of these environmental problems in Africa's urban centres vary from place to place, no place in sub Saharan Africa seem to be immunised against the tragedy. As one of the few countries in sub Saharan Africa, which had many large pre-industrial cities, Nigeria cannot possibly be an exception to the African urban environmental decay.[10]

But what is the root of the prevailing urban crisis in most African cities especially in Nigeria? Since the primary focus of this study is to proffer possible solutions to this problem, it is necessary to look critically at the environmental history of Nigeria because as Abraham Lincoln rightly submitted "If we could first know where we are, and whither we are tending, we could better judge what to do and how to do it."[11] Due to the enduring impact of colonial rule on environmental history of Nigerian cities, any attempt to proffer solution to these urban crises (such as the current effort) must include an in-depth analysis of the influence of colonial urban policies and programmes on the evolution and development in those areas. This is why this chapter (with Ibadan as the primary focus) critically examines the impact of British colonial urban environmental policies on development of Nigerian cities.

As discussed in the preceding chapter, history of Ibadan reveals that Ibadan is one of the few large pre-colonial cities in Africa. Indeed, according to Mabogunje "in a very real sense, it is regarded as the pinnacle of pre-European urbanism in Nigeria, the largest purely African City...which has attracted various epithets such as 'Black Metropolis.'[12] He further described Ibadan as 'the largest city in black Africa'....[13]As the city was expanding, a new protective wall known as 'odi Ibikunle' was built in 1858. As reported by Millson, around 120,000 out of the then

---

[9]   Ilda Lourenco Lindell, 2002:9....
[10] Laurent Fourchard, Urban Slums Reports: The Case of Ibadan, Nigeria....
[11] Carl Sandburg, Abraham Lincoln; The Prairie Years and The War Years, (New York: Dell Publishing Co. Incorporation, 1939) 13
[12] Mabogunje, Ubanisation in Nigeria....186
[13] Mabogunje, Ubanisation in Nigeria....186

population of 200,000 in Ibadan lived within this wall.[14] To further demonstrate the existence of pre-European urbanisation of Ibadan, it was said that before 1893 when it came under the colonial, the Ibadan covered about 20,000 square miles and have an estimated tributary area of 50,000 square miles consisting of most of Yorubaland.[15]

However, despite the pre-colonial pre-eminence of the city, the arrival of the British in 1893 opened a completely new epoch in the environmental history and development of Ibadan. Although, the City witnessed a number of unprecedented urban growth and expansion during this period so much that it could be concluded that modern Ibadan was a colonial creation but the haphazard nature of the growth seriously negated the outcomes of the colonial urbanism in Ibadan. The most enduring debilitating effect was the evolution of a segregated society resulting into polarisation of Ibadan into two unequalled worlds (just as in other colonies). While reflecting on the absolute inequalities that characterised the colonial cities in Africa (colonial world), Frantz Fanon observed that:

> The colonial world is a world cut in to two-(the settler's zone-where Europeans and other foreigners lived and the native's zone, where indigenes resided; which is diametrically opposed to each other)....No conciliation is possible...The settler zone is strongly built, made of stone and steel; brightly lit; the streets are covered with asphalt.....The settler town is a well fed town, an easy going town; its belly is always full of good things....The part of the town belonging to the colonised people is a place of ill-fame, peopled by men of evil repute. They are born there, they die there....It is a world without spaciousness; men live there on top of each other....The native town is a hungry town, starved of bread, of

---

[14] A. Millson, "The Yoruba Country, West Africa", *Proceeding of the Royal Geographical Society,* 13 583
[15] Mabogunje, *Urbanisation in Nigeria....* 191 and Bola Ayeni, "The Metropolitan Area of Ibadan: Its growth and Structure......" 73

meat, of shoes, of coal, of light. The native town is crouching village, a town on its kneels, a town wallowing in the mire....[16]

This description is also in line with William Bissell's view that:

Colonial designs on the city, rather than successfully reworking space, repeatedly failed to rationalize the urban sphere. These schemes, sponsored by an overextended and disjunctive state apparatus, foundered precisely in the gap between intention and implementation, hindered by internal disarray as well as the incapacity of legal and bureaucratic instruments to reorder to totality of the everyday...."[17]

He argued rightly that "this bureaucratic disarray had profound consequences for African residents in the colonial city. Sanitation was political, social, and profoundly personal; the failed struggles in implementing effective sanitation policies or building new infrastructural systems...."[18]

The choice of Ibadan is premised on the fact that cities serve as images of their societies.[19] Thus, Ibadan as a micro society within a larger entity called Nigeria can be said to visibly depict the convoluted developmental processes of the country. The general signs include but not limited to: incessant and chaotic political situation, mass illiteracy, endemic poverty, admixture of traditionalism and modernism, unregulated industriali- sation, inadequate town planning and lack of efficient urban environment management strategies. Ibadan no doubt has been a great cosmopolitan city in Nigeria and the history of its environmental decay actually reflects the history of Nigeria's ecological conundrum.[20] This chapter focuses on

---

[16] Frantz Fanon, *The Wretched of the Earth*, (Great Britain: Penguins Books, 1983). 29-30.

[17] William Bissel, *Urban design, chaos, and colonial power in Zanzibar* (Bloomington, IN: Indiana University Press, 2010) 2-3

[18] William Bissel, *Urban design, chaos, and colonial power in Zanzibar*....4. See also Jennifer Hart "Fruity" Smells, City Streets, and the Politics of Sanitation in Colonial Accra...."111

[19] Justin Labinjoh, "Ibadan and the Phenomenon of Urbanism", in G.O. Ogunremi ed., *A Historical and Socio-Cultural Study of an African City*, (Ibadan: Oluyole Club, 1999). 234.

[20] A.G. Onibokun....1987; S.I. Abumere, "Urbanization and Urban Decay in Nigeria", in P. Onibokun, F. Olokesusi and L. Egunjobi eds., *Urban Renewal in Nigeria*, (Ibadan:

the impact of colonial government's environmental policies on the unsustainable development of Ibadan in the areas of: Urban planning, Waste Management, Sanitation and Impact of Poverty during the colonial period in Ibadan.

Urban Planning

The British colonial Government assumed full control of the administration of both Northern and Southern Nigeria on January 1, 1900 and the subsequent sixty years were to have remarkable impact on urban growth and development of the country.[21] Though the interactions have started earlier, it was the cession of Lagos in 1861 that brought about British direct involvement in the transformation of what later became Nigeria in 1914. The British Crown acquired limited rights of administrative interference over land, first over the colony of Lagos and later extended to hinterland to cover all territories comprised in the Southern Protectorate.[22]In the Northern Protectorate, the colonial administration had obtained the control of the land from the Fulani under the Proclamation of 1900, culminating in the Proclamation Acts of 1910 and 1916. British suzerainty over the territory comprising modern Nigeria became complete with the 1914 amalgamation of Lagos colony with the Northern and Southern Protectorates. The colonial administration embarked on the systematic control of land use by series of statutory measures on ad hoc basis having established political and administrative control over the land.[23]

NISER, 1987); Labinjoh, "Ibadan and the Phenomenon of Urbanism…." (1999); Laurent Fourchard, *Urban Slums Reports: The Case of Ibadan, Nigeria*….

[21] NAI Oyo Prof. 811, File No 2663 Vol. I and II, "Colonial Development Schemes". See also A.L. Mabogunje, *Urbanization in Nigeria*, (London: University of London Press, 1968) 111-112

[22] A. Olukoju, "The New Lagos Town Council and Urban Administration, 1950-1953", in T. Falola and S. Salm eds., *Nigerian Cities*, (Trenton: Africa World Press, 2003). 8-46; 255-269 respectively. See also NAI Oyo Prof. 2/2, File No 676, Vol. I & II, "System of Land Tenure in Yorubaland" and NAI Oyo Prof. 2/2 File No 968 Vol I & II, "Land in Colonial Nigeria"

[23] A. Olukoju, "The New Lagos Town Council and Urban Administration, 1950-1953…."170

In 1945, the Nigeria and Country Planning Act, was enacted, fashioned after the English Town and Country Planning Act of 1932.[24] The Law had a nationwide application. According to Olukoju, the title of the Act is misleading.[25] It gives the impression that the Act is a unified Federal code covering all aspects of land usage within the federation of Nigeria, and the utilisation of land by Town Planning Authorities. However, the provision of the Act did not seek to regulate and control the use of land, but rather they spell out the type of interest obtainable in land, the mode of acquisition and quantum of rights exercisable therein. By virtue of section 4 of the Act, Governor-general was empowered to appoint a Planning Authority in any part of the country that in his opinion required a scheme.[26] An example of such planning authority was the Lagos Executive Development Board made up of the chairman and members appointed by the governor-general. The authority was the executive organ for the planning and carrying out of any scheme on the area to which it was applied. Although town planning had been conceived as a central matter at the initial stage, with the adoption of the federal constitution in 1954, it became regionalised. The emergent regions thereafter reenacted the 1946 Act as their respective regional law.

As mentioned earlier, the nucleus of a British government in Nigeria was established with the annexation of Lagos in 1861.[27] From then on, Lagos became the base from where the British operated in Southwestern Nigeria. Right from the beginning, it became obvious that the overriding and primary aim of the British was not to develop these areas. Rather their aims and goals were to prevent interest of other Europeans, extending their influence and establishing control in the area brought them into contacts with other Yoruba groups. As demonstrated in this chapter, this is why much of the taunted and documented 'developments' that occurred in most cities in Nigeria (especially in Ibadan) during the colonial rule were accidental rather than design. The environmental

---

[24] NAI Oyo Prof 21/1, File No 1566, Colonial Development Fund.
[25] Olukoju, A., "The New Lagos Town Council and Urban Administration, 1950-1953", 170
[26] A. Olukoju....170
[27] Toyin Falola, *Politics and Economy of Ibadan, 1893-1945,*(Lagos: Modelor Design Aids Limited, 1989) 16

problems associated with the nature of urban land use pattern in Nigerian cities became aggravated during the period of British colonial administration which was also the beginning of modern urban development in the country.[28] One of the major types of environmental degradation associated with land use development in Nigerian cities was the uncontrolled intensification of land use within the built up areas of pre-colonial cities. Generally, the land use development in these areas took place with little attempt at regulating it.

Indeed, the so-called development was done primarily for the convenience of British colonial administration instead of focusing on the peculiar needs of the natives. This haphazard improvement manifested visibly in the urban policies and strategies adopted by the British more than any other aspects of urban environmental governance in Nigeria at this period. In the case of Ibadan, relations with Britain were, for most of the 19th century, tied up with military and security needs. Both were aware of each other's presence and interests. The British knew that Ibadan was a major Yoruba power, "the flower of the Yoruba army" as Glover described her in 1859 in reference to her military might.[29] The signing of 1893 agreement between Ibadan chiefs Denton (who represented Mr Carter-the Governor of Lagos) ushered in colonial take-over of Ibadan.[30] Consequently, in December 1893, Captain R.L. Bower was as appointed as the first British Resident in Ibadan.[31] His first major assignment was the setting up of the nucleus of a colonial administration by recruiting the first set of clerks, messengers and interpreters. As a reflection of the British Indirect Rule system, Bower also involved native chiefs in the administration of the city (who were placed under the supervision of British officials) by inaugurating the council of chiefs otherwise known as Ibadan Town Council.[32] As part of the general introduction of local administration in Yorubaland during colonial era, the council served as

---

[28] Falola Toyin, *Politics and Economy of Ibadan*....16
[29] Falola....16
[30] Falolo 25-26
[31] S.A. Ajayi, "Bower's Tower, A Historical Monument in Ibadan", in G.O. Ogunremi ed., Ibadan....70
[32] Falola Toyin, *Politics and Economy of Ibadan*....32

Native Government Ibadan which comprised the Resident as the president, the Baale and other chiefs.[33]

Ibadan like many other pre-colonial urban centres in Nigeria witnessed a number of remarkable/ unprecedented changes during colonial era. It must, however, be clearly stated that most of these changes occurred through much coordinated efforts of the British colonialists who utlilised the instrumentality of law. As Falola once observed, law under the colonial system was an integral aspect of British administration.[34] Omoniyi Adewoye equally described law as an agency of socio-political ordering, an agency that could be applied for several ends such as 'keeping of peace and order, , the maintenance of the social status quo, the furtherance of economic interests of the dominant class or...the stabilisation of a colonial regime by one people over the other. In the case of colonial southern Nigeria, he asserted that the primary function of law was to stabilise colonial regime (mostly at the expense the colonised people).[35] It was therefore not surprising that the colonial masters enacted series of ordinances and laws in colonial Nigeria ostensibly to ensure peoples' adherence and enforcement of 'safe and adequate' urban environmental management strategies (most importantly in area of urban planning).

There were several of such ordinances and laws. For instances, there were Town Improvement Act of 1863; Swamp Improvement Act,1877, the Waterworks Ordinance, 1915, Railway Provident Fund Ordinance, 1915,Criminal Code Act, 1916, Public Health, 1917,Electricity Supply Ordinance,1917, Public Ordinance, 1917 and many more.[36] Between 1863

---

[33] C.H. Elegee in his book, *The Evolution of Ibadan* maintained that Ibadan City Council originally have the Resident, *Baale, Otun Baale, Osi Baale, Balogun* and twelve minor chiefs as members. See also, Toyin Falola 33

[34] Toyin Falola 33

[35] O. Adewoye, *The Judicial System in Southern Nigeria, 1854-1954: Law and Justice in Dependency,* (London: Longman, 1977).

[36] The relevant information about most of these Laws and Ordinances were discovered in many archival files at the National Archives, Ibadan, Nigeria. Some of them include NAI Oyo Prof. 2/2 File No 556 Vol. I-IV, "Nigerian Ordinances"; NAI Oyo Prof. 21/1, File No 1566, 'Colonial Development Fund'; Oyo Prof. 21/1, File No 4504 Vol. I and II, 'General Progress of Development' and Community Development Progress Reports; NAI Oyo Prof. 811, File No 811, Vol. I and II, 'Colonial Development

and 1900, the British acquired the whole of Southern Nigeria and introduced development control through the promulgation of the 1863 Act. In 1928, another Act to enhance development control was promulgated which established the Lagos Executive Development Board (LEDB), presently renamed as Lagos State Development and Property Corporation (LSDPC).[37] Ibadan came under firm legal control of the British in 1901 when Governor William MacGregor enacted the Native Council Ordinance which brought about some changes. With this reform, the Baale replaced Resident as the President of the council but the former continued to consult the latter for advice when necessary. Apart from the fact, the council was empowered to deliberate on internal administrative issues and matters affecting the people within its area of jurisdiction; the Ordinance also initiated the idea of including some educated men in the Council.[38]

It must reiterated here that, the 1901 reform which assigned administrative and legislative functions to the Council, instead of serving the interest of the natives, essentially entrenched colonial rule in Ibadan District. Though the impression of self-government was created by making the Baale the president and expanding the membership; all legislative and administrative functions of the Council were in conformity with the British colonial mindset, philosophy and dictates. Indeed, in line with the intricacies of the Indirect Rule system, Ibadan City Council and its subsequent Committees on health, works, education, road and agriculture were used to propagate and implement colonial social, economic, political and environmental policies.[39] Apart from these administrative reforms, railway also reached Ibadan and this later proved

---

Scheme, 1940-1941'; NAI Oyo Prof. 811, File No 4104, Provincial Geography, 1944. See also, Donald Cameron, *The Principles of Native Administration and their Application,* (Lagos, 1934) and Margery Perham, *Native Administration in Nigeria,* (London, 1962)

[37] F. O. Ogundele, O. Ayo, S. G. Odewumi and G. O. Aigbe, "Challenges and Prospects of Physical Development Control: A Case Study of Festac Town, Lagos, Nigeria", in *African Journal of Political Science and International Relations,* Vol. 5, No. 4, (April 2011). 174–178

[38] Ibadan City Council Minute Book, Minutes of Meeting for 10 October, 1903; See also, J.A. Atanda, *The New Oyo Empire: Indirect and Change in Western Nigeria, 1894-1934,* (Ibadan: Longman, 1973) 83

[39] Toyin Falola, *Politics and Economy of Ibadan*....39

very significant in the evolution of Ibadan as a prominent urban/commercial centre during colonial era in Nigeria.

In continuation of the colonial agenda and administrative convenience, by 1912, management of Ibadan was put under the Baale and chiefs. Thus, the city was divided into 18 quarters with each under the control of a chief who was allowed to be protected by one Akoda (native police).[40] The growth in importance and changes experienced in the city due to colonial urbanisation also affected the status of the traditional institutions and the structure of governance in the city. For instance, in 1936, the title of Baale was changed to Olubadan and this led to formation of Olubadan in Council.[41] Several laws such the Local government law of 1950 were also promulgated to further entrench colonial induced changes in Ibadan. This law made provision for the Regional Authority to conduct inquiries in the localities for the purpose of determining the units of local government. This development led to the replacement of the Native Authorities representative local authorities.[42]Interestingly, all through this period of changes in Ibadan City Council, the administrative centre remained at Mapo hall.[43]

As we have established earlier, there were hundreds of laws that governed British urban planning adventures in colonial Nigeria but this study argues that none seemed to have far reaching and enduring impact on Ibadan and other cities like the Township Ordinance of 1917. Sadly, its legacy outlasted the British colonial rule in Nigeria and even persists till today. Details of the impact of the 1917 Township Ordinance and other related colonial legislations of Ibadan are discussed below:

Several records and authors have reflected on the negative (especially the segregating) impact of the 1917 Township Ordinance in most Nigerian cities and especially Ibadan during and after colonial rule. For examples: Ogundele et-al, observed that the British Town and Country Planners Act

---

[40] Toyin Falola, *Politics and Economy of Ibadan*....39
[41] These changes were contained in a letter from Resident's Office to Baale's Office dated 9th July, 1936. See NAI Oyo Prof. C42
[42] NAI Iba Div1/1 2642/S.3
[43] Ruth Watson, "Change in the Symbolic Meaning of Mapo Hall, 1925-1945", in O. G. Ogunremi ed., *Ibadan*....84

of 1917 defines development "as the carrying out of building operations engineering, mining and other operations in, on, under or over land; or other land."[44] In the opinion of Olukoju:

> The British adopted an urban policy of segregation and separation in Nigeria. The Governor-General is empowered to zone each of the townships into "European", "Non-European" Reservations and open spaces for the urban physical development. Mixed racial residences were rendered unlawful in a reservation area under the 1917 Act (section 66 of the Act, 170).[45]

In the same token, Seyi Fabiyi noted that during the colonial period Nigerian cities went through tremendous changes in every aspect of their construction and physical organisation and many of these changes began in Lagos. He stressed further that the enactment of Township Ordinance in 1917 was very significant in this transformation process. In his words:

> The act provided for the creation, constitution and the administration of all towns and municipalities in Nigeria. The act also classified cities in Nigeria into three, which were first class towns, second class and third class. The criteria for this classification was not explained in the act and Lagos (derived from Portuguese Lagoon de Curamo) was the only first class town in this classification being the seat of colonial government and a port city. It was so popular among the colonialists. The first class town was administered by a Town council while second and third class towns had appointed officers. The second class towns were mainly trade centres where they served as European depots and in 1919, there were 18 of them. Most towns developed or parts of the existing cities that developed during the colonial rule were patterned

---

[44] F. O. Ogundele, O. Ayo, S. G. Odewumi and G. O. Aigbe, "Challenges and Prospects of Physical Development Control: A Case Study of Festac Town, Lagos, Nigeria....." 174

[45] A. Olukoju, "Nigerian Cities in Historical Perspectives...." 29-31

according to the planning concepts prevalent in UK at the time. Many of the road layouts were crisscross pattern.[46]

This account also revealed the fact that segregation both in terms urban planning and income were fashionable at that time. For example, the author observed that the European Reservation Area (ERAs) was residential scheme developed to house the colonial rulers and their enclaves.[47] They were marked differences between the European settlements and Nigerian or African settlements in terms of quality of housing and urban infrastructures. While the layout in the ERAs had adequate accessibility, the African settlements especially the traditional parts of the city were in the process of fragmentations of large compound into smaller units to house single or two family members. The consequence was that the accessibility into traditional part of the city was mainly narrow lanes or footpaths.[48] Evidence was also provided that since the colonial times there has been a deliberate tradition of not providing adequate resources for certain professions, skills, and technology of importance to the issues of: environmental protection, forestry, surveying and environmental health engineering and especially- town planning.[49]

In his own account, Akin Mabogunje posited that the British policy on urban planning and development was contained in the Township Act of 1917. According to him, the Act provided for the creation, constitution, and administration of all towns and municipalities in Nigeria with the exception of those native towns where the population was sufficiently homogeneous for it to be administered by Native Administration.[50] He further stated that the Township Ordinance operated until the end of the

---

[46] Seyi Fabiyi, "Colonial and Postcolonial Architecture and Urbanism", in Hakeem Ibikunle Tijani ed., *Nigeria's Urban History, Past and Present*, (Maryland: University of America Press, 1984) 148

[47] Seyi Fabiyi....

[48] See NAI IBA Prof. Vol. 1-4: Ibadan Provincial Office, 1897-1960; NAI IBA DIV Vol. 1-4: Ibadan Divisional Office, 1893-1957) NAI Oyo Prof 811, File No 2919, 'European Reservation Areas', 1949; NAI Oyo Prof. 811, File No 2080, "European Reservation Ibadan, 1937-1950"

[49] Andrew G. Onokerhoraye, "Urbanisation and Environment in Nigeria: Implications for Sustainable Development", (Benin: *The Benin Social Sciences Series for Africa*, 1995) 10

[50] A. L. Mabogunje, *Urbanization in Nigeria*.... 111-112. See also NAI Oyo Prof. 2/2, File No 556, Vol. I, "Nigerian Ordinances"

Second World War before it was then overtaken by series of constitutional changes which marked the post-war period.[51] Throughout the period of its operations, Lagos was the only first class township in Nigeria. Second-class township were about 12 in the Southern provinces (Aba, Itu, Abeokuta, Calabar, Onitsha, Enugu-Ngwo, Opoko, Port Harcourt, Forcados, Sapele, Ibadan and Warri) and Six in the Northern Provinces (Ilorin, Kaduna, Kano, Lokoja, Minna and Zaria) in1919.[52] The Act empowered the Governor General to declare and administer any place or area as first, second or third class Township (section 3 of the Act). In pursuant to this provision, Lagos was declared as First Class Town. According to Mabogunje, the selection and classification of townships did not depend on the size of the population or the traditional importance of these towns; rather it was based on the level of the presence of colonial activities of one form or the other in the area.

Interestingly among all scholars who have written on the implications of 'the Township Ordinance of 1917' on urban planning in Nigerian cities only Akin Mabogunje, P C Lloyd and Toyin Falola analysed the particular effects it had on Ibadan. Although the three authors came to similar conclusions about the ordinance as the most powerful weapon of segregation and the real instrument for the eventual polarisation of Ibadan into two unequal worlds during the colonial rule (the legacy persists till date); it was Falola's account that offered a more elaborate and much expository analysis of the heinous act at both colonial and post colonial times. According to Falola, the Ibadan Township was located in parts of Southwest and Southeast.[53] It encompassed part of what later became known as the New Reservation (or Jericho); the Railway Station, Iyaganku, Iddo gate, and Lebanon street. It was conceived to be autonomous in administration, and chiefs were not supposed to exercise power in the township. The Local authority was the station Magistrate (or any other officer) assisted by an advisory board of members chosen by the governor. The station Magistrate relieved the district officer of those activities connected with trade, European community, housing, land leases, sanitation, railway matters, road maintenance and of other

---

[51] A.L. Mabogunje, *Urbanization in Nigeria....* 111-112
[52] *The Nigerian Handbook*, (Lagos, 1919) 98
[53] Toyin Falola, *Politics and Economy in Ibadan....* 116

important affairs in the township. Like other second class townships, Ibadan was within the authority of the Supreme Court and the station Magistrate. For its finance, it had a 'township fund'.[54] It relied on the Public Works Department to assist in the construction and maintenance of roads and other works, and on Medical/Health Officer for sanitation and other aspects of public health.

The period of the First World War, 1914 to 1919 also witnessed several changes in Ibadan and most of these changes emanated from Lagos as by-products of the First World war.[55] One of the major changes that came as a result of policy directive from Lagos was the establishment of a Township area as contained in the 1917 Township Ordinance. According to Lugard, the main purpose of this Ordinance was to establish the broad principle of municipal responsibility, highlight the importance of each community and measure its ability to accept and discharge satisfactorily the conduct of its own affairs.[56] Townships were graded into three classes, from first to third, 'according to the degree of control and responsibility vested in the governing authority, and they include practically all centres where Europeans reside.[57] Only Lagos was a first class township, and Ibadan was one of the eighteen second class townships.[58]

Falola stressed further that, in order to promote a deliberate policy of segregation, the British created the Ibadan Township area with the promulgation of The Townships Ordinance of 1917.[59] According to him, this was at two levels. The first was the separation between indigenes of a town and Nigerians from other places (the so-called alien natives). Indigenes were discouraged from settling in the township; but the law was stricter on preventing the poor- labourers and others engaged by Europeans- than the elite who found the township a better place to live. The Township Ordinance also contained a provision which empowered the government to expel any native from the area. The second was to

---

[54] Falola, *Politics and Economy in Ibadan*....116
[55] Falola, *Politics and Economy in Ibadan*....117
[56] NAI Oyo Prof. 2/2, File No. 556, Vol. I, "Nigerian Ordinances"
[57] NAI Oyo Prof 2/2, File No. 556.133.
[58] N.A.I. Iba. Prof. 253/1/G. N.A.I. Iba, Div. 1/1 2642/S.3, 2642/S.3 Vol.2, and 2994 Vol. 1.
[59] Toyin Falola, *Politics and Economy in Ibadan*....135

segregate Europeans from Nigerians, irrespective of the status of the latter. Europeans lived in reservations where they have access to the best medical attention, strong security, adequate water supply, good road and other social amenities. Although the reservations were close to the townships, but the distinction was so visible as follows:

> All the townships are divided into European and a native quarter, separated by a non-residential area of a quarter of a mile in breadth, which extends round the former. This belt is kept clear of undergrowth, and may be used for recreation, and parade grounds, and even for garden allotments, in which high-growing crops are not allowed. Non-residential buildings may be erected upon it, such as churches, court-houses, stores etc. provided they do not impair its utility as a fire-break, on the side of the native quarter. Europeans may not reside in the close vicinity of a township, but must live in the European reservation, here the amenities of a pure water supply, and a police protection are as possible available.[60]

One other excuse given for this segregation was to protect Europeans from being contaminated by the allegedly yellow-fever infected 'natives'. It was believed that the segregation would enable Europeans to live more comfortable and healthier lives.[61] But there was a major problem of how to restrict the movements of those natives who must have direct contacts with the Europeans such as the customers and domestic servants who by virtue of their engagement must come in contact with the white men. There was a great anxiety that these seeming unavoidable contacts could constitute health hazards.[62] A number of suggestions were made by sanitary and medical officers in Ibadan between 1919 and 1920. One of these was that part of the township in the city be reserved for all merchants who should be compelled to trade and live there. This had two visible advantages of not allowing Europeans to go to the 'native town' or the traditional part of the city where the risk of contamination is greater and also made it easier to clean and sanitise the trading centres.[63] In the

---

[60] Toyin Falola, *Politics and Economy in Ibadan*.... 72
[61] Toyin Falola, *Politics and Economy in Ibadan*....135
[62] NAI, Oyo Prof. I., 895, Vol. IV, Sanitation-Oyo Province
[63] Toyin Falola, *Politics and Economy in Ibadan*....135

case of Ibadan where some firms had established business and residential quarters within the town for some years and had obtained land leases for long period; they were advised to stay together in an areas considered large enough for building of shops and houses. New firms were also compelled to stay in the segregated areas while the older ones were induced to join them. When their leases expired, they, too, were mandated to live in the segregated areas.[64]

With this lopsided arrangement, racial segregation and ethnic disaffection created by colonial urbanism became more drastic and damaging in Ibadan and other Nigerian cities. It also fostered class consciousness which was anchored on the concept of modernisation especially among the few educated elite. These "civilised" few felt that they were privileged to imitate western culture and who also benefited from the colonial state began to see themselves as *olaju* (the civilised) to distinguish themselves from the *ara oko* (the rural, uncivilised).[65] As a result of this, the British colonial urbanism created difference and inequalities in the intra-city relations and growth in Ibadan.[66] The problem included discrimination against the multitude who found themselves in the underprivileged class of "*ara-oko*" (the rural, civilised). Throughout the 1920s nearly all indigenes of Ibadan with the exception of a handful were "*ara-oko*" while the successful ones among the strangers were "*olaju*."[67]

However, as the colonial administration began to bring-home more of the western urban culture, the number of the "*olaju*" began to increase and each began to develop its own culture in Ibadan. For the "*olaju*' to practice their newly found culture, they needed to live in an area separate from the "*ara-oko*". The township provided this opportunity; which was seen an ideal settlement for the '*olaju*'. The British administration in the city protected their interest by supplying them with necessary amenities. The administration was of the view that segregation would discourage hostile relations with indigenes, and enabled strangers to participate fully in the

---

[64] Toyin Falola, *Politics and Economy in Ibadan*.... 73
[65] Toyin Falola, *Politics and Economy in Ibadan*....73
[66] Toyin Falola, *Politics and Economy in Ibadan*.... 220
[67] Toyin Falola, *Politics and Economy in Ibadan*....220

economy and have access to land. Unfortunately, this was to become a marked dividing line between and among cultures within the city.

The township enjoyed many privileges and opportunities. First, its members controlled the wealth of the import and export trade. The leading firms were allocated choice areas where they eventually built their shops at the expense of the natives. In 1922 for instance, it was almost impossible to secure one in the trading sites in the business area. High rents were paid in subsequent years. These shops dominated wholesale trade in the city. On March 2, 1922, the Bank of British West Africa was opened to serve the interest of the commercial elite.[68] Secondly, houses in the township were Brazilian type bungalows and one storey buildings were common. As earlier stated attempts were made to have clean environment, some even had gardens. There were also the existence of public utilities and basic infrastructure. In 1928, among the new facilities provided were four incinerators, three butcher stalls, a parking ground, four dust-bins, one slaughter slab and two public latrines at Gbagi Business District in Ibadan. In fact it was reported in 1930 that the European township also had eight incinerators, twenty-three labour and one headman for refuse disposal.[69]

The best road in Ibadan which traversed the township was tarred. On the other hand, the larger part of the town where the indigenes lived did not have as many opportunities and privileges. The administration did not concern itself with the outlying villages, except of course to provide 'cocoa roads' to evacuate their products and force them to pay the annual tax. To further show the disparity in the way the colonial government administers the 'two towns' whenever cases of illness among members of the township, cases affecting Europeans were usually taken more seriously, while little concern was shown for the indigenes. For instance in 1922, government was more concerned with the six cases of 'black water fever' among the Europeans than in the influenza epidemic in the

---

[68] Toyin Falola, *Politics and Economy in Ibadan*....220
[69] NAI Oyo Prof. File No 1323, Vol. I-III, "Town Planning Committee for the Southern Provinces, 1930-1956"

larger parts of the city[70]. When an interest was shown in epidemic it was because of the fear that it could affect indigenes and non-indigenes. Also to buttress this point, while the whole of the Oyo province had only a single Medical officer, the small township (for Europeans), in contrast, had a senior Medical officer with other supporting staff.[71] In 1924 when the larger town lacked western medical facilities, the colonial government was lamenting that the one in the township for the Europeans needed an urgent attention because of its location close to the railway.[72] Also supervision of health conditions was better in the township where the government paid for five sanitary inspectors, while the rest of the city were left in the hands of the Akoda or native police for each quarter of the city. More so, while interest was shown in Dugbe market where the members of the township usually bought beef and food stuff, the other markets were ignored.[73]

The township was organised as stipulated by the Township and Public Health Ordinance which mandated the concentration of foreign firms and with this, the place acquired a distinct outlook. The properties under the control of Native Authority were to be acquired. In 1919, the first major step was taken to acquire the Jubilee market from the Native Administration so as to incorporate it into the township. In October, 315 pounds was paid into the N.A. funds and the control of the market changed hands. In the same year, plans were completed to mark out the area between Government land (i.e. the neighbourhood of Railway Station) and the Ogunpa stream 'into trading plots with suitable dividing roads...in such a way that firms can acquire more than a plot if so desired.[74]

The trading area, that is the neighbourhood of Iddo gate, became part of the township. Europeans could only trade in this area (called Gbagi by

---

[70] NAI Oyo Prof. 21/1 File No 885, Vol. III, "Sanitation Oyo Province matters, 1921-1922"

[71] NAI Oyo Prof. 21/1 File No 1340, Vol. II, "Medical & Sanitary Department Oyo province"

[72] NAI Oyo Prof. 21/1 File No 1340, Vol. II,....a

[73] NAI Oyo Prof. File No. 808, Residential Layout at Ibadan,1932-1935

[74] NAI Iba Prof. File No. 32, "Land rules etc Acquisition of land for by Nature Authorities" 1923-1938

indigenes) and must reside elsewhere, in order to control the spread of diseases arising from contacts with the 'natives'. This area became (and up till date) the major commercial centre in Ibadan. All leading traders and the branches of European firms acquired plots of the township.[75] In no time, the township became the important part of the city. It controlled commerce and recreation; enjoyed better municipal facilities and accommodated 'the men of substance and power in the hinterland'.[76]

The diversities in the population in the opinion of Mabogunje also created cross-currents of segregation which underlie some of the complexities of the morphology of Ibadan. For instance, there was a rough type of segregation based on ethnic and racial considerations, which distinguished the areas occupied by the 'indigenous' Ibadan from those occupied by other Yoruba people. Areas occupied by other Yoruba were also separated from where non-Yoruba were living. And all of these areas were also segregated from the European reservations. But socio-economic realities in the city at this period seriously countered these divisive provisions. Thus, an indigenous Ibadan man, having attained certain level of income and social status, will move away from the family house where he pays no rent to live in another part of the city, to be among those of comparable status and income level even though he has to pay rent there.[77] With this development, the basis for the complexity of morphological differentiation in Ibadan was laid. The convergence in the city of traditional and modern economic, social and political institutions, and of the classes of people required to maintain these institutions explained much of the character of the city. In Ibadan, the new has failed to swallow up the old, both continue to exist, strangely juxtaposed, maintaining rather complex relations with each other, and functioning almost despite each other.[78]

It was therefore not surprising that the internal structure of the city showed the existence of two city centres- the older centre around Oja-Iba

---

[75] NAI Oyo Prof. File No 1187, Vol. I and II, "Ibadan Business Area Allocation intelligence Report, 1933-1937"
[76] Toyin Falola, *Politics and Economy in Ibadan,* ... 76
[77] Mabogunje.... 43
[78] Mabogunje.... 43

(now Oja Oba) and the new one around the railway station situated around an area now known as Dugbe. The characters of these centres gave a clue of the development to be expected around them during colonial rule and even beyond. Oja Oba was an area of open ground without any vegetal cover (plant or grass) owing to the lateritic (hard) nature of the rocks that form it. It served as an everyday market of the week but usually at its peak in the evenings between 7 pm and 8 pm.[79]Here, large number of small scale traders usually converged to peddle a wide diversity of commodities which ranged from home made *eko*, ready for immediate consumption, to livestock and even textiles and small hardware. In nearly all cases, minimal capital was usually invested thereby resulting in negligible and inconsequential realisable profit. Largely for this reason, such traders usually bore limited overhead charges and were also excluded from payment of rents and rates. But the scale of their trade did not limit the importance of Oja Oba during colonial era. These small scale traders usually served as essential link in the distributive chain, selling in units which were small enough for the poorest to afford. They also provided economic succour to the majority of the low income groups and especially in the older part of the Ibadan, where the new colonial economy had undermined the traditional economic base. Indeed, these traders formed a very important class in colonial Ibadan.

However, Oja Oba has other, and perhaps more compelling attractions for the majority of people in the older part of the city. Beyond being a forum for economic interactions, it also served as centre for social intercourse. In the evenings, young men attended the market with the hope of meeting their future brides. They equally visited the market to gather latest news and discuss politics of the city and that of the country as a whole; share comradeship/camaraderie with their friends or even just to feel a sense of belonging.[80] Most importantly, the market as the (highest point of socialisation) was always the terminal point for many festivities in the older part of the city. It was therefore not surprising for the market to

---

[79] Mabogunje.... 44
[80] Mabogunje.... 44

serve as a civic centre of a sort where people usually share their happiness as they were being entertained by a number of happy dancing groups.[81]

As it was at the pre-colonial times, the morphology of the older section of Ibadan was based on traditional social structure and on traditional technology with regard to building, economy and social conditions for the better part of the colonial period. Due to the circumstances of the foundation of the city, very little attempt seemed to have been made to plan the city. This was largely because the land was given out as block grants to chiefs who proceeded with their relations and followers to occupy them as they thought fit. Such blocks of buildings came to form a 'quarter' of the town and often bore the name of the chief prefixed by the significant topographic element such as hills (*Oke*), roads (*Opo* or *Popo*) or market–place (*Ita*).[82]

Being the oldest settled part of the city, the area around Mapo Hill including *Oja Iba* was the most densely occupied portion of the city. Even in 1856, Hinderer described it as 'the most populous part in contrast to the more widely built and cleaner parts occupying extensive portions of the places below the hills on every side.'[83] Thus, congestion, which according to Mabogunje defies description' became one of the most visible feature of the older part of the city. The inability of this area of Ibadan to grow beyond its 1900 boundary and some significant changes during colonial rule was given as the main reasons for this.[84]

To understand the significance of these changes, it is necessary to stress the nature of traditional social and economic conditions which sustained compound life in the past. The mutual relationship between the compound land in the city and the farmland in the rural area was a fundamental factor. Usually, the status of a lineage in a community implied rights over both types of land. Consequently, in a society where the main source of income was from agriculture, much power resided in

---

[81] A.L. Mabogunje, "The Problems of a Metropolis", in Lloyd, Mabogunje and Awe eds., *The City of Ibadan*, (London: Cambridge University Press, 1967) 261

[82] Mabogunje, *Urbanisation in Nigeria*.....46

[83] CMS C/A2/049/103, D. Hinderer, Journal Entry, June 2,1856.

[84] Mabogunje, *Urbanisation in Nigeria*.... 47

the hands of the head of a lineage, who decided on issues affecting both town residence and the use of farmlands, and who had powers of dispensing hitherto unused land for the use of the members of the lineage. The concentration of such economic powers in the hands of the head of the lineage (apart from other social and political powers attached to his position within the community) tended to encourage greater solidarity within the lineage and preserve the compound as a single residential entity. But certain developments especially during the colonial rule drastically reduced the power of the heads of lineages.

In the first place, many young members of a lineage became aware of wider range of opportunities to acquire either new skill unconnected with family membership or new employment that did not require the use of land. Even where an individual remained a farmer, the rise cocoa production for export, with its unprecedented opportunities for wealth accumulation within relatively short period meant that the hold of the head of lineage over his juniors was greatly reduced. As a corollary, lineages began to witness increased differentiation in the range of income and a stronger tendencies towards individualism occasioned principally by differences in social expectations. Therefore, as chances presented themselves, members of a lineage separated their portions of the compound and transformed into single-family houses. This was a much more manageable unit on which to concentrate the improvement they desired for themselves and their immediate family. Compounds then not only began to break up but to start depicting differing standards, some of which survive till today. Some of the houses have adequate ventilation, others do not; some are plastered with cement, others remain of mud; some distempered with paint, others retain the reddish-brown colour of local clay; some have electricity, others continue to depend on oil or kerosene lamps.[85]

Religion also played a role in this 'transformation'. The two dominant religions in the city are Islam and Christianity. In spite of their differences, both religions emphasise that the basic unit of the society is the family, a man, his wife or wives and his children, rather than the lineage. Converts

---

[85] Mabogunje.... 47

to these two religions thus had little compunction in concentrating their income on the improvement of living conditions for themselves and their immediate dependants though the latter group was often extended to include a man's mother, his brothers and sisters.[86]

There was other element that particularly encouraged the congestion which today characterises this part of the city, more probably than it did in the past. It may be described as the pull of newer central business districts. Many of the city's indigenous inhabitants sought non-farming work, the location of which was usually on the farther side of older city. Considering the relative poverty of these people, a further expansion of the older city in the direction away from the new business district would mean greater cost of transportation either in terms of the effort needed to walk the distance, or hard-won cash. For this reason it was a matter of great convenience not to have to move much further than was essential. As long as there was vacant land within the compound, buildings continued to be crowded onto the available space.

In consequence, the built-up area in this eastern part of the city did not increase for decades. But growth began again only the 1950s.With the Free Primary Education Programme introduced in 1955, numerous new schools were built on the extensive vacant land in the area that was mostly located at the edge of the city. As the schools grew up in ring in this area, there was a spate of housing development to provide improved accommodation to lease to the numerous teachers and other workers required to serve these schools.

In summary, the major features of the morphology of the older section of the city are its amorphous lay-out, its high density of housing and population which only results into some order in terms of 'quarters', the indifferent, though highly varied, quality of housing, its sprawl of rusty brown roofs with the towers of churches and mosques providing the only breaks on the skyline, and its 'green-belt' eastern margin, made up largely

---

[86] M.O. Filani, F.O. Akintola and C.O. lkporukpo eds., *Ibadan Region*, (lbadan: Rex Charles Publication, 1994)

of the football fields of the numerous schools, colleges and hospitals, which have grown up here only recently.[87]

The 'newer' part of Ibadan is largely the immigrant section of the city. Its development and differentiation in many ways epitomises the history of human contact in Nigeria-the mixing of the people, the colonial struggle for internal autonomy and the independence of Nigeria. Although all of these events have remarkable repercussions on the townscape of Ibadan but none (as it be demonstrated in this chapter) affected the environmental history of the city like the colonial rule, which of course greatly aided the influx of immigrants into the city. The immigrant areas of Ibadan lie to the west of the central range of hills and especially to the west of 'ogunpa stream'. The morphology of this half of Ibadan reflects the ethnic and racial differences among the immigrant groups; the date of arrival of earliest substantial section of each group; the prevalent idea of housing and architectural design; the incidence of land apportionment and sales; and increasing social and economic differentiation in the city.

Ibadan also witnessed some notable positive changes during the colonial rule in terms of urban services and planning. With the establishment of colonial administrative machinery as argued by Abayomi Makinde, attention of the Native Authority shifted from war to maintenance of law and order which aided development.[88] The new administration took measures that operated to the benefit of Ibadan's development in the spheres of physical infrastructure (railway and road health, education, abolition of tolls, tapping of forest products and farming. There were changes in health habits especially the prevention and the treatment of diseases. As far as prevention of diseases was concerned, sanitation was encouraged by the enforcement of certain rules of health. Vaccination was undertaken to prevent smallpox. For example, in 1920 the Native Authorities in the Province employed fifty vaccinators with the financial cost of One thousand five hundred and eighty-nine Pounds.[89] The digging

---

[87] M.O. Filani, F.O. Akintola and C.O. lkporukpo eds.....

[88] Makinde Abayomi, "Local Administration in Ibadan City, 1893-1934", Undergraduate Long Essay, Department of History, University of Ibadan, Ibadan, Nigeria, 1974

[89] NAI, Oyo Province Annual Report, 1920, (para 176)

of pit toilets was enforced and building of public latrines was also undertaken. On water supply, digging of wells was encouraged to provide useable water.[90] This was improved upon in Ibadan by the provision of pipe-borne water in the 1920s.[91] Dispensaries and a few hospitals were also provided for the treatment of diseases. The first of such dispensaries was opened in Ibadan in 1902[92] and the Adeoyo Hospital as the first Native Authority Hospital in Ibadan was opened to the public in 1927.[93]

At the Adeoyo Hospital alone, in-patients between 1929 and 1930 were about 400 and 458 respectively, while out-patients for the period were 2,831 and 7,554 respectively[94]. In spite of this, there can be no doubt that those who took advantage of the newly introduced western method of curing diseases or delivering health care services were in the minority. The majority of the people still relied on traditional or indigenous medicine.[95] This lack of faith in orthodox medicine by the natives during colonial era was further engendered by the grossly inadequate provision of health care facilities.

On road construction, the new administration built some roads within Ibadan and the ones linking the city with the neighbouring towns and villages. This began in August, 1897 by the order of the Governor of Lagos, Sir Henny Mc Mcllum. Generally, as opined by Makinde, both the British administrators and Yoruba chiefs deserved some credit for the building of roads in the Province. The former usually gave direction and supervision while the later ensured the adequate supply of labour. For the greater part of this period, these labourers were neither paid nor fed.[96] However, as records showed, payment for labour commenced from the

---

[90] NAI Oyo Prof. 2/1, File no 772, Ibadan Water Right Schemes, 1921-1929
[91] NAI, CSO. 26/2, File 12723, Vol. V, Oyo Province Annual Report 1927, (para 138) NAI Oyo Prof. 2/1, File No 773 Vol. III, Ogunpa Water Supply Ibadan, 1827-1932
[92] C.H. Elgee, *Evolution of Ibadan*....11
[93] NAI CSO 26/2, (para. 11)
[94] NAI Oyo Province Annual Report, 1930, (para. 98)
[95] Makinde......33
[96] H.W.W. Pollitt, *Colonial Road Problems: Impressions from Visits to Nigeria*, (London: His Majesty Stationery Office, 1950)

1920s. In 1924 for example, it was confirmed that there was no longer 'free labour for making roads.'[97]

It was not too long before these efforts started yielding results. By October 1906 when Governor Egerton of Lagos brought the first Motor car to Ibadan, the city and its neighbourhood could boast of twenty miles of roads 'sufficiently good to support wheeled traffic.'[98] By 1907, Ibadan and Oyo which were three miles apart, had been linked by road, which made it possible to start a motor transport services between the two towns.[99] The construction of roads, the establishment of motor transport services and railway service brought a great change in the means of transportation. By the 1920s, Engineer Robert Jones who also must be credited for designing and construction of Mapo Hall between 1925 and 1932 already started transport business in Ibadan.[100]

Perhaps, one of the most noticeable features of colonial urbanism in Ibadan was the rise in the commercial importance of Ibadan. This was further enhanced when foreign companies already operating in Lagos opened their branches in Ibadan. Some these companies included: Miller Brothers, G.B. Ollivant, African and Eastern Trade Corporation, John Holt, Patterson and Zochonis (PZ), United African Company (UAC) and the likes. Apart from the fact that this attracted more people into the city; it also encouraged further diversification in the character of its population in terms of ethnic origin, education and levels of income.[101]

Over the years, as the railway linked diverse parts of the country together and as the road linkages wore – extended and improved, more and more people speaking different languages and with varying educational qualification migrated into Ibadan. By 1924, the population of the city had become more diverse and heterogeneous in its socio-economic characteristic. This diversity however, was not only apparent in terms of

---

[97] NAI Oyo Prof. 4/12, File No 59/1923, Memo No. 713/92/1924 of 14 March, 1924, D.O. Ibadan to Resident Oyo Province, Handing Over Notes by H. L.Ward Price
[98] Elgee.... 20
[99] 1907 Annual Report for the Colony of Southern Nigeria 85
[100] Elgee ...21 and Makinde 35
[101] A. L. Mabogunje, *Urbanization in Nigeria*, (London: University of London Press, 1968) 202

occupation or of educational status.As the city offered an increasingly wider range of employment opportunities, it at attracted people from other parts of Nigeria and from different ethnic groups. Apart from other Yoruba people, notably the Ijebu, Ijesha and Egba, who were some of the earliest to move into Ibadan, there were substantial members of peoples from nearly every ethnic group within Nigeria who equally migrated to the city around this period. The Hausa (possibly due to their long trading link with the Yoruba) were the earliest to come to Ibadan. The Ibo, Efik and Ibibio came particularly after 1927- with the connection of the eastern line of the railway with the Lagos –Kano line. The number of Edo, Urohobo and others from the mid-west tended to increase with the improvement of the road links, but particularly with the rise of Ibadan as the headquarters of the western province in 1946 and capital of the western region in 1952.[102] With this development, the city began to attract more of the clerical and executive types as well as non-Nigerians such the Europeans, Syrians, Lebanese and Indians.

This influx of immigrants to Ibadan also led to evolution of new settlements reflecting the distinguished characters of the settlers. For instance, the labourers from Lagos who worked with the British on the rail construction settled in an area known as 'Ekotedo' (the settlers from Lagos); the Hausa and Fulani settled in Sabo; the Nupe, Ebira, Ibo, Efik, Ibiobio, Urhrobo and the likes resided in Mokola area. Lebanon Street (which still exists till date) was also named after the Lebanese shopkeepers due to their prominence during the colonial period.[103]

The British also deserve some credit for their attempt to entrench a semblance of merit in the pursuit of urban expansion in Ibadan. While the status in the new urban tradition was based on achievement (initially anchored on acquisition of reading and writing skills); the advancement within the system was based on standards of efficiency, higher education and higher output of works.[104] As soon as colonial residency was opened

---

[102] NAI File No PR/E5 "Western Government on Survey-Mother of all Developments, 1953 and NAI File No PX/F2 "Reports on Colonial Nigeria Federal Development, 1949-1952"

[103] Toyin Falola, *Lebanese Trader in Southwestern Nigeria....*

[104] P.C. Lloyd, A.L. Mabogunje and B. Awe eds., *The City of Ibadan.....* 41-42

at Ibadan, it attracted a group of Western educated Africans from Lagos. With the arrival of new institutions such as: the railway, the commercial houses, the banks, the agents of the central government, the numbers of these people continued to increase and they eventually settled around the new institutions which attracted their services.[105] With the above, it was to be expected that the growth of population during colonial rule was remarkable. Between 1911 and 1921, the population increased at about 3.1 per cent per annum to 238,075; 5.0 per cent per annum between 1921 and 1931; 0.8 per cent between 1931 and 1952 when the population rose from 387,133 to 459,196.[106]

Within the colonial set-up, this earlier development was not of dramatic significance. The whole administrative organisation was for instance contained within a single building called the secretariat. The change in 1952 involved substantial transfer of the political power from the British colonial office to the nationals of the country and began the process of ministerial appointments and cabinet rule. In terms of the administrative set-up this meant rapid expansion of staff and building in Ibadan where before a single block was adequate to house all the officials and clerks, now some dozen or more blocks had to be built to house the various ministries.

More than ever before, Ibadan became the focal point of political and economic activities for a region with some 42,000 square miles in area. Its importance was further enhanced by the presence since 1948 of what was then the only University College in Nigeria and also the most modern and best equipped teaching hospital (the University College Teaching Hospital) in the whole country.[107]

The increasing concentration of so many people (senior civil servant, University teachers, professionals technicians e.t.c) earning income well above the average for the region, substantially increased purchasing power in the city. This remarkable concentration of purchasing power in the city stimulated rapid growth in commerce and in employment

---

105 P.C. Lloyd, A.L. Mabogunje and B. Awe eds.....
106 Bola Ayeni, "The Metropolitan Area of Ibadan....." 75
107 A.L. Mabogunje, *Urbanization in Nigeria* 199

opportunities. Numerous new firms were established and many of the older ones expanded. But as we have demonstrated earlier in this chapter, much of the taunted and documented 'positive developments' that occurred in most cities in Nigeria (especially in Ibadan) during the colonial rule were accidental rather than by design.

That Ibadan witnessed a period of remarkable changes during the colonial period would be stating the obvious. This period was very important in the evolution of the city. The convergence in this city of traditional and modern economic, social and political institutions, and of the classes of people required to maintain these institutions explains much of the character of the city. In Ibadan, the new has failed to swallow up the old, both exist, strangely juxtaposed maintaining rather complex relations with each other, and functioning almost despite each other. Thus, the internal structure of the city shows the existence of two city centers, the older can be around Oja Iba and the new one around the railway station situated around present day Dugbe in Ibadan.[108] The character of these centres gave a clue to the type of development to be expected around them and much of this was the creation of the colonial administration in Ibadan. As mentioned earlier, the colonial urban policies created and consolidated the existence of this dual personality inherent in the city and this engendered lopsidedness and arbitrariness which later had tremendous and enduring negative effects on the environmental and physical development of the city.

Indeed, Ibadan has a long history of urban arbitrariness, which was complicated rather than eradicated by the colonial masters. For example, a 1945 colonial report revealed that though Ibadan was founded in 1829, it took successive administrators a century to commence a comprehensive planning for the city.[109] Consequently, it could be argued that development had proceeded in Ibadan without any comprehensive master plan or effective urban planning and environmental management policies or strategies that could standardize the street systems, parks, buildings and so on. Invariably, Ibadan since beginning was what

---

[108] A.L. Mabogunje, *Urbanization in Nigeria* …. 199
[109] NAI Iba Div 1 File No 1400 Vol. I-VI, "Town Planning: Ibadan"

Labinjoh called an "epitome of planlessness...."[110] Fourchard corroborated the above, nothing among other things that "after taken a deep historical search, what Ibadan reflects is a near total absence of urban management and urban planning."[111]As it has been demonstrated in this chapter, the dual personality created in Ibadan by colonial urbanism led to a lot of enduring environmental, social, economic and cultural complicated changes which sharpened and dampened almost in the same manner the general character and patterns of growth and development of Ibadan from colonial era and subsist till post independent period. Lack of adequate waste management strategies and precarious sanitation systems were also fundamental features of the period under review.

Waste Management

As reflected in the opinions of experts on environmental management, wastes constitute a serious problem in most cities of the third world, especially in Africa.[112] Their management is one of the most intractable problems facing city administrators and sanitary/environmental officers. Thus, wastes have generated considerable research interest especially as they have affected and still affecting various aspects of life in African and Nigerian cities. For instances: waste magnitude has been handled by Filani and Abumere; waste characteristics by Adedibu; disposal problems by Oduola and Adedibu and factors affecting waste generation by Onidundu-Amao and Ibitoye.[113]

Femi Ayorinde sees environmental degradation as one of improper waste disposal methods through indiscriminate generation and disposal of both industrial and domestic waste.[114]The implication is reflected in high

---

[110] Labinjoh, "Ibadan and the Phenomenon of Urbanism...." 19

[111] Laurent Fourchard, *Urban Slums Reports: The Case of Ibadan, Nigeria*.....237

[112] A.S. Adeyemi, J.F. Olorunfemi and Adewoye, "Waste Scavenging in Third World Cities: A Case Study in Ilorin, Nigeria", in *The Environmentalist*, 21, (2001). 93–96

[113] A.G. Onibokun and A.J. Kumuyi, "Ibadan, Nigeria", in A.G. Onibokun ed., *Managing the Monster: Urban Waste and Governance in Africa*. Available online at http://web.idrc.ca/en/ev-9402-201-1-DO_TOPIC.html

[114] F. Ayorinde, *Solid Waste Management in Commercial Area*, (Lagos: Bambee Press, 2001)

mortality, outbreak of epidemic, lowlife expectancy, development of slums and its attendant problems as seen in most cities in Africa. Many of the waste generated are from the goods sold in the market composes such as peels, shaft, suckers, leaves, stalk etc. To him, it is rather unfortunate that no external body takes the management of this waste in the market into cognizance despite its importance to the national economy through revenue generation. He contended that the issue needs to be addressed collectively by the traders, all tiers of government, non-governmental agencies to ensure a clean environment. Public and private agencies should be solved in piece meal. In his opinion, a comprehensive approach should be considered to include changes in population, commercial activities, types of waste generated, and infrastructural development comprehensive planning policy will equally help to reduce the extent of the problem both in Lagos and Nigeria as a whole.

In the same vein, Glencoe McGraw-Hill (based on the United States experience) advocates recycling as one of the best waste management options. He posited that paper makes up more that 40 percent of our trash and if recycled, it save lots of landfill space because making brand new paper from trees uses lots of water and pollutes the air. But recycled paper takes 61 percents less water and produces 70 percent fewer air pollutants. He, however, maintains that paper is not the only thing that can be recycled. There are other things like aluminum, plastic and many more can be recycled with the energy needed to produce a single brand-new can from ore.

In McGraw-Hill's account, while recycling is yet to have firm root in some countries for various reasons; some have made it mandatory. For examples, in the United States, only 10 percent of garbage are now promoting and even requiring recycling while in Japan and Germany about 50 percent of the garbage is recycled. But there are ongoing efforts to promote recycle culture in the United States. For instances, in Seattle, Washington, people who recycle pay lower trash collection fees. Some states even require local businesses to use recycled products for instance, the Ohio congress proposed a law to require newspaper publishers in Ohio to use some recycled paper. Federal government in Washington, DC as at 1993 was just becoming involved in recycling. A good example was the container law, which requires at least a five-cent refundable deposit on most beverage container nationwide. This means paying five cents extra at the store for a drink, but getting your nickel back if you take the container back to the store. If everyone in the nation would participate in this programme, the nation could save enough energy to light a large city for four years.[115]

Daniel B. Bolkin *et-al* focus on the different types of waste generated by human societies. These include nuclear wastes, medical wastes, industrial hazardous wastes, household hazardous waste, mining wastes and municipal solid waste (M.S.W). Much emphasis was laid on municipal solid waste as it consist of materials people no longer want because they are broken, spoilt or have no further use. Methods of waste disposal such as landfills, incinerations, source reduction, compositing and recycling and their inadequacies were discussed in the last part of the work.[116]

In his contribution, Eldon O. Enger introduces the concept of integrated waste management and applies the concepts to urban waste, hazardous, chemical waste and disposal of waste in the marine environment. He also reveals that there are still inadequate environmental management in some developed countries, where open drums and illegal roadside dumping are leading to problems such as dumping spoils scenic resources, pollution of soil and water resources and potential health hazard to plants, animals and people. Emphasis was further placed on how this misappropriate dumping of waste which affects the health of people and causing air and water pollution remain a serious challenge to environmental managers in the affected societies.[117]

---

[115] G.L. Glencoe and M.C. McGraw-Hill, *Merrill Earth Science,* (USA: Macmillan Press, 1993). In Nigeria, the most popular techniques of solid waste management include, open dump sites, land filling, open burning, reuse/recycling and conversion. Among these management techniques, reuse, recycling and conversion have generated relatively little research attention. Refuse collection in Nigeria is generally believed to be the responsibility of the Municipal Council or the State Environmental Protection Agency

[116] D.B. Bolkin and F.A. Keller, *Environmental Science, Earth as a Living Planet,* (Boston Von. Hoffmann Press, 1995)

[117] E.D. Enger, B.F. Smith and A.Y. Bockarie, *Environmental Science: A Study of Interrelationship,* (New York: McGraw Hill Companies, 2006)

Gordon and Breach identified three systems required in Waste management, namely analysis of waste the natural system, waste system and social system. It also mentions eight steps to which waste management can be controlled which include: waste generation, waste treatment, waste transport, waste interaction, waste decay, waste impact, social assessment and social response. All these are interlinked. Each system requires information from other systems and in turn produces information for other system. The components which make up the waste management cycle have been described as the way in which they may be quantified. The first five steps describe the selection of the water quality level appropriate to the prevailing condition and the eight describes the selection of the most acceptable and efficient waste control strategy. It was concluded that all waste control management decisions are made on the basic of the waste management cycle.[118]

In the opinion of Aguwanba, J.C., the environment of man lies at the mercy of both natural disaster and negligence on the part of man in the course of controlling the gifts of nature. The later takes the form of dumping solid waste in an uncompromising pattern, that can cause: desert encroachment, erosion, depletion of ozone layer, depletion of natural resources, pollution of land, rivers, the air and generally the environment.[119] Sadly, inefficient management of solid waste, perhaps, stands as the most visible environmental problem facing most communities and cities in Nigeria.[120] According to Chukwuemeka Emma, the problem is growing daily as a result of increasing urbanisation and the solid waste problem is visible in most parts of the communities within

---

[118] O. Gordon and C. Breach, *"The International Journal on Environmental Studies"*, 8, (1976) 227-233

[119] J. C Aguwanba, "Solid Waste Management in Nigeria: Problems and Issues", *Environmental Management,* Vol. 22, No. 6. 849-856

[120] Emma O. Chukwuemeka, B. C. Osisioma, and Joy Ugwu, "The Challenges of Waste Management to Nigeria Sustainable Development: A Study of Enugu State", *International Journal of Research Studies in Management*, Vol. 1, No. 2, (October, 2012). 79-87

most Nigerian cities: on the roads, within the neighbourhoods and around residential buildings.[121] Ibadan is not an exception.

Although it is generally agreed that enormous quantities of wastes are generated in Ibadan daily, the exact figures have not been determined, probably owing to the use of diverse methods of calculation. Maclaren International Ltd during a survey conducted in 1970 found that the average per capita quantity of solid waste generated was 0.37–0.5 kg/day for the traditional areas of the city and 0.53 kg/day for the newer areas.[122] Oluwande estimated the average solid waste generated and its mean production rates per head for three distinguished areas of Ibadan: 0.420 kg/day in the GRA; 0.377 kg/day in outlying areas; and 0.35 kg/day in the old city.[123] According to Egunjobi, 38 million kg of solid waste was collected in the suburbs of Ibadan in 1986. The suburbs constitute about 21% of the city. On this basis, it can be estimated that 181 million kg of solid waste was generated in the city as a whole in 1986. This gives a per capita waste-generation rate of 0.31 kg/day, using the 1986 estimated population of 1.6 million for the city.

The solid-waste composition in Ibadan comprises leaves, paper, food waste, tins, glass, and rags. This is because Ibadan is located at the heart of a rich agricultural land and has a large old and unplanned section. PAI Associates made a comparative analysis of the composition of solid waste from two acres of Ibadan in 1970, which showed that residential land use accounted for 70.1% of the waste generated, followed by commercial land use (18.8%) and industrial land use (9.7%). Institutional and other land use accounted for 0.7% each.

Apart from the challenges of solid waste management, the Ibadan metropolis also has a lot of problems with the management of its liquid and industrial wastes. Liquid waste in Ibadan contains tins, sticks, excreta, oil, pieces of iron scrap, and refuse. Outside of large institutions,

---

[121] Emma O. Chukwuemeka, B. C. Osisioma, and Joy Ugwu, "The Challenges of Waste Management to Nigeria Sustainable Development: A Study of Enugu State...."

[122] A.G. Onibokun ed., *Managing the Monster: Urban Waste and Governance in Africa*, http://web.idrc.ca/en/ev-9402-201-1-DO_TOPIC.html

[123] A.G. Onibokun ed

such as the University of Ibadan's Teaching Hospital and the International Institute of Tropical Agriculture, Ibadan has no sewage system. The city's human waste is disposed of largely by means of septic tanks, pit latrines, and buckets. PAI Associates in 1983 estimated the magnitude of liquid waste within Ibadan at 22 650 million L (an average of 6.2 L per household), and Akintola and Agboola projected the amounts of liquid waste for 1990 and 1995 at 113.7 million and 126.5 million, respectively.[124]The industries in Ibadan also generate a lot of waste, particularly chemical and toxic waste, explosives, and ash, but the exact quantities have not been measured. The industries make private arrangements for disposal of their waste, with little or no monitoring. Groundwater pollution is a possibility, as companies do not take precautions at disposal sites to supervise and ensure proper sanitary conditions.[125]

The uncontrolled disposal of liquid waste into open gutters, open spaces, along roads, etc., poses serious health hazards. Bodies of stagnant water produce bad odour, breed mosquitoes, and sometimes obstruct the movement of people and goods. For instance, the 1983 study by PAI Associates revealed that 50% of the stagnant pools emitted bad odour, 70% bred mosquitoes, 24% obstructed the movement of people, and 12% bred worms and other germ-breeding pests.[126] Poor practices for liquid-waste disposal are responsible for waterborne diseases that are common in the city, particularly in its inner core. The unwholesome environment forces the populace to spend appreciable portions of their low income and time on improving their personal health, with adverse consequences for general economic well-being.[127]

---

[124] A.G. Onibokun ed

[125] L. Egunjobi, "Human Elements in Urban Planning and Development: Ibadan", in *Habitat International*, Vol. 10, No. 4, (1986).147-153

[126] A.G. Onibokun and A.J. Kumuyi, "Ibadan, Nigeria", in A.G. Onibokun ed., *Managing the Monster*....

[127] T.O. Lawoyin et. al, "Outbreak of Cholera in Ibadan, Nigeria", *European Journal of Epidemiology*, Vol. 15, No. 4, (1999), 369

Although waste management is one of the most pressing problems in Nigerian cities, it is not a new problem.[128] According to Egunlobi in the early times (pre- colonial days) up till 1970s, the disposal of refuse and other waste did not pose any significant problem.[129] This was because the population was small and enough land was available for assimilation of waste. He stressed further that solid waste problem started with urban growth, resulted partly from national increase in population and more importantly from immigration (especially with the advent of colonial rule).[130] While it may be absurd to blame the British Imperialists solely for the complications occasioned by colonial urbanism (including the grossly inadequate waste management strategies); but with the unprecedented changes that accompanied their urban environmental policies; it is not also possible to exonerate them from the current crises confronting most cities in Nigeria

In the case of Ibadan, traditional solid waste management practices such as waste burning, indiscriminate open dumping of waste and so on that were prevalent in the pre-colonial period also subsisted during colonial era. As an area with substantial urban population, the city had a hierarchy of chiefs, community heads, area heads, and compound heads, with defined areas of jurisdiction for the administration of each community. The inhabitants of these communities lived by a system of well-defined rules and functional differentiations. Public places were swept in rotation by groups of women; household and other refuse was deposited in surrounding bushes, where it decomposed. Traditional system of environmental regulation also existed in forms of customary laws and practices peculiar to the Yoruba people were enforced. This fact was attested to by another scholar who observed that "long before the advent of the colonial laws, each clan and village had different laws for the protection of the environment. Some of these unwritten laws are in the

---

[128] L. Egunjobi and A. Oladoja, "Administrative Constraints in Urban Planning and Development. The Case of Ibadan Metropolis", in *Habitat International*, Vol. 11, No. 4, (1987). 87-94

[129] J. K. Egunjobi, *Solid Waste Management in an Increasingly Urbanized Nigeria*, (Ado Ekiti: Proceedings of the National Practical Training Workshop, 2004)

[130] J.K. Egunjobi...

form of societal norms; others are in the form of adages."[131] Thus, the village or communal head or king and his council of elders and chiefs evolved and enforced these unwritten policies and laws respectively. Their pronouncements, affected all sanitation issues including refuse collection and disposal.

However, with the incursion of the British imperialists, specific laws on the management of wastes in Nigeria gradually emerged from solely focusing on basic environmental sanitation regulation and were processed into a more comprehensive body of legislations that address other environmental management issues.[132]The Colonial period witnessed greater impetus in the formal establishment of public health policies. At some point, public health policies became a priority of the British Colonial Government in its attempt to protect the health of its officers and troops from prevalent local diseases; specific environment sanitation programmes were introduced and were to be implemented by Native Authorities such as anti-leprosy campaigns, control of mosquito breeding areas which were backed by law.[133] The Public Health Ordinance of 1917 which initially applied only to Lagos Township and later to the whole country (including Ibadan) was aimed at abating identified nuisances. Such nuisances include the existence of "any pool, ditch gutter watercourse cesspool, drain, ashpit, refuse pit, latrine, dust-bin or other thing in such a state or condition as to be injurious to health."[134]

Thus, any improperly managed refuse pits and dust-bins were regarded as depicting nuisances and constituting offences for which various fines and court orders were issued. Noxious matters discharged into public streets or channels also constituted nuisances and overcrowding of premises which tended to contribute danger to the health of residents were also address by the legislation. Authorised health officers issued abatement notices in respect of such nuisances. Another pertinent legislation during this period was the Town Improvement Ordinance of

---

[131] A.L. Mabogunje, "Ibadan-Black Metropolis", *Nigeria Magazine,* No. 68, (March, 1961)
[132] Onibokun, Managing the Monster....20
[133] NAI Oyo Prof. 2/2 File No 556 Vol. I-VI, "Nigerian Ordinances",
[134] A.G. Onibokun, *Public Utilities and Social Services in Nigerian Urban Centres: Problems and Guides for Action.* Canada: IDRC and Ibadan: NISER, 1987

1863 which aimed at controlling urban development and waste management (applicable in Lagos and later in the rest of the country). The Lagos Town Planning Ordinance of 1928 was also passed as part of effort to combat the outbreak of bubonic plague in Lagos.[135]

In the 1950s, management of waste in Ibadan was in the hands of Western Region Government and Ibadan City Council. Sadly, this period did not witness stable and uniformed waste management strategies due to mainly to the unhealthy rivalry between the two authorities in charge of the exercise (ICC and WRG).[136] Evidently, the absence of cooperation between these two bodies greatly hampered government efforts at getting rid of filth in the city at this period. For example, it was reported that at a point when the Western Regional Government proposed an idea of urban renewal scheme in Ibadan; Ibadan City Council vehemently resisted it.[137]

Another colonial legislation seemingly on waste management/sanitation was the Public Health Ordinance (cap.125 of 1958).[138] This was enacted preparatory to Nigeria's independence from British colonial rule concentrated mainly on environmentally related aspects of domestic waste disposal. This legislation was applicable in the country until 1990 when it was replaced by a national law.[139] The law contained haphazard provisions on controlling infectious diseases, slaughter of animals, night soil disposal. The Criminal Code Act of 1958 also made provisions to prevent and control threats to public health through the vitiation of the atmosphere. Offences in this respect resulted in imprisonment for six months or small fines. The role of colonial laws thus primarily focused on the management of domestic wastes. The implication of the observation was that hardly any effort appears to have been to regulate the activities of emerging industries during the colonial era, particularly in terms of enacting legislations to manage waste. A number of reasons have been

---

[135] N.A.I. Oyo Prof. 1, 895 Vol. IV, Sanitation - Oyo Province: Ibadan Sanitary Committee
[136] Julius Oluyitan, "Sanitation in Ibadan, 1942-1999", M.A. Dissertation, Department of History, University of Ibadan, 2005. 26-27
[137] Ibid. 27
[138] NAI Oyo Prof. 2/2, File No 556, Vol. I, "Nigerian Ordinances", (Section 13, 14,30, 34, 44, 45, 45, 47, 49 of the 1917 Township Ordinances)
[139] Onibokun, Managing the Monster.....23

adduced for the apparent ineffectiveness of the waste management strategies in most of Nigerian cities (especially Ibadan) during the colonial rule.

According Folake Okediran, despite several laudable attempts at specifically regulating waste disposal activities during this period, most of the governing laws fell short of their objectives due to various factors which included: lack of awareness by the populace, poor enforcement of laws, institutional inconsistencies, public apathy and lack of legal provisions for industrial and domestic waste management in Nigeria.[140]

As mentioned earlier in this chapter, the British deliberate policy of urban segregation led to convoluted development (waste management inclusive) in most of Nigerian cities and the legacy of the urban crisis still persists. The resultant creation of the GRAs, which have remained the best parts of Nigerian cities, reflected the major shortcoming of the colonial era. As argued by Onibokun, to promote their heinous agenda, the British colonial masters treated Nigeria as primarily a rural country and regarded the urban centres as accidents of area development.[141] They therefore made no concrete effort to solve the emerging urban problems, particularly of sanitation and waste management. Wraith observed that in the west, which had six cities of more than 100 000 inhabitants, "the chaotic urban communities presented a challenge which should have been met. They now present almost insoluble problems of planning and sewage and lack any normal civilised amenities."[142]

Another cause of the ineffectiveness of city administration during the colonial period was that the towns administered by town councils were subordinate to the native authorities, who were constituted and controlled by obas and chiefs, people who were mostly "old men by the time they attained office, with interest in another age".[143] Their main interest was "ancient law and custom," and it was left to the Health and

---

[140] This was extracted online at http://www.folakeokediran.tripod.com/id1.htm
[141] Onibokun, Managing the Monster.....25
[142] R. Wraith, *Local Government in West Africa*, (London: George Allen and Unwin, 1964) 68
[143] R. Wraith 94

Public Works Department of government to deal with the most urgent and compelling of municipal problems"[144] especially waste management and Sanitation.

Sanitary Institutions, Legislations and Administration in Colonial Ibadan

As discussed in the previous chapter, prior to the arrival of the colonial administration, the inner part of Ibadan was organised in such a way that the traditional ruler's compound was at the center and the chiefs' compounds in order of their seniority and status.[145] The result of this was an unplanned spatial growth and development evident by congestion occasioned by the cluster of family compounds and quarters that led to a number of general environmental problems especially the deplorable sanitary conditions in Ibadan. As demonstrated earlier, the morphology of the inner city which is characterised by the compounds and houses system developed in such a way that the houses were so congested that there were hardly any efficient road network linking the streets and proper drainage and sewage facilities to prevent environmental problems like floods and erosion, sanitation and health problems.

Sadly, the bad situation of sanitation in Ibadan did not change much for the better even with incursion of the British colonialists. Hence, a number of health and environmental problems created by the peculiarity of the spatial growth of Ibadan continued almost unabatedly during the colonial period. The houses did not have proper means of waste disposal like quitters and sceptic tanks. Also most of the houses lacked proper sanitary facilities, as most had only pit latrine and holes and swaps for the disposal of the city's waste. Also it was not uncommon to see numerous houses within a compound with little room or space for adequate ventilation.

However, this is not to say that colonial administration outrightly neglected environmental sanitation and health issues in Ibadan. Indeed,

---

[144] R. Wraith 95

[145] P.C. Lloyd, Bolanle Awe and Akin Mabogunje eds., *The City of Ibadan*, (Ibadan: Institute of African Studies, University of Ibadan, 1967).

various environmental and sanitary programmes were initiated ostensibly to help improve not only the environmental condition of its inhabitants but also water supply and the establishment of dispensaries and hospitals.[146] As early as 1900, attempts were made to drain the swamps in the town, notably the Oranyan Swamp. In June 1901, the first dispensary was built in the town and 1903 marked the beginning of smallpox vaccination which was made compulsory in 1905.[147]

Part of the measures taken by the colonialist to ensure proper sanitation in Ibadan and the entire province was the establishment of Ibadan Training School for the native administration Sanitary Inspectors and School of Hygiene essentially for the training of fresh sanitary overseers/inspectors and refresher courses for the senior ones. Bicycles were also bought for the sanitary Inspectors to enhance their mobility as well as the discharge of their responsibilities.[148] The first training centre for African Sanitary Inspectors in Ibadan was built in 1932 with money provided by the Colonial Development Fund and one thousand Pounds as the take off grant. Twenty Youths were admitted as first set of intakes (12 from Ibadan Division, 4 from Oyo and 4 from Ife-Ilesa Divisions. They were given monthly allowance of one pound each for the training which lasted between one and two years and receives fifty pound as monthly salary after the training.[149] The Sanitary Inspectors were mandated to educate the public on sanitary matters especially in towns without sanitary staff at the period.

The breakdown of the course contents reveals the following:

1. **Elementary Parasitology (two weeks):** The trainers recognised the presence of lice and guinea worm and the students are to be

---

[146] A.L. Mabogunje, "Ibadan-Black Metropolis", *Nigeria Magazine,* No. 68, (March 1961). See also: NAI Iba Div 1/1 File No 355-306, Acquisition of Land for Ibadan Waterworks, 1938-53, NAI Iba Prof. 1/3 File No IB.7/1951-1954, Materials for Ibadan Water Works Major Scheme and NAI Iba Div. 1/1, 1978 Vol. 1, Health Committee, Ibadan: Annual Sanitation Report — Oyo Province, 1941, I

[147] Oyo Prof/460/40.....
[148] Oyo Prof/1/2/699 10,23
[149] NAI Oyo Prof. 21/1, File No 1566, Colonial Development Fund

taught some practical measures (such as protection of wells, disinfectitation etc) to be taken against such parasitic elements.

2. **Refuse/Waste management (5 weeks):** Trainers were taught the following: types of refuse, dangers, practical methods of collection in the bush, dumping, reclamation, incineration, construction of incinerators with special attention to the economy; dumping sites advantages and disadvantages etc.

3. **Night Soil (8 weeks):** They were trained about: the dangers from the job, practical economic methods of collection and disposal, construction of trenching grounds, disposal areas, portable latrines for private and public use, disposal of urine, water supplies and so on.

4. **Elementary Entomology (7 weeks):** This aspect of the training educated the attendees about species, breeding habits, life history, diseases and dangers from mosquitoes, house-flies, latrine flies, market-flies, tsetse flies etc. They were also trained on the methods of controlling these dangerous species.

5. **Infectious Diseases (6 weeks):** The participants were also taught some practical preventive measures to be taken against communicable diseases. These steps include: disinfestations, de-ratting, vaccination, fumigation and construction of working models moist and dry heat, disinfection pots, disinfestations and fumigation stations etc.

6. **Sanitation of Villages (2 weeks):** In order to cater for the sanitation and general hygiene of the villages, this course was meant to train those involved in sanitary rules, types of houses, planning of streets, how to prevent congestion, how to ensure ventilation and proper sitting positions for latrines, incinerators and markets.

7. **Sources of Water Supply:** The content of this course consist of methods of collection, clarification and sterilisation of small and large quantities of water; practical methods of purification/ protection of wells, streams, rivers etc; prevention of pollution; preventive measures against guinea worm and bilharzias. It also contained training on the construction of wells, collecting areas, filters and so on.

8. **Food and Meat Inspection:** Methods of inspection of food and meat, causation of diseases (air borne, water borne etc), detection of micro organisms etc.[150]

With the above actions, it was evident that the British invested in the institutions of health and sanitary inspection as measures to implement a wide spread programme of environmental sanitation. As far back as the 1940s, schools of hygiene in Ibadan, Kano and Aba. Although the sanitary Inspectors have numerous functions, the major ones for which they were popularly identified were to enforce refuse management in residential premises and house to house check on the local people's sources of water supply. They usually impose drastic and immediate penalty for infraction of sanitary regulations. By the introduction of health and sanitation officers, the British attempted to stimulate public awareness and actions about environment, sanitation and hygiene.

Setting up of sanitation and health Committees was another measure by the colonialist to improve environmental situation in Ibadan at this period. As early as 1903 when Captain Elgee became the Resident and it is on record that he sought the advice of the Council on most of his projects designed to raise Ibadan up to the higher standard of municipal life'.[151] He encouraged the formation of Committees for Health, Education and Agriculture. The Health Board was established in August, and it comprised the Resident, the Medical officer and three chiefs appointed by the Council. The Board was to make suggestions on sanitation and discuss the reasons for reforms so that it would be easy to pass the laws in Council and implement them among the people. In 1934, through the initiative of the District Officer, Ibadan Sanitation Committee, was formed.[152] Lack of support and interest for the organisation by the

---

[150] N.A.I., Iba Div. 1/1, File 1978, Vol. 1, Health Committee, Ibadan: Annual Sanitation Report, Oyo Province, 1941.10

[151] NAI Iba, Div. 1/1 2642/S.3, 2642/S.3 Vol.2, and 2994 Vol. 1

[152] NAI Oyo Prof. I., 895, Vol. IV, Sanitation-Oyo Province: Sanitation Committee, Appointment of, 133

Medical Officer rendered the body ineffective.[153] Consequently, it died naturally.

The year 1942 was also significant in the environmental history of Ibadan. It witnessed the resuscitation of Ibadan Health Committee responsible for making recommendations against poor sanitation.[154]This committee emerged as a Follow up to the receipt of unfavourable reports on Ibadan sanitation in 1941. The then Resident of Oyo Province instructed the District Officer of the Ibadan Division to form a Health Committee that would be working in alliance with the Health Department with a view of making recommendations for the improvement of Sanitation in Ibadan.[155] Subsequently, the Inner Council made up of the Olubadan and the senior chiefs appointed some of their members to serve on the committee. Amongst the chiefs appointed were the following: the Ashipa Olubadan, the Ashipa Balogun, the Ashipa Ekerin Olubadan, the Ashipa Ekerin Balogun and Councillor J.L. Lasebikan. Equally, Dr. A.S. Agbaje and Messrs S.T. Omikunle and S.A. Oloko were appointed by the Native Administration Advisory Board to serve on the Committee. [156]

Several laws/Acts were also enacted by the British against sanitation problems in order to rejuvenate Ibadan environment. For example in 1945, Sanitary Rule No. 9 rule specified that 'no house should be built in Ibadan without the formal approval of the Health Authorities" was enacted.[157] It was alleged that the rule was very unpopular as people often refused to comply with it. This was not unconnected with the fact that the committee on health was not consulted before its enactment. The fact was that people had overheard that the rule did not receive the consent of the majority of the councilors with the Ibadan Native Council. In any case, some of the members of the Health Committee were also part of the

---

[153] NAI Oyo Prof. I., 895, Vol. IV, Sanitation-Oyo Province: Sanitation Committee, Appointment of, 133

[154] NAI Iba Div. 1/1, 1978 Vol. 1, Health Committee, Ibadan: Annual Sanitation Report — Oyo Province, 1941, I

[155] NAI Iba Div. 1/1, op.cit, 14

[156] NAI Iba Div. 1/1, File 1978, Vol. 1, Health Committee, Ibadan: Annual Sanitation Report, Oyo Province, 1941. 10

[157] NAI Oyo Prof.1895, Vol. IV, Extract from the minutes of the I.N.A. Inner Council Meeting held 13 August, 1945. 390

council. Consequently, the rule was restricted to strangers' settlements like Ekotedo, Oke-Bola and Gbagi. Prior to 1946, most of the houses in Ibadan were roofed with thatch and the incidence of fire was very common. In that year, the Ibadan Native Authority (as the Ibadan City Council was then known) passed a bye-law abolishing the use of thatched roofs and ordered that only incombustible roofing materials should be used. The result was the promulgation of the building rules and Sanitary Rule 9 which mandated the universal use of corrugated iron sheets instead of the old thatched roofs.[158]

By 1948, the lbadan and District Native Authority Drainage Rules had been enacted.[159] Section 2 of the Rules stipulated that, No person shall deposit any matter whatsoever in any stream or water course nor dig or otherwise interfere with any land lying within 6 feet of either bank of any stream or water course.[160] Another portion of the law relevant to this discussion is Section 3 which states: The Ibadan and District Native Authority or any duly Authorised person shall at all time have a right of entry into or upon any land lying within 25 feet of either bank of any of the stream or water course set out in the schedule hereto, for the purpose of improving and/or maintaining the flow of water in the stream.[161]

The public health Acts of the 1950s were also aimed at controlling the slaughtering of animals, handling of night soils and general waste disposal. The major laws on water pollution include Criminal Code of 1958 with section 246 aimed at controlling burial in houses and the Public Health Act of 1958 which aims to control the spread of diseases, slaughtering of animals and disposal of night soil and refuse.[162] On land pollution, the criminal Code Act of 1958 was designed to regulate burial

---

[158] NAI Oyo Prof.1895, Vol. IV
[159] NAI, lba Div. 1/1, File 1978, Vol. 1, Health Committee, lbadan: Annual Sanitation Report, Oyo Province, 1948
[160] NAI, lba Div. 1/1, File 1978, Vol. 1
[161] NAI, lba Div. 1/1, File 1978, Vol. 1
[162] NAI AD 110/28, "IBADAN Health Committee Minutes, 1952-1958" and NAI Oyo Prof. 2/2, File No 799 Vol. II, "Slaughter House in Ibadan".

within houses while the Noxious Act of 1958 made it an offence to produce noise or ordour that constitutes nuisance to others.[163]

In 1952, Ibadan became the headquarters of the Western Region and the presence of medical, educational and commercial facilities attracted many immigrants to Ibadan. There was also a larger and more rapid expansion in the indigenous areas like Ode-Aje, Agugu, Elekuro, Oke-Aremo, Oke-Offa and Oniyanrin.[164] The result of these developments was a high density of population which created vast problems of sanitation. In order to address the problems, the Western Region Public Health Law of 1957 gave more control regarding sanitation to local authorities, and the results were disappointing. The recommendations of the Medical Officer of Health (M.O.H.) were often neglected, partly because he was employed by the Region and not directly by Ibadan City Council. The law had in mind, the employment of more Medical Officers of Health by the City Council, but Lagos was the only city which did in fact employ any doctor at all until 1964. In that year, Dr. T.B. Adesina was employed by the Ibadan City Council as Assistant Medical Officer of Health to Dr. R.S. Sethi. The responsibility for sanitation was often moved from the Health Office to the Town Engineer and back again, and much the same process occurred with the supervision of markets.[165]

Apart from the Public 'Health Law of 1957, there was the Night Soil Adoptive Bye Laws Order, 1959, enacted by the lbadan City Council and other regulatory legislations against sanitation problems. It was evidently clear from the available records that the laws were not enforceable due to inadequate sanitary facilities that could provide legal alternatives to illicit practices. As a result of congestion and overcrowding, the Western Region Government made attempt to introduce a slum clearance scheme to develop and resettle people of the indigenous areas similar to what was done in Lagos in 1928 and 1956.[166] This was resisted by the City Council

---

[163] Adebisi Adedayo, "Environment Sanitation and Waste Management at the Local Level in Nigeria", *Geo-Studies Forum* 1 and 2, 2000

[164] NAI AD 110/28, Ibadan Health Committee Minutes....

[165] NAI AD 110/28, Ibadan Health Committee Minutes....

[166] Adebisi Adedayo, "Environment Sanitation and Waste Management at the Local Level in Nigeria....." 5

because it would mean displacement of the indigenous inhabitants, perhaps to the advantage of the immigrant groups. Nonetheless, the Regional Government in 1959 went ahead with the development of a housing estate at Bodija. This was designed to provide moderately – priced houses, particularly for people of the low-income groups desirous of escaping congestion in the centre of the town.[167]

Furthermore, efforts were made by colonial administration to provide electricity for the city and establish a water supply scheme.[168] In the case of water supply the colonial administration saw the improvement of domestic water supplies as a necessary and vital component of the general effort to improve environmental sanitation; and health and social welfare of the people. In the interior area of the core, the colonial administration took certain measures to protect and sustain the people's traditional sources of drinking water. Some of these measures included the deepening and the concreting of the walls of existing wells, the fencing of spring heads and the construction of concrete structures to store and protect spring water.[169]

But at the same time the colonial administration embarked on schemes to provide pipe borne water.[170] The same factor of township status seemed to have underlined the provision of public utilities like water supply and electricity. For many years Lagos, a first class township was the only centre with pipe borne water supply. The Lagos water works were completed in 1914 on the Iju River. Subsequently other areas, mainly second class and third class towns developed smaller scheme.[171] Except in Lagos however, no water rates was paid initially in many of these centres. In 1930, a new services reservoir with a capacity of 44,000 gallons was constructed at Ogunpa Hill again to serve the commercial area in the

---

[167] NAI IBADIV 1/489/Vol.XIX, Ibadan Division Annual Report, 1959
[168] NAI Iba. Div. 1/1 File No. 937, Vol. 3, "Combined Waterworks and Electric Lights Scheme for Ibadan" and NAI Oyo Prof. 21/1 File No 1918, Vol. II, "Electricity Schemes, 1937-1950".
[169] NAI Oyo Prof 21/1, File No 394, Vol II-IV, "Water Works II Urban and Rural Water Supply: Development, 1937-1957."
[170] NAI Oyo Prof 21/1, File No 394, Vol. I Water Works Schemes Oyo Province, 1927-1937
[171] NAI Oyo Prof 21/1, File No 4730, Colonial Economic Research, 1941

township. The plan for the larger town did not come up until 1938 when a proposal was made for a water pipe from government Hill through Oje, Adeoyo and Oja-Oba. This was to supply the prison, the hospital at Adeoyo and council office at Mapo. The twenty-one and half miles pipeline was indeed well conceived, but it was never achieved during the decade under consideration. The first modern water supply system for Ibadan city became fully operational in 1942 when the construction of Eleyele reservoir on River Ona was completed.[172] The reservoir has a catchments area of about 323.8km [2], an impoundment area of 156.2 hectares and a storage capacity of 29.5 million liters of water. The water is usually treated and pumped-out at the rate of about 13.6 million liters a day.

The Eleyele water supply scheme was built partly with funds from the Native Authority as was the case with most of the water supply schemes established by the colonial administration. During the period the scheme was under the management of the Public Works Department (PWD) which was in general supervision of all waterworks in Nigeria until 1953.[173] With the establishment of regional government 1953, the management of the scheme came under the in ministry of works. An offshoot of the ministry, the Western Nigeria Water Corporation was established in 1959 to see to the development and maintenance of water supply schemes in the region.[174] The Eleyele waterworks served Ibadan city throughout the colonial period and until 1972 when the Asejire water supply was completed.[175]

The provision of electricity services followed the same pattern. Electricity was first provided for Lagos residents in 1896 from a small central power station operated by the electricity branch of the Public Works Department. The discovery of coal in Udi in 1916 led to the construction in 1923 of more modern power stations having a plant capacity of 3,000kw. By 1950,

---

[172] Report on the Ibadan Water Supply.... See also, NAI Iba. Div. 1/1, File No. 2997, 'Waterworks Undertaking'

[173] NAI Oyo Prof 21/1, File No 1566, Colonial Development Fund

[174] Report on the Ibadan Water Supply....

[175] Report on the Ibadan Water Supply Expansion Project, Western State Ministry of Home Affairs and Information, Ibadan, 1972

twenty towns had electricity installations in Nigeria, and although by then the distinction into grades of township was no longer in operation, no less than twelve of these centers were either first or second class Township. In that year, all electricity installations were brought under the unified control of the Electricity Corporation of Nigeria.[176]

However, in 1953 the question of a combined water and electricity scheme was raised and a preliminary estimate of cost of the scheme was 60,000 pounds. The scheme provided for a steam power station situated near the railway station with high and low tension overhead lines for giving supplies for waterworks pumping purposes and the general supplies in the Ibadan city.[177] Probably due to its efficiency and cheaper cost, general approval to a combined water and electricity scheme was later consented to by the colonial government when steps were taken to obtain prices of plant and equipment for all purposes of the preparation of a report with firm estimates of cost and revenue. With the provisions of these social amenities, the population of the city grew very rapidly at this period. For instance the 1952 population census gave the population of Ibadan city as 459,196. This had increased to 627,379 by 1963 (a 36.6% increase) and to an estimated 783,511 by 1972.

Expectedly, the transformation of Ibadan into an urban centre also had negative effects on urban environmental management and development of the city. Urbanisation and its attendant environmental issues such as sanitation, proper water channel, sewage disposal, drainage scheme, public health among others posed daunting challenges to the development of the city of Ibadan. It must, however, be noted that various social and sanitary programmes discussed above were initiated which helped to improve the sanitary conditions, reduced mortality and stimulated a rise in the total population during the early decades of the twentieth century. But unfortunately, these colonial programmes which aimed at improving the environmental sanitation in Ibadan only achieved minimal positive fruits. This was due to a number of reasons, namely:

---

[176] NAI Oyo Prof. 811, File No 2663, Vol. I, Colonial Development Scheme
[177] NAI Oyo Prof. 21/1, File No 1918, Vol II, "Electricity Schemes, 1937-1950"

The disruption of traditional resource management systems marked the beginning the crisis. This was orchestrated by the British colonial regime in Nigeria with the establishment of central government which initiated the integration of the Nigerian economy into the world capitalist system. By doing so, environmental issues were based on western models and public policy on environmental sanitation and management were fashioned to serve the political and economic interest of Britain, rather the traditional communities were being handled by bureaucratic organisations such as Ministry of Health or Livestock Department.[178]

In fact they had more interest in this because of the need to protect administrators, officials and other Europeans residing in Ibadan. Thus, the colonial administration concentrated on protecting the township and preventing the spread of diseases to the township areas. With this Euro centric approach, it became apparent that colonial urbanism was not meant to correct any inherited urban malaise but rather to protect the British colonial interest at the expense of the natives. It was therefore not a surprise that the British colonial regime introduced and entrenched a deliberate policy of segregation in their urbanisation which entrenched dual personalities in Ibadan and other Nigerian cities at this period.

As mentioned earlier, in order to promote a deliberate policy of segregation, the British created the Ibadan Township area with the promulgation of the "The Townships Ordinance of 1917."[179] This was at two levels. The first was the between indigenes of a town and Nigerians from other places (the so-called alien natives).The second segregated Europeans from Nigerians, irrespective of the status of the latter. Europeans lived in reservations where they have access to the best medical attention, super security, adequate water supply, good road and adequate sanitation amenities.[180] With this ugly arrangement, it was not surprising that colonialism initiated unsustainable development in

---

[178] C.S. Ola, *Town and Country Planning and Environmental Laws in Nigeria*, (Ibadan: Ibadan University Press, 1984) and B.A. Chokor, "Government Policy and Environment Protection in Developing World: The Example of Nigeria", *Environment Management*, Vol. 17, No. 1, (1993), 15-30

[179] Falola Toyin, *Politics and Economy of Ibadan, 1893-1945*....135

[180] Oyo Prof 811, File No 2919, 'European Reservation Areas'

Nigerian cities. Mabogunje stated that one of the major problems of Ibadan was the pre-European foundation and colonial complications which resulted in "its almost unbelievable density of buildings, their spectacular deterioration, and virtual absence of adequate sanitation.[181]

One of the reasons given for the segregation of Europeans in the township was to protect them from yellow fever and other contagious diseases which were prevalent among the natives. This seemed to be plausible but the near absolute neglect of non-European part the Township and the inner part (which constituted the greater part) of the city demonstrated the ill motivation of the British urban environmental policies. For examples, while the European part of the Township at the inception had a government hospital with a senior medical officer and sanitary inspectors who used to supervise the area, the other parts were left to grapple with negligible attention from the colonial government and acute shortage of health and sanitary officers. By 1951, the number of Europeans had reached more than two thousand and three new reservations had to be established – the Commercial, the Links and the New Reservations. It must be noted that when proper town planning practice started in Ibadan, the area benefited more from most of the colonial urban planning policies and were the first beneficiaries of most amenities provided by the colonial administration in Ibadan. In fact these areas remain the only part of Ibadan in which planning laws were executed and enforced.[182] Due to this reason the water supply scheme was able to provide for limited areas. Hence, the European reservation and Agodi Hills were the main beneficiaries of the water supply scheme during the colonial period. The improvement in water supply was felt more in the township. More wells were also constructed for the township.

For the rest of the city, the administration also took some negligible steps to solve some of the sanitation and health problems. For instance: Government established the health board whose main concern again was to serve the few Europeans in the city by seeking means to eradicate

---

[181] A.L. Mabogunje, "The Problems of a Metropolis", in Lloyd, Mabogunje and Awe eds., *The City of Ibadan*, (London: Cambridge University Press, 1967), 261
[182] IBADIV 1/895/Vol.II: Europeans Resident in Ibadan, 1947 and NAI Oyo Prof. 811 File No 2080 "European Reservation Ibadan"

mosquitoes. The health board co-operated with the resident to clear some swamps in the town in 1900 and opened a small dispensary opened in 1902. The council was particularly useful in the fight against small pox which was prevalent at the period. Also in 1921, with advice from the British officials, the Board proclaimed that burials in private houses constituted a health hazard and decided to establish two burial grounds, both outside the town. In the following year, a leaper's village was created to isolate people afflicted with leprosy.[183]

Other measures during this period were the campaigns for environmental sanitation, and a search for a better way of disposing faeces. During the period measures on the collection of faeces and refuse were grossly inadequate.[184] On health matters for the larger city, a medical officer was provided for the whole of Oyo province. The Native Authority hospital at Adeoyo, Ibadan was established in 1927, but it was until 1929 that the hospital became busy with 420 in-patients and 2,831 out-patients. Also in 1927, two additional small dispensaries were proposed for the larger town.[185] A year later, it had become necessary to build additional ward and extend the out-patient waiting room. Also in the large town sanitation was left in the care of Akoda-the traditional police.[186]

Housing facilities in the increasingly populated inner parts of the city were also overstretched and consequently, there was an outbreak of small pox. Correspondence between the senior Resident Oyo province and the Secretary, Southern province between 1925 and 1926 showed the rapid rate at which the epidemic spread in Ibadan.[187] In addition, there was an acute shortage of water supply. There was also the indiscriminate dumping of refuse and ugly sights of human faeces in the city. In order to address the deplorable condition of the city, the authorities between 1930 and 1931 provided sanitary facilities like latrines and public dustbins and

---

[183] NAI CSO 15965, "Infectious Diseases treatment, Ibadan: Memorandum from Senior Resident, Oyo to the Secretary, Southern Provinces", 8

[184] NAI Oyo Prof. 1, File No 895, Vol. IV, Sanitation — Oyo Province: Ibadan Sanitary Committee, 137

[185] NAI, Iba Div. 1/1, File No 1978, Vol. 1, Health Committee, Ibadan: Minutes of Meeting of Ibadan Health Committee

[186] NAI, Iba Div. 1/1, File No. 1978, Vol. 1....

[187] NAI CSO 15965, "Senior Resident Oyo to the Secretary Southern Provinces...."

incinerators for areas such as Iba market, Agbeni, Ekotedo, Sabo, Oje, Ayeye, Gbenla. Policemen were also stationed in the areas to arrest culprits for trial at the Native Courts.[188]

Part of the problem was also the near absence of stable/uniform sanitary policies and half-hearted law enforcement. The colonial authorities believed strongly that education of the people was more beneficial than prosecution. As a result of this, community participation in form of Health Committee was introduced and encouraged. Much could not be done in terms of provision of sanitary facilities. As demonstrated earlier in this chapter, because the primary interest of British Rule in Nigeria was economic; the colonial development policies and plans contain few stringent rules to preserve the environment. Consequently, the breach of major environmental laws on sanitation and public health usually attract liberal penalties and the laws often poorly enforced.[189]

The above imposition of light punishment on the sanitation offenders and ineffective law enforcement mechanisms naturally led to incessant and flagrant flouting of environmental regulations among the residents of Ibadan at this period. This had entrenched some grievous consequences on Ibadan and its environment (some of which still persist till date). This was one of the reasons for incessant destructive floods in Ibadan since 1933.[190] In that year, Gege River drowned the houses of those living on its bank. This unfortunate incident marked the beginning of destructive flood disasters in Ibadan. For example, in 1951, a two-day heavy downpour between July 9 and 10 caused considerable damage to property along the banks of the major rivers that passed through the city. Trees, Vehicles, and houses were swept away in June, 1955 by the flood that followed a two-day heavy rainfall. On 17 August, 1960, the city was again ravaged by the flood waters of swollen rivers and streams which killed hundreds of people and over 1,000 people were rendered homeless and property estimated at over N100, 000 damaged. In addition, roads, railways, bridges, were damaged and washed away. Again, in late

---

[188] N.A.I., Iba Div. 1/1, File No. 1978, Vol. 1....

[189] NAI Oyo Prof. 1 File 895, Vol. IV Sanitation....

[190] Nigerian Environmental Study/Action Team, *Nigeria's Threatened Environment: A National Profile*, (Ibadan: Intec Printers Limited, 1991) 107

August, 1963, a more devastating flood occurred causing damage to property worth over N200, 000. The story was the same in May, 1969 with floods damaging property worth over N 100,000. All these floods and subsequent ones were largely as a result of violation of the rules mentioned above.[191]

As argued by Ola, C.S. and Chokor B.A., despite its limitations public policy existed at the local level during the colonial era.[192] The first development planning exercise in Nigeria which came into operation in 1946 focused on environmental sanitation and hygiene. Emphasis was placed on refuse disposal and management of liquid and solid wastes in abattoirs, residential areas and streets. Andrew Onokerhoraye also corroborated this when he submitted that "during the colonial period, the environment was conceived of as essentially a sanitation and health issue by the colonial administrators."[193] But they could not achieve much for Nigerians and their cities. This was due to the "more deep-seated racist fears and sentiments in colonial physical planning, legislation and urban development policies,"[194] which culminated in palpable poverty in Nigerian cities (especially among the natives).

The Interface between Poverty and Environmental Crisis

The continued prevalence of poverty throughout the World keeps its alleviation as a central objective of global economic development agenda.[195] Strategies for reducing poverty have begun to pay more attention to the relationship between environmental degradation and poverty. This two-way relationship is a significant one. Environmental degradation contributes to poverty through worsened health and by

---

[191] Nigerian Environmental Study/Action Team, *Nigeria's Threatened Environment*....
[192] C.S. Ola, *Town and Country Planning and Environmental Laws in Nigeria*, (Ibadan: Ibadan University Press, 2000); B.A. Chokor, "Government Policy and Environment Protection in Developing World: The Example of Nigeria", *Environment Management*, Vol. 17, No. 1, (1993). 15-30
[193] Andrew G. Onokerhoraye, *Urbanisation and Environment in Nigeria, Implication for Sustainable Development*, (Benin: University of Benin City Press, 1995) 6
[194] Ibid 6
[195] Stephen D. Mink, "Poverty, Population and the Environment", *World Bank Discussion Paper*, No. 189

constraining the productivity of those resources upon which the poor rely, and poverty restricts the poor to act in ways that are damaging to the environment. In addition, demographic factors can be involved in complex ways; high growth rates are associated with poverty, and directly exacerbate problems of environmental degradation. The poor's exposure to environmental degradation is distinctive for two reasons. First, locations inhabited by the poor are often environmentally vulnerable or degraded. Whether erosion-prone hillsides in rural areas or urban neighborhoods with inadequate water and sanitation infrastructure, the areas to which the poor can gain access are often the riskiest for health and income generation. Second, being poor entails lacking the means to avoid the impacts of environmental degradation. A lack of resources makes it difficult for the poor to buy out of exposure to environmental risks, or to invest in alleviating the causes of environmental degradation.[196] Sadly, most of the cities and their inhabitants in sub-Saharan Africa have not escaped this tragedy of poverty.

Indeed, the prevalence of poverty in Africa is a key factor in the environmental crisis in most of her urban centers. As early as 1960s, this has engaged the attention of some scholars. For instance, P.C. Lloyd observed that "most African dwellers suffer from extreme poverty."[197] He argued that this is traceable to a number of interrelated factors such as absence of basic infrastructure-(road, electricity, toilets, portable water, health care facility etc); mass unemployment, under employment, under-nourishment and so on. This would be particularly true of the illiterates who constitute more than 80% of the population in most African cities. This gave rise to deep sense of frustration among the millions of these poor urban dwellers who are mostly under-nourished, ill educated, poorly housed, uninformed and constantly exploited by others. Due to the fact that most of them usually rely on nature for bare existence, keeping basic environmental laws become unthinkable or out-rightly impossible. In establishing the unbreakable nexus between poverty and environment degradation, the World Bank in a 1993 report on Jamaica submitted:

---

[196] Stephen D. Mink

[197] P.C. Lloyd, Bolanle Awe and Akin Mabogunje eds., *The City of Ibadan*, (Ibadan: Institute of African Studies, University of Ibadan). 1967.

Though many environmental problems are general in their effects
....the poor in many countries bear more of the immediate pressure
of environmental degradation. They are the least to be served with
public water, sewage and garbage services; tend to live in crowded
homes where indoor pollution is high and may live in close
proximity to disposal sites for solid and hazardous waste.... the
poor are not able to escape these areas.[198]

The report concluded that poverty is a cause of environmental
degradation as the "poor must meet their urgent survival needs."[199] This
is especially true of rural, semi-urban dwellers and the urban dwellers in
the informal sector that operate without the benefit of health, safety or
environmental regulations.

The social tension engendered by the continued and worsening poverty
always has negative effects on the country's ability to attract long term
investment (domestic and foreign).[200] Rapid urban growth concomitant
with continued poverty and economic stagnation is a major problem in
developing countries, Nigeria inclusive. The population of some cities is
increasingly beyond the capacity of the society to cope with the human
influx. The political forces and elite continue to funnel resources towards
cities to the neglect of the impoverished rural areas. The depletion on the
quality of life in the rural areas leads to the congestion of the urban
centres thereby making them socially unhealthy. The poverty stricken
condition of our rural areas is an elective mechanism, which motivates
urban migration. The urban centres are mainly incapable of absorbing the
rural migrants who are usually lacking in essential training to make them
employable.[201] Accordingly, the inevitable consequences of this vicious
circle are: universal poverty, housing shortage, urban slum, environ-

---

[198] World Bank, "Poverty Report on Jamaica", (1993)...
[199] World Bank, "Poverty Report on Jamaica", (1993)....
[200] Mosope Fagbongbe and Hakeem I. Tijani, "Legal Aspect of Urban Development", in
Hakeem I. Tijani ed.,...
[201] O. Adigun, "The Problem of Housing in Nigeria", in A.O. Obilade ed., *A Blueprint
for Nigerian Law*, (Lagos: Faculty of Law, University of Lagos, 1995) 155

mental squalor and a host of other environmental and human problems.[202]

According to Pelling, there is a direct correlation between poverty and vulnerability to environmental risks.[203] Low-income groups in African cities are relatively disenfranchised from decision-making, having the least resources at their disposal to meet lifestyle challenges, even less during times of change or disaster.[204] The urban poor, especially women and the very young are shown to be most at risk from disease, pollution and disasters which might all be exacerbated by climate change.[205]

Opinions are divided over the contribution of the colonial rule to the environmental crises and entrenchment of poverty in most African cities. In the opinion of Peter Ekeh, colonialism destroyed and supplanted African ways of life and beliefs to create, colonial social structures which were eventually left behind in Africa.[206] These social structures which involved models of development, system of education, Christianity, capitalism, urbanisation, Western medicine, and industrialisation, constitute enduring social structures and legacies that are active in Africa today. Some colonial social structures were based on the modern European view of ontology, scientific method, and utilitarian ethics. The European model of economic development imposed on Africa, which was left as a legacy, culminated in Africans destroying the jungles and its animals for economic gains. Africans had been indoctrinated by

---

[202] P. Onibokun ed., *Housing in Nigeria: A Book of Reading*, (Ibadan: Nigerian Institute of Social and Economic Research, NISER) 8

[203] M. Pelling, "Toward a Political Ecology of Environmental Risk: The Case of Guyana", in Karl Zimmerer and Thomas Bassett eds., *Political Ecology: An Integrative Approach to Geography and Environment-Development Studies*, (The Guildford Press, 2003)

[204] G. McGranahan, D. Mitlin, D. Satterthwaite, C. Tacoli and I. Turok, "Africa's Urban Transition and the Role of Regional Collaboration", *Human Settlements Working Paper Series*, (Theme: Urban Change-5, IIED, London., 2009) 58

[205] D. Schwela, "Review of Urban Air Quality in Sub-Saharan Africa", *Clean Air Initiative in Sub-Saharan African Cities*, (Washington DC: The World Bank, 2007), 64; World Bank, "Climate Change, Disaster Risk, and the Urban Poor", (World Bank: Washington 2012); S. Bartlett, "Climate Change and Children: Impacts and Implications for Adaptation in Low to Middle Income Countries", *Environment and Urbanization*, Vol. 20, No. 2, (2008). 501–519

[206] Peter P. Ekeh, *Colonialism and Social Structure: An Inaugural Lecture.* (Ibadan: University of Ibadan Press, 1983)

Europeans and forced to accept this model of development. The image of real estate or industrial complexes in Europe, among other things, represent in the mind of many Africans, the idea of development; this was an idea or ethos that colonialism brought to Africa and left behind as a legacy.[207]

Polycarp Ikuenobe also argued that the activities that have raised environmental concerns in Africa did not exist prior to colonialism because Africans had conservationist values, practices, and ways of life. To him, African views and thoughts on ontology, cosmology, medicine and healing, and religious practices supported their moral attitudes toward the conservation and preservation of nature.[208] Traditional African thought sees nature as holistic and as an interconnected continuum of humans and all natural objects which exist in harmony. People's actions and ways of life reflected the efforts to exist in harmony with nature. These efforts led to the preservation of environment. Many of these traditional African values, ways of life, and the moral attitudes of conservation were destroyed by the exploitative ethos of European colonialism and modernity.[209]

While recognising the universality of the environmental crisis, Ogungbemi noted that in understanding the nature of the environmental crisis within the context of sub-Saharan Africa, three points are sacrosanct: 1) Ignorance and poverty 2) Science and technology 3) Political conflict, including international economic pressures.[210] He contended further that in order to properly understand the nature of the environmental crisis in Africa, we need to understand the ways in which both traditional and modern social structures have led to environmental degradation. Concerning the factors of ignorance and poverty, Ogungbemi explained that the majority of traditional Africans lived in

---

[207] Ekeh, Peter P., *Colonialism and Social Structure: an Inaugural Lecture*....13

[208] Polycarp A. Ikuenobe, "Traditional African Environmental Ethics and Colonial Legacy International", *Journal of Philosophy and Theology*, Vol. 2, No. 4, (December 2014). 1-21

[209] Polycarp A. Ikuenobe, "Traditional African Environmental Ethic...."

[210] S. Ogungbemi, "An African Perspective on the Environmental Crisis", in Louis J. Pojman ed., *Environmental Ethics: Readings in Theory and Application*, (Belmont, CA: Wadsworth Publishing House, 1997) 204

rural areas where the people wallowed in poverty, and lacked basic amenities such as good water supplies, adequate lavatories and proper energy use. As a consequence, the rivers were polluted with human waste, exposing the people to avoidable water-borne diseases such as dysentery, typhoid and cholera. Also the excessive use of fuel wood and constant bush burning which is a predominant practice in traditional Africa, increased air pollution. This affected air quality adversely and depleted the forest and other natural habitats. This factor of poverty cum ignorance on the part of traditional Africans, Ogungbemi argued, "does not necessarily exonerate our people from their contribution to environmental hazards."[211] This is particularly so given that the relevant patterns of behaviour may come, at least in part, from an inability to exploit nature because of low levels of economic and technological development.

In the case of Ibadan, the failure of environmental management strategies; near collapse of urban infrastructure and poor sanitary conditions had been a recurring decimal since the height of colonial era in the 19th Century. This has been exposing the city to constant environmental hazards which still persist till date. A close examination of the contribution of colonial urbanism to the prevalence of poverty in colonial Ibadan reveals four categories viz: poverty of philosophy; poverty of planning; poverty of personnel; and poverty of the people.

Arguably, the fundamental foundation for the poverty that eventually ravaged Ibadan and other Nigerian cities during colonial rule was the policy of urban segregation, which served as the underpinning philosophy of the British colonial urbanism. This was exemplified by the promulgation and implementation of "The Townships Ordinance of 1917," (already discussed fully above). The most noticeable debilitating effect of this poverty of urban philosophy was the evolution of a segregated society resulting into polarisation of Ibadan into two unequalled worlds. And the two are diametrically opposed to each other. The first prosperous part which harboured the Europeans and very few Nigerian elite was small but well regulated. This zone was strongly built,

---

[211] *Ibid*

made of stone and steel; brightly lit; the streets are covered with asphalt. Indeed, the place was always full of good things; as they had access to the best medical attention, super security, adequate water supply, good road and other social amenities. The other larger part of the town belonging to the natives was poverty stricken and largely unregulated. It was a place of ill-fame; without spaciousness; where men live there on top of each other; a place starved of all good things including bread, meat, shoes, coal, or light. To borrow from Fannon, the native town is crouching village, a town on its kneels, a town wallowing in the mire....[212]

Although, Ibadan was said to have a long history of urban arbitrariness, but this was further complicated and multiplied rather than eradicated by the colonial masters. As a corollary, colonial Ibadan (like most Nigerian urban centres) became the most conspicuous environment in which economic capabilities were impeded rather than expanded and social qualities of life frustrated instead of being fulfilled. Hence, colonial urbanisation became as dangerous as it was important for the development of Ibadan. This was because in their efforts to create 'a new Ibadan', the British colonialists became more pre-occupied with the protection of their interest. In the same manner, they neglected the need for adequate and qualitative urban environmental management strategies for sustainable development; which could have stimulated progress in education, transportation and communication. Consequently, Ibadan like most African urban areas grew more rapidly than its sustainable capacities. At the same time, colonial urban authorities were unable to cope with the housing, educational, healthy welfare and recreational needs of the expanded population of colonial Ibadan. The result was the evolution of what Blackwell called the "Parasitic City" or unproductive city. Rather than providing the basis for sustained economic growth, Ibadan (like most cities in Nigeria) has become serious impediment to sustainable development. Sadly, the problem persists till date.

Closely related to the issue of acute philosophical deficiency in the content of colonial urbanism were the mind boggling perfidious planning strategies of the era. Right from the pre-colonial era, poverty of planning

---

[212] Frantz Fanon, *Wretched of the Earth* 30

had been entrenched in Ibadan. It is relevant to recall here that Ibadan evolved as a military camp and as such, the soldiers had no time to settle and plan for the city upon establishment. Rather, they engaged in waging wars and accumulating war booties. This did not help Ibadan at all in its development. It had impeded the city's progress due to lack of sound town planning strategies which had affected the physical development of the city and resulted into seemingly intractable problems of urban development. The planlessness of the city unfortunately and quickly manifested in its slum outlook especially in the traditional core areas of the city.

From the above, it could be inferred that no town planning existed before the colonial invasions. Therefore, it was natural that the preparation of plans began with the colonial regime under the apparent influence of Euro centric philosophy. Though there were reasons that prompted the British colonial government to embark on planning for the colonies (Ibadan inclusive), it has been asserted that these plans were defective — there was no consistent development strategy, in short, the plans were no more than lists of projects.

The Nigerian Town and Country Planning Ordinance of 1945 heralded Town Planning in Ibadan.[213] In section 4 of the ordinance, the Governor appointed a planning authority to carry out planning schemes. Consequently, the Ibadan Town Planning Authority was inaugurated and was gazetted on the 9th of June, 1949. However, following the enactment of the ordinance in 1945, a town planner, Mr. Farm, an engineer, was brought from the United Kingdom, and was commissioned to study Ibadan and produce a plan for the city. Farm was a friend' to Maxwell Fry, who was also a town planner and the duo from a six weeks study of Ibadan produced, the plan. This was the famous Farm Fry Plan report of 1945.[214]The most striking part of the report demonstrated a long history of urban arbitrariness in Ibadan.

The report specifically revealed that though Ibadan was founded in 1829, it took successive administrators a century to commence a comprehensive

---

213 N.A.I. Iba. Div. 1/1, File No. 1400, Vol. IV
214 N.A.I. Iba. Div. 1/1, File No. 1400, Vol. IV

planning for the city.[215] Consequently, it can be argued that development
had proceeded in Ibadan without urban planning and management or
sound environmental policies that could standardize the street systems,
parks, buildings and so on. Invariably, Ibadan since beginning was what
Labinjoh called an "epitome of planlessness...."[216] Fourchard
corroborated the above noting, among other things that "after taken a
deep historical search, what Ibadan reflects is a near total absence of
urban management and urban planning"[217] While these views generalize
on Ibadan, it is difficult, if not impossible to deny the enormous
challenges present generation of Ibadan dwellers are facing with the
poverty of planning bequeathed by the British colonialists. Suffice it to say
that informal settlements outnumber legally planned developments and
are increasing more rapidly. This was further complicated by high
demand for land, housing shortages and inability of the majority of
Ibadan residents to afford other urban services as a result of high rates
imposed by urban authorities. The situation was not made better by the
fact that planning and building standards, regulations and procedures
were based on European norms rather than local circumstances. Attempts
at Victorian model of physical urban development in Nigeria, as in other
non-English areas under the British, did not help matter. Rather, it created
other problems such as maintenance, life span, and suitability to the
environment.[218]

Poverty of personnel also contributed seriously to the failure of colonial
urbanism to initiate and entrench sustainable development in Ibadan.
Evidence provided by Onokerhoraye shows that since the colonial times
there has been a deliberate tradition of not providing adequate resources
for certain professions, skills, and technology of importance to the issue of
environmental protection.[219] One major profession which has been
affected in this way is town planning. Others that come to mind include
forestry, surveying and environmental health engineering. During the

---

[215] NAI Iba Div File No 1400, (1945) 496
[216] Labinjoh 1999
[217] Laurent Fourchard, *Urban Slums Reports: The Case of Ibadan, Nigeria*.... 237
[218] Ibid. 168-169.
[219] Andrew G. Onokerhoraye, *Urbanisation and Environment in Nigeria: Implications for Sustainable Development*....10

colonial era critical environmental sectors such as the collection of faeces and disposal of refuse suffered greatly due to grossly inadequate availability of personnel.

The poor residents in Ibadan also did not enjoy good medical services during the colonial rule due to acute shortage of medical personnel and hospitals at this period. As a reflection of urban segregation policy of the British, while the European part of the Township at the inception had a government hospital with a senior medical officer and sanitary inspectors who used to supervise the area, the other parts were left to grapple with negligible attention from the colonial government and acute shortage of health and sanitary officers .For instance, a medical officer was provided for the whole of Oyo province.[220] Furthermore, the Native Authority hospital at Adeoyo, Ibadan was established in 1927.[221] As the only General hospital in Ibadan as at 1929, the hospital was overwhelmed with 420 in-patients and 2,831 out-patients.[222] Unfortunately, the problem of inadequate provision of hospitals persisted till post-colonial era. For instance, it was reported that in 1983, "not a single hospital was located in the traditional slums of Ibadan and out of the 21 hospitals only four were located in the periphery of the slums. Also due to inadequate number of trained sanitary inspectors in core areas of Ibadan (where most poor natives lived); enforcement of sanitation laws was left in the hands non-professionals called 'Akoda' (palace guards).

With the above discussion, it has become obvious that rapid urban growth concomitant with continued poverty and economic stagnation has been a major problem in Ibadan just like other third world cities.[223] Since, colonial times, the population of Ibadan has been increasing largely beyond the capacity of the society to cope with the human influx. The political forces and elite continue to funnel resources towards city centres to the neglect of the impoverished traditional/rural areas. The depletion

---

[220] NAI Oyo Prof 21/1, File No 1566, Colonial Development Fund.
[221] F. Lyun, "Hospital Service areas in Ibadan City", in *Social Science and Medicine*, Vol. 17, No. 9, (1983) 601
[222] F. Lyun, "Hospital Service areas in Ibadan City"....601
[223] O. Adigun, "The Problem of Housing in Nigeria", in A.O. Obilade ed., *A Blueprint for Nigerian Law*, (Lagos: Faculty of Law, University of Lagos, 1995) 155

on the quality of life in the poor areas led to the congestion of the traditional core areas of the city; thereby making them socially unhealthy. The poverty stricken condition of the poor areas is an elective mechanism, which motivates urban migration.[224] The urban centres are mainly incapable of absorbing the rural migrants who are usually lacking in essential training to make them employable. Consequently, prevalence of poverty has been the most constant denominator in Ibadan and other cities in Nigeria.

In the opinion of Justin Labinjoh, the character of a city's economic activities determines the nature of its domination.[225] Ibadan case is no exception. The city grew as a garrison and fortress; it soon became a center of administration and like all cities, a market. It was never dominated by a real bourgeoisie interested in production; rather it was dominated, first by an indigenous aristocracy who were mainly consumers, then by middlemen merchants and much later by a stratum whose common link was (mere) literacy (i.e. education but not necessarily erudition) and whose primary concern was not the modernisation of the city. As a result, industrial capitalism never developed there. Ibadan has therefore remained underdeveloped till today-like the rest of the Nigerian society. Accordingly, the inevitable consequences of these vicious circles are: universal poverty among the people, housing shortage, urban slum, environmental squalor and a host of other environmental and human problems.

Indeed, areas that harboured the poorest of the poor in colonial Ibadan have no building plans, no pipe borne water, no toilet, no electricity, no proper roads, no adequate waste disposal system, and no drainage. The whole environment generally reflected the elements of poverty.[226] These places are the real definition of slums. The areas are also full of old houses, which are often overcrowded, with mounds of refuse; rampant diseases (but limited health centres) because of the poor sanitary state of the environment. To worsen the situation, upper mobility among the multitude of slum dwellers in colonial Ibadan was near impossibility due

---

[224] O. Adigun, "The Problem of Housing in Nigeria".....155
[225] Justin Labinjoh, Modernity....3
[226] Fourchard, Ibadan....12

to extreme poverty. Although many of them occupied this part of Ibadan due their cultural heritage or familial links (family houses, grandfather's land etc) as most of them were born there; but the main reason for living in the squalor in spite of the extremely limited facilities for human welfare was their economic penury. For instance, apart from the fact that most of the residents lived in family houses free of charge, the areas were the cheapest in Ibadan in terms of house rent payment. So, the people with a very limited income could not afford very expensive house rent in other part of Ibadan Township which was very neat and well organised. Unfortunately, as regards poverty of the city, little has really changed for Ibadan even with the so-called independence. This clearly reflected in the overall morphology of the city.

Conclusion

According to Mabogunje the morphology of Ibadan exhibits three dominants characteristics, which were entrenched by colonial urbanism but are still present at the post colonial era. The most obvious is the extensive confused mass of housing which occupies a large proportion of the city, where poor people occupied.[227] Although, there are other challenges created by this such as impossibility of orderly house numbering and difficulty in mail delivery; the environmental implications were so enormous. For example, the situation presented an almost intractable problem of organising a proper sewage disposal or refuse collection. Waste household water from the back of one house fouls the frontage of the adjoining houses. Apart from creating a nuisance, it represents a serious danger to people's health. The position is made more serious by the fact that channel cut by this flow of water is deepened and widened by the torrential rains, and sometimes undermines the foundation of houses.

The second feature of the morphology is the small proportion of land devoted to roads, especially in the older part of the city which contains vast majority of poor, low- income unskilled workers and 'non-income' persons who provided cheap and low grade labour for the city's

---

[227] Mabogunje, *Urbanisation in Nigeria*.....55

industries and businesses. Due to the awkward transport situation, most of this people wake up as early as 4 a.m. mostly to trek long distance to work places.

The third feature is the persistence of uncontrolled growth. Certainly, there was town Planning Authorities with responsibility of promoting development. But since they have no overriding powers over land development of the city, their role has been purely permissive. The so-called Authorities even after colonial era tried not to impose a lay-out but rather to make sense of the jig-saw of small-lay-outs of individual landowners. They are also involved in approving, and thereby ensuring, a certain basic standard of construction and amenities in many of the new houses. But the Authorities had tended not to use these powers with respect to the indigenous parts of the city.[228]

Overall evaluation of the impact Colonial urban environmental policies and programmes on development of Ibadan reveals a near-absolute paradox. On one hand, the colonial period reinforced the position of the city in the Yoruba urban network. After a small boom in rubber business (1901-1913), cocoa became the main produce of the region and attracted European and Levantine firms, as well as southern and northern traders from Lagos, Ijebu-Ode and Kano among others. On the other hand the reproduction of the colonial model of urban development initiated and entrenched a convoluted and lopsided development in Ibadan. Thus, in the words of Mabogunje, Ibadan remains as perplexing a phenomenon as it is a perplexing problem. The city's terrible environmental features are only the physical reflection of a vast array of other problems of a social, economic and political dimension. It is therefore not surprising the city continues to grow as 'a vast, untidy, amorphous aggregation of rusty tin-roofed shacks.'[229]

As discussed earlier, the consequences of the near-absolute disequilibrium of colonial urbanism are most obvious for the poor people in Ibadan. In most areas, there is a near-total absence of urban management and urban

---

[228] Mabogunje, *Urbanisation in Nigeria*.....56
[229] Mabogunje, *Urbanisation in Nigeria*.....56

planning; no standard waste disposal system, no drainages, a limited number of roads. In others, there was a near total lack of basic facilities like water and electricity supply. Access to health centres was also limited in most places. Sadly, these identified environmental crises which negated the principles of sustainable development endured from colonial times to post-colonial era in Ibadan as in most cities in the Global South (especially in sub-Saharan Africa where Nigeria is located).

Chapter Six

Environmental Management and Sustainable Development in Postcolonial Ibadan

Preamble

It is incontestable that postcolonial Ibadan presents a despicable environmental picture. In fact, Francis Egbhokare's revised Standard version of Professor J.P Clark's 'Ibadan': "Ibadan/oozing blob of rot and mold squashed and splattered among human wastes/ like Bodija market in the rain"[1] vividly captured the filthy nature of the 21st century Ibadan. Indeed, like most of the Nigerian urban areas, there is little evidence of any realistic physical planning. Visits to major streets and residential areas especially the core city centers show a prevalence of uncontrolled heap of refuse in open spaces and all pervasive repulsive odour of open sewers. There is also infrastructural decay: deplorable roads, pitifully inadequate water supply, erratic electricity supply and acute shelter shortage. The picture is generally that of urban disarray. At the heart of problem is the ubiquitousness of non-industrial and industrial pollutants. Individual and corporate bodies are reckless in waste storage and disposal often resulting in environmental hazards inimical to animal and human health. Despite the above environmental challenges, little attention is currently given to the problems. Indeed, many have concluded that filthiness in Ibadan is cultural in origin as the people are dirty by nature. This ugly situation presents grievous threats to the achievement of sustainable development in the city.

As demonstrated in the preceding chapter, the impact of British colonial rule on the growth and expansion of Ibadan (like most of Nigerian cities) was a near-absolute paradox. On one side, the colonial urban

---

[1] Quoted in Soji Oyeranmi, "A Civilization without Toilets? Ibadan and her Environment in the Postcolonial Era" *Sociology and Anthropology*, 6(2): 187 DOI: 10.13189/sa.2018.060201, 2018 available online at https://www.hrpub.org/journals/article_info.php?aid=6771

environmental policies engendered unprecedented growth in Ibadan. On the other hand, it initiated enduring environmental crises which negated the much touted positive contributions of colonial urbanism. Sadly, these identified environmental crises endured from colonial times to post-colonial era in Ibadan. Expectedly, as the city continues to experience series of development activities since independence in 1960, environmental hazards kept arising from a wide array of sources which included the issues discussed earlier in this study. These problems have more direct and immediate negative impact on urban planning, waste management and sanitation with horrendous implications for human health and safety especially for the poor. The urban poor basically housed in slums or squatter settlements often have to contend with appalling overcrowding, bad sanitation and contaminated water. The sites are often illegal and dangerous, so, forcible evictions, floods and road slides and pollution are constant threats.

However, it is important to note that factors responsible for the dire challenges of environmental management in Ibadan during the post-colonial era are in two-folds. First, efforts made to remedy the existing environmental problems especially urban planning, waste management and sanitation in Ibadan have been largely inadequate and so ineffective. Second, for more than six decades after independence from colonial rule, the city administrators did not recognise the need to focus on sustainable development, which is the only way to sustain the process of development for the future generations. It was therefore not surprising that they failed to entrench effective environmental management strategies into the developmental agenda of Oyo State as a whole and Ibadan particularly. What was required therefore during the early decades of the post-colonial era was a balance between development aimed at upgrading the quality of life and the conservation of the city's environmental quality. However, the failure to successfully tackle these environmental challenges over the years is responsible for the worsening environmental situations in Ibadan. This has resulted in some of the most severe negative implications for the development of the city.

In view of the above, with a primary focus on Ibadan, this chapter is to examine: City and Sustainable Development; economy and the

Environment, the Stakeholders and the Environment vis-à-vis the various urban environmental problems that are confronting the city in post-colonial period.

City, Environment and Sustainable Development

The drive to ensure that the sporadic urbanisation which is a revelation of rapidity of human civilisation is sustainable has led to the creation of the concept of "Sustainable Cities."[2] As usual the Sustainable Cities Project (SCP) has impacted positively in cities in developed parts of the World while vast majority of cities in the Third World still largely remain huge squalors and mostly unsustainable. Thus, most First World Cities have become more sustainable through: taking in to account economic development and environmental costs of urbanisation; self-reliance in terms of resource production and waste absorption; cities become compact and energy sufficient; and the needs and rights of all are well balanced.[3] According to Darshini Mahadevia, cities in the underdeveloped parts of the World could only be sustainable if: there is environmental sustainability; social equity; economic growth and redistribution and; the empowerment of the disempowered.[4] This clearly shows the close link between environment and development.

Across ages, the inseparable link between environmental sustainability and human development actually authenticates the fact that existence of Flora and Fauna has been central to human existence on earth. Indeed, the totality of man depends on the environment; he walks on land, works on it, builds on it; might die and be buried on it. While Commenting on the link between environment and human development, Mabogunje observed that apart from the fact that all development takes place within an environment, it often gives direction to development in cities. According to him, cities have special role to play in this linkage as they

---

[2] Darshini Mahadevia, "Sustainable Urban Development in India, An Inclusive Perspective", in David Westerndorff ed., *From Unsustainable to Inclusive Cities*, (Geneva: An UNRISD Publication in Collaboration with Swiss Agency for Development Cooperation, October 2002, 2-19)

[3] Darshini Mahadevia, "Sustainable Urban Development in India.... 4

[4] Darshini Mahadevia, "Sustainable Urban Development in India.... 4

are primary to development. To drive home his points, Mabogunje further argued that:

Cities are engines of growth and development. Because production mostly takes place in cities, we usually have chains of demand and supply. On the supply side, you have people who are making or producing things such as agric implements or trading. On the demand side, the workers will need food, so, their presence encourages rural areas to grow. Sometimes, people have funny idea that the only way to encourage Africa's development is to concentrate solely on agriculture but agriculture cannot grow without markets. Cities naturally provide these markets. And as the cities continue to provide these markets, the more they enhance the growth and productivity of the rural areas. This may be achieved by bringing machinery to the rural areas. This enhanced productivity will in turn be reducing the number rural dwellers; as many of them would have to move to the urban centres, which are presenting more economic opportunities for them. And as cities continue to grow, the numbers of people in the rural areas continue to reduce. This is why we have countries like the USA with just 3% rural population.[5]

In demonstrating the real link between environment and development in cities, the erudite urban geographer observed further that:

The environment provides the raw materials which urban centres usually process for an increased productivity. Man has always depended on his environment for survival. From the earliest times to the 1700s (an era characterised by low productivity), the relationship between man and environment was governed by mutual respect. But with industrial revolution which brought about mass production and consumption (in cities) with a very high impact on the environment, things started changing for the worse. Due to technological driven and energy sapping industrialisation, we have depleted ozone layer globally which has led to climate

---

[5] Interview with Professor Akin Mabogunje - The globally renown doyen of African/Nigerian Urban geography at his Bodija, Ibadan residence

change. With this, environment seemed to have gotten to full circle, thus turning against our excesses. There are also other lesser problems such as poor urban planning, sanitation and waste management. Cities do import a lot, thereby generating enormous waste through consumption.[6]

The close relationship between Development and environment in Cities was also recognised at the UN conference on human settlements in Vancouver June 1976.[7] The forum concluded that problems of uncontrolled population growth, rural stagnation, migration, the inability of urban centers to cope with the rate of population increase and environmental deterioration demand corrective action at both the national and international levels. This is why Borofice rightly linked the current global environment crises to the development crisis, both of which are the result of unsustainable system of production and consumption in the North, inappropriate development models in the south and a fundamentally inequitable world order.[8] Interestingly, as environmental crises in cities have assumed universal phenomenon in the 21st Century so is the global response especially under the frameworks provided by the United Nations' systems.

As it has been established earlier, cities are often characterized by stark socioeconomic inequalities, social exclusion, extreme poverty, unemployment, poor environmental conditions, and high production of greenhouse gas emissions; their potential for growth and development makes them strong drivers for positive change and sustainable development . Their density and economies of agglomeration act as strings that connect all Sustainable Development Goals together, linking economy, energy, environment, science, technology and social and economic outcomes. According to the UN, nearly half of humanity—(3.5 billion people) are already living in cities today, and this number will continue to grow; indeed it is projected that urban population will get to 5

---

[6] Interview with Professor Akin Mabogunje

[7] the UN Conference on Human Settlements in Vancouver June 1976

[8] R.A. Boroffice, "Environment and Development", in S. Otokiti and S.G. Odewunmi eds., *Issues in Management and Development*, (Ibadan, Rex Charles Publication, 2001). 587 – 596

billion by 2030.[9] Because the future will be urban for a majority of people, the solutions to some of the greatest issues facing humans—poverty, climate change healthcare, education—must be found in city life.[10]

Considering the enormity and global ramifications of the crises, it is therefore not surprising that they attracted much recent global policy discourses orchestrated under the aegis of the United Nations, such as the Agenda for Sustainable Development (2030) and the New Urban Agenda of UN Habitat.[11] They advocated for urgent concerted focus at the city and the community scale – not only to achieve long-term developmental objectives but also to make direct tangible benefits to the quality of lives of the people. To address this specific issue, Sustainable Development Goal 11 is to make cities and human settlements inclusive, safe, resilient and sustainable provides an unparalleled opportunity for the attainment of collective and inclusive progress, and for the achievement of sustainable development in the world.[12]

While launching Sustainable Development Goals (SDGs), Ban Ki Moon, former UN Secretary General noted that cities are where the battle for sustainable development will be won or lost. This is because urbanisation has become the defining phenomenon of the 21st century as we are increasingly living in an urban world.[13] For instance, in 1950, the world was predominantly rural as global urbanisation level at mere 30%.

---

[9] The United Nations Organisation, *High Level Political Forum on Sustainable Development SDG 11 Sustainable Cities and Communities: SDG 11 and the New Urban Agenda: Global Sustainability Frameworks for Local Action*, 2018, Available online at https://www.un.org/sustainabledevelopment/wp-content/uploads/2018/09/Goal-11.pdf

[10] visithttps://sustainabledevelopment.un.org/content/documents/18785E_HLPF_2018_2_Add.4_ECAadvanceduneditedversion.pdf

[11] Hitesh Vaidya and Tathagata Chatterji, "SDG 11 Sustainable Cities and Communities: SDG 11 and the New Urban Agenda: Global Sustainability Frameworks for Local Action" in I. B. Franco *et-al* (eds.) *Actioning the Global Goals for Local Impact, Science for Sustainable Societies*, Singapore, Springer Nature .January 2020 DOI: 10.1007/978-981-32-9927-6_12

[12] The United Nations Organisation, *High Level Political Forum on Sustainable Development SDG 11 Sustainable Cities and Communities: SDG 11 and the New Urban Agenda: Global Sustainability Frameworks for Local Action*, 2018....

[13] Hitesh Vaidya and Tathagata Chatterji, "SDG 11 Sustainable Cities and Communities: SDG 11 and the New Urban Agenda: Global Sustainability Frameworks for Local Action...." 174

According to the UN, by 2050, the scenario is expected to reach 70%. Most importantly, cities are hubs of innovation, employment and wealth creation; urban areas already account for 55% of the global population and produce 85% of the global GDP with concomitantly grievous environmental cost, wastes, pollution and other sustainability crises.[14]

Since 2015, SDG 11 has catalyzed collaboration and partnerships between diverse groups of stakeholders at the local level, and between local, regional, and national governments. The breakdown shows that: 11.1 focuses on safe and affordable housing and basic services; 11.2 on safe, sustainable transport systems; 11.3 on inclusive urbanisation and participatory, integrated planning; 11.4 on cultural and natural heritage;11.5 on resilience to disasters; 11.6 on reducing environmental impact of cities; 11.7 on green and public spaces; 11.a on rural-urban linkages; 11.b on integrated policies and plans; 11.c financial and technical support for sustainable and resilient buildings.[15]

However, cities and regions continue to struggle with providing adequate housing, services, and infrastructure, especially in light of the increasing global incidence of natural disasters. A number of challenges in data availability to track progress towards implementation of SDG 11 also present significant barriers to assessing global progress on the goal. Inequality remains a big concern. 833 million people live in slums and this number keeps rising. The levels of urban energy consumption and pollution are also worrying. Cities occupy just 3 per cent of the Earth's land, but account for 60-80 per cent of energy consumption and 75 percent of carbon emissions. Many cities are also more vulnerable to climate change and natural disasters due to their high concentration of people and location so building urban resilience is crucial to avoid human, social and economic losses. The situation in most African/Nigerian cities is so pathetic in this regard.

---

[14] The United Nations Organisation, *High Level Political Forum on Sustainable Development SDG 11 Sustainable Cities and Communities: SDG 11 and the New Urban Agenda: Global Sustainability Frameworks for Local Action*, 2018....
[15] The United Nations Organisation, *High Level Political Forum on Sustainable Development SDG 11 Sustainable Cities and Communities: SDG 11 and the New Urban Agenda: Global Sustainability Frameworks for Local Action*, 2018

As a corollary, this study also recognizes the fact that the ecological crisis in Ibadan (as in most sub-Saharan African cities) is a reflection of overall underdevelopment of the city.[16] According to Michael Renner, it is unfair to blame series of security and developmental challenges in African cities on the persistent primordial hatred among the people because the underlying cause is more likely to involve environmental degradation, which in turn led to depletion of natural endowment.[17] Consequently, cities rather than providing the basis for sustained economic growth therefore have become serious impediments to development in sub Saharan Africa and disputes are often sharpened or even triggered by glaring social and economic inequalities. Regrettably, no place in sub Saharan Africa seemed to be immune against this tremendous tragedy, though, the impact varies from place to place. Nigeria as one of the few countries in sub Saharan Africa, which had many large pre-industrial cities could not possibly be an exception to the African urban environmental decay.[18] While the world is full of both urban environmental "successes" and "failures", most African cities are examples of urban decadence; Ibadan is particularly notorious in this regard.

At the heart of the crises is the fact while most cities in the developed World have taken concrete actions to entrench sustainable development plans; most cities in the underdeveloped countries, especially in sub-Saharan Africa, are yet to have effective programmes or sustainable national action plans to ensure provision of the basic necessities of life and maintaining decent and healthy environment. For this reason, most cities in post- colonial sub Saharan African that ought to be "epic centres of development" are bogged down with severe environmental problems. Ibadan is not an exception! For this reason, post colonial Ibadan like most urban centres in sub-Saharan Africa has become a "sick city" overwhelmed by air pollution, noise, traffic, waste, racial tension, slum

---

[16] Abiodun Areola, Environmental Justice and Green Spaces in Ibadan Metropolis, Nigeria: Implications on Sustainable Development in Urban Construction *Environ. Sci. Proc.* 2022, 15(1), 57; https://doi.org/10.3390/environsciproc2022015057 24 May 2022
[17] See Aidan Campbell, *Western Primitivism: African Ethnicity - A Study in Cultural Relation*, (London, Cassell Press, 1997). 86
[18] Laurent Fourchard, *Urban Slums Reports: The Case of Ibadan, Nigeria......*

conditions, maladministration and many other urban malaises.[19] Onibokun equally concluded that "today, the hearts of many cities in Africa are like islands of poverty in the seas of relative affluence."[20] Other challenges include: the deterioration of basic services, housing and environment, mass unemployment and underemployment, the virtual absence of State welfare and many more. All of this culminated in what Lourenco Lindell called "urban crisis."[21] Although the impact of environmental problems in Africa's urban centres vary from place to place, no place in sub Saharan Africa seem to be immune against the tragedy. As one of the many cities in sub Saharan Africa, which inherited grievous damages from colonial urbanism, Ibadan cannot possibly be an exception to the African urban environmental decay.

Ibadan was described by P.C Llyod as a "city- village."[22] This description truly reflects the frenetic mixture of traditionalism and modernism in most African "modern cities". Ibadan as the study shows continues to draw more people to her expansive land and over the years has developed a paradoxical character. Like other pre-industrial societies, Ibadan, even after independence presents a sprawling agglomeration of buildings, spread out in numerous directions without any coherent order. At the same time, it demonstrates the frantic pressure and restless energy associated with modern metropolis.[23]Unfortunately, the environmental disorderliness, which has been ravaging this huge urban space has become dreadful. Yet, both government and the people are seemingly growing more apathetic about environmental issues. It is therefore not surprising that Ibadan appears today as "a crippled city" in terms of environmental management and development.

This study fully recognises the fact that a lush/healthy environment is not a luxury and that environmental sustainability is at the heart of a

---

[19] Herbert Werlin, *Governing an African City, A Study of Nairobi*, (New York: Routledge, 1974) 12

[20] A. G. Onibokun et.-al, eds., *Urban Renewal in Nigeria*, (NISER, 1987). 10.

[21] I. L. Lindell, *Walking The Tight Rope: Informal Livelihood and Social Networks in a West African City*, (Stockholm, Stockholm University Press, 2003)

[22] P. C. Lloyd, *Africa in Social Change*, (Baltimore: Penguin, 1967) 3

[23] See the prologue in Dapo Adelugba ed., *Ibadan MESIOGO: A Celebration of a City, its History and People*, (Ibadan: Bookcraft Ltd, 2002) 2

meaningful development through effective environmental techniques. Thus, reinforcing the unbreakable link between environment and development. Unfortunately, failure of environmental management strategies has been a recurring decimal in the history of Ibadan. Things have become worse and almost unbearable at the post colonial times. Consequently, the city is experiencing constant environmental hazards such as: poor sanitation, inefficient waste management, outbreak of epidemics, incessant flooding, unregulated urban planning and many more. Some experts and non experts alike are blaming some of the crises on the negative impact of the economy on the environment.

Ibadan Economy and the Environment

It must be reiterated here that adequate urban environmental management is essential for sustainable development in any country.[24] As a corollary, urban centres (especially cities) have become the most conspicuous environments in which economic capabilities are expanded or impeded and social qualities of life are fulfilled or frustrated. However, urbanisation can be as dangerous as it is important for the development of most African countries. This is because the urban centers' crucial role as major media of development has not been sufficiently realised by development planners. In the struggle to create new capacities, African leaders tend to be more pre-occupied with economics while neglecting the need for adequate and qualitative urban environmental management strategies for sustainable development. With the stimulated progress in education, transportation and communication, most African urban areas are growing more rapidly than their sustainable capacities. At the same time, urban authorities are unable to cope with the housing, educational, healthy welfare and recreational needs of the population. The result is the evolution of what could be called the "Parasitic City" or unproductive city.[25] Rather than providing the basis for sustained economic growth, cities have become serious impediments to sustainable development, especially in sub Saharan Africa. Ibadan is no exception.

---

[24] Werlin Herbert, *Governing an African City: A Study of Nairobi*....6
[25] Werlin Herbert, *Governing an African City: A Study of Nairobi*....6

Before delving into the details of the interactions between Ibadan economy and the environment; it is essential to briefly examine the general links between economy and environment through the relevant views of some experts and scholars on the subject matter.

Obviously, the environment is a significant component of the economic system. In the opinion of a scholar without the natural environment the economic system would not be able to function.[26] Hence, Environmental economics posits that we need to treat the natural environment in the same way as we treat labour and capital; that is, as an asset and a resource.[27] Environmental economics is a branch of economics concerned with environmental issues. It also involves theoretical and empirical studies of the economic effects of national or local environmental policies around the world. Particular issues in environmental economics include the costs and benefits of alternative environmental policies to deal with air and water pollution, toxic substances, solid waste, and global warming. Thus the discipline addresses environmental problems and valuation of non-market environmental services. In general, environmental economics focuses on efficient allocation and accepts the assumption of neoclassical economics that the economic system is the whole and not a subsystem of the global ecosystem.

According to environmental economists, environmental degradation is the result of the failure of the market system to put the deserving value on the environment, even as the environment serves economic functions and provides economic and other benefits.[28] It is argued that, because environmental assets are free or underpriced, they tend to be overused and abused, resulting in environmental damage. The solution offered to the above problem is to put a price on the environment so that it can be incorporated into the economic system and taken seriously by those who

---

[26] Jim MacNeill, "Strategies for Sustainable Economic Development", *Scientific American*, (September 1989) 105

[27] D. J. Thampapillai, *Environmental Economics,* (Melbourne: Oxford University Press, 1991)

[28] Peter Self, "Market Ideology and Good Government", *Current Affairs Bulletin,* (September 1990). 4-10; J. Seneca and M. Taussig, *Environmental Economics,* (New Jersey, Prentice-Hall, 1984)

make decisions.[29] However, one of the major problems in putting a price on the environment is that it is highly objected by many as it is similar to putting a price tag on your family and friendship.[30] Another problem with valuing the environment is that the preferences of future generations and other species are not taken into account.[31]

To counter these criticisms, the proponents of environmental economics frantically explain the real nature of the interaction between economy and environment. It is a general belief that we can't have both economic development and environmental quality simultaneously, that if we want to improve economically we must sacrifice the environment. This is why often in the past economic development has been given preference over the environment and society. But the environmental economists believe that there is a mutual connection between environment and economy that is often not recognised. There is a widely held theory that resource management practices and policies which protect the environment are most likely to harm the economy and reduce employment opportunities. However, empirical data supporting this theory are rare. In recent years, economists and ecologists have increasingly begun to use quantitative methods to test this theory. Studies examining industrial emissions, endangered species, air quality and other issues have found no evidence that economies suffer as environmental policy strength increases. On the contrary, numerous researchers have reported slight positive correlations between environmental and economic indices, suggesting that environmental health may help to improve the economy.[32]

Environmental economists argued further that in order to prevent or reduce environmental degradation that usually accompanies economic

---

[29] David Ehrenfeld, "Why Put a Value on Biodiversity?" in E. O.Wilson *Biodiversity*, (Washington DC: National Academy Press, 1988)

[30] Paul Barkley and David Seckler, *Economic Growth and Environmental Decay: The Solution Becomes the Problem*, (New York: Harcourt Brace Jovanovich, 1972)

[31] Paul Barkley and David Seckler

[32] J. Sachs, *The End of Poverty: Economic Possibilities for Our Time*, (New York: Penguin); Herman E. Daly and John B. Cobb Jr., *For the Common Good: Redirecting the Economy Toward Community, the Environment, and a Sustainable Future*, (Boston: Beacon Press, 1989); World Commission on Environment and Development (WCED), *Our Common Future*, (Oxford: Oxford University Press), 1987

activities; we must start putting the deserved value on the environment. With this, the environment will continue to serve economic functions and provide other benefits without necessarily suffering any form of devastation. Therefore putting a price on the environment is required to mitigate the negative impact of the economy on the environment. It is argued that, because environmental assets are free or underpriced, they tend to be overused and abused, resulting in environmental damage. Because they are not owned and do not have price tags then there is no incentive to protect them.

The solution of putting a price on the environment (as recommended by environmental economists) will lead to its incorporation into the economic system and attracts serious attention from decision makers.[33] As a result of this, environmental values will then be integrated into economic decisions, market failures will be repaired and sustainable development assured. In the views of environmental economists, Cost-benefit analysis (CBA) is one of the key ways in which environmental values are incorporated into economic activities. Another is through economic instruments. Economic instruments include taxes and charges on polluters that aim to internalise environmental costs into the decisions of companies and individuals and therefore provide an incentive to curtail behavior that is destructive to the environment.[34]

A major problem with valuing the environment according to individual willingness to pay is that the preferences of future generations (and indeed other species) are not taken into account. For this reason, according to critics of this school of thought, the market value might not be consistent with long-term welfare or survival. Individuals might prefer to continue adding to the greenhouse emissions rather than cutback on energy use even though this might threaten future generations.[35] Gladly, the concept of sustainable development (which is a paramount guide for this study) has filled this lacuna with the focus on taking care of the

---

[33] D. J. Thampapillai, *Environmental Economics....*
[34] *Ibid*
[35] Paul Barkley and David Seckler, *Economic Growth and Environmental Decay: The Solution Becomes the Problem...*

economic and other needs of the present without compromising the capacity of the future generations to meet their own needs[36].

Sharon Beder and other advocates of sustainable development strongly opined that the market or economics cannot resolve environmental problems and "that there is a need to find a solution that embraces the ethical dimension of environmental protection in the sustainable development debate."[37] Sustainable development recognises that economic growth can harm the environment but argues that it does not need to. The sustainable development approach claims to be able to avoid the environmental degradation that has previously accompanied economic growth by integrating economic and environmental decisions. For most governments this means incorporating the environment into the economic system. For more conservative environmentalists, economists, politicians, business people and others, the concept of sustainable development offers the opportunity to overcome previous differences and conflicts, and to work together towards achieving common goals rather than confronting each other over whether economic growth should be encouraged or discouraged.[38]

For instance, it was reported that David Pearce and his colleagues, in their report on sustainable development to Margaret Thatcher, the then British Prime Minister, submitted that the principles of sustainable development meant recognizing that 'resources and environments serve economic functions and have positive economic value.[39] As a component of the economic system, the environment is seen to provide raw materials for production and to be a receptacle for its wastes. One of the most significant arguments for continued sustainable or environmental friendly economic growth, is that it is necessary to meet the needs of poor people. In line with this, a report concluded that poverty is a cause of environmental degradation as the "poor must meet their urgent survival

---

[36] World Commission on Environment and Development, *Our Common Future*....

[37] Sharon Beder, "Economy and Environment: Competitors or Partners?" *Pacific Ecologist* 3, (Spring 2002) 50-56

[38] Sharon Beder, *The Nature of Sustainable Development*, (Melbourne: Scribe Publications, 1996) 7

[39] David Pearce, Anil Markandya and Edward Barbier, *Blueprint for a Green Economy*, (London: Earthscan, 1989) 5

needs."[40] And the surest way to rescue people from extreme poverty is to stimulate economic growth. Jim MacNeill, the Secretary-General to the 1987 Brundtland Commission, (the most renowned advocates of sustainable development) made this clear when he argued that:

> The most urgent imperative of the next few decades is further rapid growth. A fivefold to tenfold increase in economic activity would be required over the next 50 years in order to meet the needs and aspirations of a burgeoning world population, as well as to begin to reduce mass poverty. If such poverty is not reduced significantly and soon, there really is no way to stop the accelerating decline in the planet's stocks of basic capital: its forests, soils, species, fisheries, waters and atmosphere.[41]

The above submission from the most recognised advocates of sustainable development could be the reason why the term 'sustainability' still angers some radical environmentalists, such as Mohamed Idris who observed that:

> The term 'sustainable' from the ecological point of view means the maintenance of the integrity of the ecology. It means a harmonious relation between humanity and nature, that is, harmony in the interaction between individual human beings and in their interaction with natural resources.' The term 'sustainable' from the point of view of non-ecological elites means 'how to continue to sustain the supply of raw materials when the existing sources of raw materials run out.[42]

The concept of sustainable development has undoubtedly gained global acceptability (despite of its vociferous critics) as no one can deny the fact that the world today still confronts the horrendous challenges on how best to combine the desired economic development with the unavoidable

---

[40] Jamaica, *Economic Issues for Environmental Management*, (Washington DC: World Bank, 1993)

[41] Jim MacNeill, "Strategies for Sustainable Economic Development", *Scientific American*, (September 1989) 106

[42] S. Mohamned Idris, "Going Green: A Third World Perspective", *Chain Reaction*, No. 62, (October, 1990) 16-17

environmental sustainability. A closer evaluation of the principles of sustainable development reveals the much needed positive nexus between advocates for economic prosperity and advocates for environmental protection. More importantly, it suggested many ways to pursue both economic prosperity and environmental protection and how to make them reinforce rather than compete with each other. Since cities are the epic centres of human development, they present gravest challenges to environmental sustainability due mostly to inadequate knowledge by urban managers (especially in poor countries) on how to ensure economic development without hurting the environment in the era of super-accelerated urbanisation.[43]

In many African countries, new environmental institutions at different administrative levels, such as environment management agencies, have been established to tackle both green and brown environmental issues. At sub-regional and regional levels, economic groupings accepted the challenges to develop interlinked and forward-looking strategies to ensure that Africa achieves some of the MDG targets. Despite such progress, the environment is yet to be fully mainstreamed in all sector-specific policies and in economic development. In particular, the relationship between the environment and continued poverty has not been fully acknowledged. The conclusion reached by the Brundtland Commission in 1987 that institutions tend to be independent, fragmented, and work "to relatively narrow mandates with closed decision-making processes," remains true several decades later.[44] In essence, today, Africa is still facing challenges of interlinking the environment and human development, understanding the causes and effects of environmental change and developing appropriate policy responses.[45] Ibadan like many

---

[43] Dennis Church, 1992, "The Economy Vs. The Environment: Is There A Conflict?" Available online at http://www.ecoiq.com/dc-products/prod_conflict.html

[44] Susan Parnell and Jenny Robinson "Development and Urban Policy: Johannesburg's City Development Strategy", *Urban Stud*, Vol. 43, No. 337, (2006)

[45] United Nations Environment Programme (Content Partner); Peter Saundry (Topic Editor), "Seizing Opportunities in Africa: Interlinkages in Environment for Development", in Cutler J. Cleveland ed., *Encyclopedia of Earth*, (Washington, D.C.: Environmental Information Coalition, National Council for Science and the Environment, 2007). Available online at http://www.eoearth.org/article/Seizing_opportunities_in_Africa:_interlinkages_in_environment_for_development

other cities in Africa is bedeviled by these problems. Some of these crises are discussed later in this chapter.

As mentioned above, cities are known to be engines of economic growth.[46] While they generate revenues for urban government, they provide income for urban residents to meet their welfare requirements. However, all over the world, these development centres are currently plagued with environmental problems of various types, and Ibadan is no exception. The fact that urbanisation in Africa is not a product of economic development further compounded the situations in the cities. Indeed, as it has been argued, urbanisation has emanated in most parts of Africa as a result of negative consequences of development policies, particularly the disarticulation of rural economies that fueled rural-urban migration.[47] In the opinion of Mabogunje:

> ...the failure of the urbanisation process in Africa to seriously improve the lot of the majority of the population either in the urban or in the rural areas calls for a re-examination and a deeper insight into the nature of the complex social forces which urbanisation represents in the particular circumstances of the African continent today. And the single most important problem facing, post-colonial Ibadan (like most other cities in Nigeria) today is the lack of administrative and revenue-raising capacity....[48]

As reflected in the opinion of Labinjoh, the character of a city's economic activity determines the nature of its domination.[49]The fact that Ibadan economy was unable to transit from its pre-industrial beginning to a real industrial one even at the post-colonial period ought to reduce its environmental degradation. But this is far from the case.

---

[46] B. Wahab, "The Institutionalisation of the Environmental Planning and Management Process", in T. Agboola et. al eds., *Environmental Planning and Management: Concepts and Application to Nigeria,* (Ibadan: Counstellation Book, 2006) 48

[47] The Cities Alliance, *Foundations for Urban Development in Africa: The Legacy of Akin Mabogunje,* (Washington, DC: The Cities Alliance Secretariat, 2006) 17-18

[48] The Cities Alliance....17

[49] J. Labinjoh, *Modernity and Tradition in the Politics of Ibadan, 1900-1975,* (Ibadan: Fountain Publication, 1991) 3

According to Oyemakinde, Ibadan economy from inception rested more on the operations of the indigenous enterprises that promoted local industrial growth.[50] This did not only assist the political preeminence enjoyed by the city in the Nineteenth Century Yorubaland but also allowed the evolution of specialisation of labour through which some critical needs of the people were met. For instance, blacksmiths were able to produce cutlasses, swords, axes and guns for the soldiers and manufactured hoes, sickles and cutlasses were for the farmers.[51] Cottage industries equally provided employment for quite a sizeable proportion of the population. Women were mostly found in the looms where textile materials were manufactured and the dyeing industry. There were also soap making, local beer, and liquor distillation industries. A number of quarters and streets that took their names after the businesses of the inhabitants also emerged in Ibadan at this period. For examples, a street that harboured the dyers became known as Idi-aro (dye center); there is also Ile-Oloyin (Honey seller's house) at Dogo's compound, Oluyoro Oke-Ofa Ibadan.[52] These various economic activities kept the people occupied such that while soldiers were fighting on the battle field, the farmers were busy on their farms and the merchants crawling from one market to another. With this economic progress, the influx of migrants into the city in the 19th Century did not create much difficult regarding provision of opportunities for economic survival of the population.[53]

The incursion of the British colonialists also brought a number of limited economic opportunities to the city. As early as 1901 when railway got to the city, Ibadan warlords were demobilised and employed in railway construction and operation and many of them were also rehabilitated by the new mercantile houses. A number of new innovations such as: cement, pre-cast blocks, corrugated iron sheets, bicycles, motor cars and Lorries were also introduced by the British. The introduction of the modern transportation enhanced the import and export trade. As a corollary, old markets were expanded and new ones emerged in order to

[50] Wale Oyemakinde, "Indigenous Enterprises in Ibadan since 1830", in G. O. Ogunremi ed., *Ibadan….* 265
[51] Wale Oyemakinde …..265
[52] Raifu Isiaka, *Urbanisation in Ibadan….*33
[53] Wale Oyemakinde, "Indigenous Enterprises in Ibadan since 1830…." 267

take care of the improved economic activities. However, despite the above limited innovations, the colonial economic policies failed to entrench sustainable development due to a number of fundamental laxities.

In the first place, rather than directing the colonial economic policies towards the revitalisation of indigenous enterprises in Ibadan, the imperialists either neglected them or discouraged their development. Thus, the colonial state as an imposed entity with its egoistic and self-centred goals and policies set serious restriction on the aspirations of local entrepreneurs which made it difficult for them to achieve the required economic self-reliance. More reflection along this line revealed that the colonial enterprise was essentially exploitative, so its goals included incorporation of the economy into that of Europe. With this, the economy of Ibadan (like most colonial cities) operated under the general condition of the economic layout which was organised with neither intention of enhancing local economic development nor with consideration for the environment. This was primarily done to serve the interest of the British colonialists. Sadly, since Nigeria's independence in 1960, all the efforts at redeeming the economic situation of the city as well as the country have achieved very insignificant success. To worsen the situation, in bids to achieve economic growth, leaders in the city have mostly paid little or no attention to the environment.

Every generation lives and works in an inherited environment shaped, in some cases, by very distant predecessors.[54] This is why is not surprising that there is a very strong link the three eras (pre-colonial, colonial and post-colonial) that shaped environmental history of Ibadan. There were also enough evidences to demonstrate constant interaction between the economy and environment during these periods. At the pre-colonial period, the ecological characteristics of the city evidently reflected enormous economic advantages derivable from the natural environment. In the opinion of T.A. Akinyele:

> The physical nodality of Ibadan (the city of sixteen gates by itself guarantees free flow of resources – human and material – and that

---

[54] Gerald Burke, *Towns in the Making of London*, (London: Edward Arnold Publishers Ltd., 1975) 41

is one of the ingredients of economic growth through trade and commerce. No wonder throughout its history, Ibadan has been an important transitional market and a gateway for the transmission of goods and services between the south and north, between the forest zones and the savannah regions.[55]

Toyin Falola has also underscored the economic pre-eminence (albeit, in a traditional form) of pre-colonial Ibadan when he concluded that "indeed, very rapidly in the 1830's the city established its politics, built a military machine, and an extensive production system based on the use of slaves and dependent labour."[56] The above historical accounts have revealed several positive characteristics of Ibadan and proved that right from the beginning, the city is destined it to become contributory to national economic development, rather than being parasitic. "It is in the combination of these natural endowments, the well-oriented system of kinship and clientage, an equitable system of government, human freedom and dignity"[57] that seemed to have enhanced 'sustainability' in the city at this period.

What was most significant about pre-colonial Ibadan's management of the environment especially land resources was the republican nature of the government of the city. This was brought to bear on the land tenure, such that the system did not allow a few people to own, control and exclude others from ownership of land.[58] A clear departure from this practice during the colonial and postcolonial eras led to a great distortion and unsustainable development in Ibadan.

As discussed earlier, the British colonial urbanism in Ibadan (like most other Nigerian cities) through the policy of deliberate segregation was

---

[55] T. A. Akinyele, "Economic Relevance of Ibadanland: Past, Present and Future", Annual Lecture of the Oluyole Club of Lagos Delivered at Kakanfo Inn, Ibadan on 9th January, 2010

[56] Toyin Falola, *The Political Economy of a Pre-Colonial African State, Ibadan, 1830-1990,* (Lagos: Modelor Design Aids Ltd., 1989)

[57] T. A. Akinyele, "Economic Relevance of Ibadanland: Past, Present and Future…."; I. B. Akinyele, *Outlines of Ibadan History,* (Lagos, 1946) Iwe Itan Ibadan, (England, 1950 2nd Edition)

[58] Babatunde Oyedeji ed., *Readings in Political Economy and Governance in Nigeria; Selected Speeches and Articles of Chief T.A. Akinyele,* (Lagos: CSS Ltd., 2002) 5

meant to serve the basic interest of the Europeans with little or no respect for the needs of the natives. Thus, the growth of the local economy was the least on the minds of the colonialists. So the question of sustainable development (which seeks to promote economic growth through positive urbanisation and ensure the balance between economic development and environmental sustainability) was not part of the agenda at this period. Hence, we have urbanisation in Ibadan without industrialisation (which ordinarily should have produced critical urban infrastructure needed for sustainable development). In any event, the direction of colonial policy was to encourage colonies to produce raw materials and not to industrialise.

Nevertheless, the economic importance of Ibadan endowed by nature and concretised by military adventurism could not be ignored by the British at a time when industrial revolution was demanding enormous raw materials for British nascent industries and at a time when the slave trade became outlawed to be replaced by legal trade in import and export of goods.[59] To accelerate development in agriculture and cottage industries, land laws were introduced to legalise land transfers and to ensure adherence to physical planning regulations. Agricultural production was encouraged by the colonial government. Three crops – cotton, cocoa and rubber – were the first to become cash crops. Sadly, post-colonial Ibadan never experience much elevation from colonial urban retardation as urbanisation still moves on rapidly without necessary industrialisation to boost economic development. To complicate matters, there is also the absence of efficient environmental management strategies. So, the concept of sustainable development remains largely a mirage.

With this, it becomes understandable why post-colonial Ibadan presents such a despicable environmental picture. In fact, Francis Egbhokare's revised Standard version of Professor J.P Clark's *Ibadan*: "Ibadan/oozing blob of rot and mold squashed and splattered among human wastes/ like Bodija market in the rain"[60] vividly captured the filthy nature of post-colonial Ibadan. The main reason for this (as in most parts of Africa) was that the governance of the city fell in to the hands of post-colonial leaders

[59] Babatunde Oyedeji ed.....
[60] Francis Egbhokare's Revised Standard Version of Professor J.P. Clark's *Ibadan*

who merely replaced the British as local imperialists. Thus, they only stepped into shoes left behind the Europeans and built on their heinous legacy instead of uprooting them. With the introduction of partisan politics and in the wake of the intense political struggle for independence, Ibadan became the headquarters of Western region and the hotbed of cut-throat politics in 1952. Prior to Nigeria's independence in 1960, there had been the effects of the Second World War in which able-bodied persons of Ibadan origin (like other Nigerians) took part at the expense of employment for local development. Secondly, the period witnessed some global economic depression, down-turn in trade and commerce and general shortages of essential commodities and food. To worsen matters, the swollen shoot disease that started to take its toll on the cocoa production and export reached its peak shortly before independence.[61]

As a consequence, there was economic retardation and massive rural depopulation. Immigration into the city of Ibadan became less orderly than even the pre-colonial era. The organised systems of kinship and clientage broke down and the farming population trooped to the city in search of better life. Expectedly, the city's infrastructure could no longer sustain the horde of city adventurers. Thus, Ibadan since the mid 1960's has turned into a conglomeration of urban filth almost totally overwhelmed by environmental decadence and ecological degradation.[62]

Sadly, this beginning of pervasive environmental decay in Ibadan coincided with the period of political crisis in the Western Region which beclouded the vision of the leaders of thought at all levels and diverted their energies to less productive pursuits. It must be admitted that environmental problems and very low industrial level of post-colonial Ibadan (especially between 1960 and 2000) never assumed top priority in the scheme of things by successive governments. To worsen the situation, the sinister political debacle forced the political leaders to redirect the establishment of some industries that ought to have been sited in and

---

[61] T. A. Akinyele, "Economic Relevance of Ibadanland: Past, Present And Future,"....12
[62] Babatunde Oyedeji ed., *Readings in Political Economy and Governance in Nigeria*.....25

around Ibadan to Ikeja, Mushin and Isolo by the Western Regional Government being controlled by the Action Group.[63]

It must also be mentioned here that the wanton distortion of the land tenure system in Ibadan since the 1960's by the political leaders did a lot of irreparable destruction to the economic fortunes and sustainable development of Ibadan. For instance, it was alleged that the objectives of the Land Use Decree of 1978 ran against the grain of Ibadan system and it has contributed to the haphazard growth of Ibadan since then. As contained in the opinion Akinyele the land system which is at the centre of Ibadan politics and well-being was bastardised by the Land Use Decree of 1978. Some of the direct consequences of this include: the decimation of a big chunk of Ibadan's fortunes; poverty of land owners; lack of private layouts; over-concentration of attention to alienation of government land in the G.R.As (Government Reservation Areas); the turning of lawyers into liars and land surveyors into land bandits through the manipulation of land deals and so on.[64] With this entire horrific situation, it is only natural that post-colonial Ibadan continues to witness the emergence of "glorified slums" everywhere. Initially, this ugly development was much more prevalent in the traditional core areas (where the poor mostly stay) but according to Afolayan, the problems are now being replicated in the so called modern parts or new areas where "the big and beautiful mansions sandwiched between the ugly and the slummiest tenements."[65] Consequently, as the economy of post-colonial Ibadan continues to witness acute stagnation, the environment occupies the least position in minds of the government (just as it did with the colonial government). The people and few available businesses (small and big) are also paying little or no attention to the environment. This is why this study tends to blame nearly all the stakeholders for the recurring and enduring environmental crises in post-colonial Ibadan.

---

[63] Babatunde Oyedeji ed.....25

[64] Babatunde Oyedeji ed....25

[65] Interview with Dr Afolayan at the Department of Geography, University of Ibadan, Ibadan, Nigeria. This fact was also attested to by other informants such as Mr Niyi Agboola, a lawyer/pastor, at Academy area, Ibadan; Mrs Nofiu Dasola, a teacher , at Sanyo area, Ibadan Mr Bisi Ilori a Civil Servant, Idi- Ayunre, Ibadan; Mrs Adetola Tanimola, a trader , Odo Ona, Elewe, Ibadan, Miss Olaide Ajunwo, a student ,New Garage area, Ibadan on 10/02/14

The Stakeholders and the Environment

Thus far, it has been established that cities are the world's greatest assets for pursuing sustainable development. This is because cities are the areas with the highest possibility and ability to create economic growth and have also continued to play an increasingly significant role in regional and spatial development policy.[66] Sadly, most cities in sub-Saharan Africa in post-colonial era are not agents of sustainable development due largely to horrendous environmental history inherited from previous generations. While it is true that every generation lives and works in an inherited environment shaped, in some cases, by very distant predecessors; but this does not foreclose environmental rejuvenation by successive generations. With a primary focus on Ibadan, this chapter intends to raise environmental consciousness of Nigerians in this direction. In order to harness Ibadan and other Nigerian cities as assets, all stakeholders must synergise and be committed to the ideals of sustainable development. The absence of this essential symbiotic relationship among the critical stakeholders in the management of environment is at the heart of ecological crisis in post-colonial Ibadan. Stakeholders in this case include: the government, the people and few companies in post-colonial Ibadan.

No doubt, cities have become dynamic centers of economic growth and development: providing jobs, education, and markets and often producing more than twice their proportional share of GNP.[67] If this growth can be sustained and if these cities can help to manage the wastes and pollution that threaten to engulf them, then, those same cities can provide one of the most important contributions to a sustainable world environment.[68] Hence, efficient environmental management has become pivotal in ensuring that cities play that historic role as drivers of balanced

---

[66] See The Executive Summary of a 2013 Special Report by The United Nations' Agency for Human Habitat : The State of European cities in transition: taking stock after 20 years of reform" available online at *www.unhabitat.org*

[67] Herman E. Daly and John B. Cobb Jr, *For the Common Good: Redirecting the Economy toward Community, the Environment, and a Sustainable Future,* (Boston: Beacon Press, 1989)

[68] Commonwealth Government, *Ecologically Sustainable Development: A Commonwealth Discussion Paper,* (Canberra: AGPS, 1990)

and equitable development. Not only can well managed environmentally sustainable –cities contribute to health, welfare and productive capacity of their own citizens but they can also make a major contribution to environment sustainability and economic development on a global scale.[69] This work denounces the hitherto widely circulated belief that urbanisation and economic development are antithetical to environmental sustainability and calls for strengthening the capacity and capability of cities across post-colonial Africa (especially, Nigerian cities) to protect the environment at this age of surging urbanisation particularly in the areas of urban planning, waste management and sanitation. This is in line with the thought of David Foster that:

> Environmental quality in rapidly growing areas (cities) is really a matter of choice; management, not chance, is the determining factor in deciding whether urban growth will help or harm the environment. Cities can capitalize on the same trends and resources which lead to economic growth and use them to invest in environmental infrastructure which will make that growth sustainable. To manage the life of a city is ultimately to choose a future: to identify priority objectives and the risks that threaten them and then to mobilise resources effectively with which to meet those threats.[70]

The issues of government, governance and Urban Environmental Management (UEM) have been closely linked to the discourse(s) of development and the role(s) of the state, the market and the private sector and the citizens.[71] But as the custodian and epitome of peoples' hope and aspiration towards personal and national development, government at all levels must ensure sustainable development of cities. Historically, the state was seen as the key promoter of development and the main

---

[69] Schaltegger Stefan et. al, *An Introduction to Corporate Environmental Management: Striving for Sustainability,* (Sheffield: Greenleaf, 2003) 4

[70] David Foster, The Role of the City in Environmental Management Regional Environmental

[71] Soeren Jeppesen, Joergen Eskemose Andersen and Peter Vangsbo Madsen, "Urban Environmental Management in Developing Countries – Land Use, Environmental Health and Pollution Management, Research Network on Environment and Development", available online at www.ReNED.dk

responsible party and a dominant perspective in international development assistance was a belief that the developing countries should establish government institutions, draft legislation and carry out enforcement similar to the North. However, with growing dissatisfaction due to limited improvements in many countries, changing international political and economic regimes, new perspectives were aired with the ambition of 'resolving state failures' or encounter 'bad governance'.[72] The 'solutions' centred on 'good governance' that will allow markets to flourish and ultimately the private sector in tandem with state institutions now seen as the key promoters of development. Lately, public-private partnerships have been further advocated as a means of sharing the responsibility and/or seeking additional funds and/or involving new – private and civil society– actors in the development efforts.[73]

UN-HABITAT's global campaign on Urban Governance argues that there has never been a more important time than now, to focus on the quality of governance at the local level.[74] The new social contract arising out of the emerging democratic dispensation, the strong re-emergence of the civil society and the expansion of the public space, foster the need for taking responsibility and accounting for outcomes and impacts.[75] In addition, the forces of globalisation and the movement towards decentralisation are putting cities and local governments under tremendous pressure to deliver an ever-expanding range of benefits.[76] The realisation of these expectations, however, is affected by several important realities related to urban governance.

Many observers have also noted that the quality of urban governance can make the difference between cities characterised by growth and

---

[72] Soeren Jeppesen, Joergen Eskemose Andersen and Peter Vangsbo Madsen....

[73] A. M. Kjaer, "Central Government Intervention as Obstacle to Local Participatory Governance: The Case of Uganda." Paper prepared for the ILO Conference on Governance, 2005. 9-10

[74] UN-HABITAT (United Nations Human Settlements Programme), "I'm a City Changer", available online at www.worldurbancampaign.org and www.imacity changer.org

[75] UN-HABITAT (United Nations Human Settlements Programme)....

[76] UN-HABITAT (United Nations Human Settlements Programme)....

prosperity and cities characterised by decline and social exclusion.[77] What is most clear is that the quality of urban governance and management is critical to gaining the benefits and reducing the negative aspects of cities of any size.[78] The rapid growth of cities will put a premium on building institutions to address the problems of those cities.

In the present context, therefore, good urban governance describes a situation in which the mechanisms, processes and instruments for decision-making and action facilitate civic engagement and accountability.[79] Regrettably, today, most cities in sub-Saharan Africa are experiencing differing levels of decline mainly due to the absence of good urban governance, Ibadan is no exception.

One of the most critical findings of this study is that there is no significant positive shift between the colonial and post colonial eras in Ibadan due to extremely poor urban governance and grossly inadequate environmental management strategies. Indeed, there is a general consensus that the major problem of Ibadan has of always been planlessness which was exemplified by the absence of master plan from the colonial period till the present.[80] There have been many urban policies and programmes in Ibadan from the colonial to postcolonial times but these have been hampered greatly by half hearted implementation and absence of master plan for Ibadan. All we have are skeletal plans for pockets of housing estates such as bodija (1970), Jericho, Oluyole, and Ring Road.[81] So, Ibadan has grown over the decades (still expanding) without a comprehensive

[77] World Bank, *Sustainable Development in a Dynamic World: Transforming Institutions, Growth and Quality of Life,* (World Bank and Oxford University Press, 2003).108-110.
[78] UNDP *Governance and Sustainable Development,* (New York, 1997) 2-3
[79] UN-HABITAT (United Nations Human Settlements Programme) and Transparency International, *Tools to Support Transparency in Local Governance: Urban Governance Tool Kit Series,* (Nairobi: UN-HABITAT, Secretariat, 2004) 1-4
[80] For instances, Mabogunje: 1968, Onibokun: 1987, Abumere: 1987, NISER: 1997, Labinjoh, 1999, Fourchard: 2003 have come to same conclusion about the enduring negative implication of absence of any credible Master Plan for urban planning and development of Ibadan. Many of experts interviewed such as: Prof Mabogunje, Dr Idowu Johnson, Dr Dickson Ajayi, Dr Femi Olaniyan, Dr Murtala Monsor, Dr Godwin Ikwuyatum, Mr S.B. Taiwo, Mr Ademola Daud Michael Adeleke also came the same conclusion
[81] Interview with Mr S.B. Taiwo (Deputy Director, Development Planning, Oyo Ministry of Urban and Physical Planning)

plans or strategies to address the identified environmental challenges and develop the city. According to Afolayan:

> Consequently,there has been growth without sustainable development; at best all we have is haphazard or convoluted development. To worsen the situation (unlike cities in the developed world) Ibadan is a contaminated city because there is no coherent city governance. For example, Ibadan metropolis has 5 LGAs and 6 at periphery which made extremely difficult if not impossible to have a comprehensive and a well coordinated plans to development the city in a sustainable manner. Although there are existing environmental management policies but due to the absence of the needed synergy, coordinated implementation become impossible as these local governments always operate regularly in isolation with different orientation.[82]

Closely related to the above is the absence of zoning arrangement in post-colonial Ibadan (in which land will be specifically allocated strictly for particular purposes and so demarcated).[83] With zoning arrangement, areas meant for residential purpose would not be used for any other purposes; same thing goes for industrial area, markets areas, school areas, churches, mosques, shrines etc. Sadly, the near-zero presence of zoning arrangement is one of the most important reasons for the squalid nature of Ibadan (just as in many other Nigerian cities).

Although, there were pockets of local plans, individual layouts but there are no general master plan. S.B. Taiwo also claimed despite the fact the first School of planning in Nigeria was founded in Ibadan; lack of development control tools and absence of development plans to guide the planning of the city had resulted into uncontrolled growth.[84] Thus, the city was just growing in all directions without any coordination leading to haphazard development.

---

[82] Interview with Dr Femi Afolayan.....
[83] Interview with Dr Femi Afolayan.....
[84] Interview with Mr S.B. Taiwo.....

This was not due to lack of planning authorities but rather because of compromising attitude of the planners and lack of political will in the enforcement of environmental rules by successive governments at all levels. Pervasive corruption, poor motivation, lack of commitment and undue familiarity between the environmental officers and Ibadan residents have also been fingered for this ineffective enforcement of planning laws and procedures.[85] It was even alleged that at a point when Local Government Areas were in charge of planning at grassroots, a commissioner of Local Government Affairs gave instruction to the enforcement agencies that the moment that any building has gotten to lintel level and such building has violated any rule, such structure should not be demolished. So, people took advantage of this by encroaching other peoples' lands; building on illegal plots of land and usually start work by Friday, by Monday the building would have gotten to lintel.[86]The massive corruption and absence of proper enforcement of environmental laws have culminated in the city's infamous status as the dirtiest city in Africa.[87]

Loss of operational autonomy by Planning Authorities in Ibadan, also created enormous during and beyond our period of study. In the 1980s, we used to have independent Ibadan Metropolitan Planning Authority that was completely autonomous but things have changed drastically in this regard (since year 2000) as everything has been taken over the government.[88] This has paved way for the undue interference from the government with serious negative implications for the growth and development of the city. This was further compounded by the problem of gross inadequate funding of environmental management particularly

[85] Interviews with: Mr Dauda Ogundeji, a trader, Iyana Agbala Itura,Old Ife Road Ibadan; Mr Olagunju Oyebajo, Furniture maker, Adegbayi area, Ibadan and Mrs Aina Iyanda, Alakia area, Ibadan; Miss Bankole Olaogun, a Secretary in a Law Firm, at Alakia area, Ibadan; and Mr Lanre Atanda, a businessman Bola Ige international market ,Ibadan on 14/01/14

[86] Interview with Mr S.B. Taiwo.....

[87] Mr. O. Oyewole, HOD or Director, Environmental Health Dept. En Mrs B.O. Areo, Principal Environmental Health Officer and Mr. K.K. Popoola (Chief Environmental Officer, Oyo Ministry of Environment and Habitat 6/1/15.

[88] Interview with Mr Ademola Daud (Senior Technical Officer, Oluyole Local Government office of Oyo State Ministry of Urban and Physical Planning on 20/01/16

urban planning, waste management and sanitation in post-colonial Ibadan. This has led to: serious shortfall in the supply of modern/ critical facilities such as: incinerators, mobile refuse collectors; logistic crises in the areas of irregular payment of salary, extremely bad road networks, problems of office spaces and accommodation, transportation, little or non-existent running cost etc.[89] However, under-staffing or shortage of man-power seems to the most vicious consequence of inadequate funding of environmental management by successive governments in post colonial Ibadan. The fact was attested to by nearly all our informants as the problem have negatively affected (in the past) and still affecting the areas of physical planning, waste management and sanitation.[90]

This definitely took a huge toll on other segments and compounding gravely other previously identified problems. For instances it was alleged that due to paucity of fund and contrary to the WHO recommendation of 1 environmental staff to 10,000 people, Oyo state within our period of study only employed less than 500 environmental officers to the population of over 6 million; less than 200 in Ibadan of close to 3 million people. This was despite to the fact that there were thousands of unemployed trained environment health personnel across the state and the city. In similar vein, with acute shortage staff (with the concomitant effect of over stretching and over stressing of few ones available) enforcement of environmental laws in the areas of waste management and urban planning became herculean (if not impossible) task. The implications of these are dire and enormous especially as the people are living in extremely poor sanitary conditions and badly managed urban environment which directly exposed them to serious health hazards, epidemics, floods etc. Although, government may take the lion share of blame for this crises (especially for near total failure to prioritise environmental sustainability in the bid to achieve economic development in Ibadan); the people (residents) could not also be exonerated from the

---

[89] Ibid
[90] Interviews with Mr. S. B. Taiwo,, Mr Daud, Mr Michael Adeleke (Deputy Director Town Planning Office, Ibadan North LGA);Mr Akinpelu Jegede (Ona Ara Local Government Area); My Yomi (Egbeda Local Government Area); Mr Onaolapo Abideen (Ibadan North-east Local Government); Mr Akinola (Ibadan South-west Local Government Area)

enduring environmental conundrum that had been the lot of Ibadan since 1960. This stems principally from the lethargic and lackadaisical attitude of the people to environmental issues such as sanitation, waste management, urban planning.

As discussed above, constraints on sustainable development in cities across Sub-Saharan in Africa are legion. Some are general and others are sectoral or specific. Some are local while others are national or regional.[91] It must also be admitted that, prior to the adoption of the current sustainable development paradigm, Sub-Saharan Africa lagged behind other regions in food security, standard of living, and various aspects of development. Consequently, the adoption of a new development paradigm that places more emphasis on combining economic development plans with environmental sustainability does not eliminate the existing constraints on sustainable development in the region.[92] This study argues that failure of governments, organisations and institutions across the continent to put their citizens at centre of the developmental efforts is the most important factor for the backward trends in the attainment of sustainable development in our cities. To worsen the situation, most peoples in these countries lack the awareness of basic ideas about the concept of sustainable development.[93] As a corollary, their participation and positive contributions towards establishing proper Urban Environment Management strategies become extremely low and near total counter-productive to sustainable development. The case of residents of post-colonial Ibadan is particularly notorious in this regard.

Indeed, some scholars like Dickson Ajayi squarely place blame for filthy nature of post-colonial Ibadan on the doorstep of the people. According to him:

There are array of environmental problems confronting the post-colonial Ibadan which arose from the points that it is a city-village

---

[91] "Constraints on Sustainable Development in Sub-Saharan Africa", http://unu.edu
[92] B. von Droste and P. Dogse, "Sustainable Development: The Role of Investment", in R. Goodland, H. Daly, S. el Seraty, and B. von Droste eds., *Environmentally Sustainable Economic Development: Building on Brundtland*, (Paris: UNESCO, 1991). 71-82
[93] B. von Droste and P. Dogse 71

and that culturally the people are dirty-sometimes they eat on leaves which they usually throw around anyhow. Their fundamental environmental problem is poor sanitation as they mostly live in unhygienic environmental without proper disposal of waste especially in the core areas.The city do generate Volumes of waste more than what the government can manage. This is why you find refuse littering the major streets and sometimes heaps of wastes all over the place. These waste often eroded into drainages and rivers channels which usually resulted in erosion and floods during raining season. This does not exclude industrial wastes from industries around Ibadan. Even when incinerators and dump sites are provided, companies and people hardly make use of them; sometimes parents do send their kids who usually dump the refuse at the feet of the incinerator because they do not have the required strength to dump the refuse properly.[94]

Godwin Ikwuyatum equally shares this belief about the people of Ibadan when he submitted that:

With few access links, evacuation during disasters and even refuse disposal became difficult; people often use that excuse to throw garbage anywhere, anyhow. There is also cultural aspect to this. People around here believe in certain delicacies which require unique packaging in leaves etc without much discipline as to dispose the waste properly. Indiscriminate dumping of refuse by market people into river channels and along the streets due to lack of monitoring and provision of needed infrastructure by the government.[95]

It was even alleged that at some point, there seemed to be resistance from the traditional core areas of Ibadan against the attempts to modernize the areas.[96] Although many of the occupants of this part of Ibadan resisted

[94] Interview with Dr Dickson Ajayi at the Department of Geography, University of Ibadan on 20/1/16
[95] Interview with Godwin Ikwuyatum at the Department of Geography, University of Ibadan on 20/1/16
[96] Interview with Dr Femi Afolayan....

change from status quo partly due their cultural affiliation or familial links (family houses, grandfather's land etc) as most of them were born there; but the main reason for 'being happy' living in the squalor in spite of the extremely limited facilities for human welfare was their economic and intellectual penury.[97] To worsen the situation, upper mobility among the multitude of slum dwellers right from colonial period to the post-colonial times in Ibadan was near impossibility due to extreme poverty- the number one enemy of sustainable development.[98]

The investigation conducted in selected core areas across the eleven Local Government Areas of Ibadan reveals that people mostly build without recourse to any plan; without regard to street layouts; encroach on roads thereby creating congestion and inaccessibility to those areas.[99] Even in some places such as markets, walk-ways and roads are often blocked; thereby creating uneasy passage for both vehicular and human movements. Majority of the houses in those areas also have no toilets. When some of the residents of these areas were asked questions on how they defecate or even dispose of their wastes, a few of them were sincere enough to open up that they usually dump their faeces (and other wastes)

---

[97] Onibokun, Adepoju "Forces Shaping the Physical Environment of Cities in the Developing Countries: The Ibadan Case", *Land Economics*, Vol. 49, No. 4, (Nov., 1973), 424-431

[98] Fourchard, Ibadan....12

[99] As mentioned in the proposal for this study, the Participant-Observation method has been extensively employed. This took the researcher to all the eleven Local Government Areas that form the study area for this research. I must state here that, series of visits to the affected areas and interviews conducted strongly confirmed what the researcher as a resident of the city had been encountering over three decades. The filthy nature of post-colonial Ibadan is staggering and simply unbelievable especially in core areas like Beere, Oje, Agugu, Idi arere. Oranmiyan, Oke-adu, Itutaba, Gege, Oke foko, Beyerunka, Isale-jebu, Orita-merin, Inalende, Omitowoju , Abebi, Bode, Itutaba and so on. In these areas, there is near- zero presence of sanity in terms of planning, waste management and sanitation. Congestion everywhere; refuse usually litter the environment and due to absent of personal or public toilets, people normally defecate at any available space (sometimes at night and many times in broad day light) without any shame. So the situation in those areas is better imagined than experienced! Unfortunately, the supposed new areas such as Mokola, Bodija, Oke –Ado, Challenge, Ring –Road, Eleyele, Sango, and Satelite towns and fringes (Ajibode, Ojoo Apete, Ido etc) are also not immuned against these environmental maladies. Both the people and the government refused to learn from the past, hence the crisis is escalating in those new areas and if care is not taken, the situation could be worse in some instances.

in channels such as gutters, canals, rivers, streams, creeks, refuse sites, the streets, even on highways and roads.[100] Even in the few houses where conveniences are available, they are pit latrines which are usually in state of dilapidation. These latrines are not well-constructed, they are roofless some of them are a sort of make-shift in structure either covered with rusted corrugated iron roofing sheets or uncovered at all; others are filled to the brim, almost spilling onto the bare ground, and one's face is greeted with maggots in an attempt to examine the state of such latrines. Sadly, some of the problems like (congestion, lack of access roads, indiscriminate dumping of refuse etc) identified in the core city centres are now being replicated in the so-called new or modern areas in post-colonial Ibadan.

Uncontrolled population growth is also a problem (up till now, there is no accurate census figures for governments to be able to plan for the citizens). The population of the city is growing beyond the capacity of managers of the environment. This is being compounded by problems unregulated human habitation, careless consumption and reckless production activities of the few available small and big businesses in post-colonial Ibadan. Godwin Ikwuyatum summed this up that:

> One major problem of Ibadan is that of urban planning which reverberates around the triparte concerns of human habitation, consumption and production. Due to absence of proper planning, the city is unable to cope with proceeds from these. Waste generated come largely from human consumption and production without effective disposal techniques (refuse always litter everywhere).[101]

If the government lacks proper planning/environmental management philosophy and tenacity to enforce environmental laws; if the people are reckless in handling environmental issues: the few industries in post-colonial Ibadan also are not helping matters. The negative impact of

---

[100] Interviews with: Miss. Oluwaseyi Adebayo, a trade apprentice, at Beere area, Ibadan; Mr. Odunayo Adigun, a bricklayer, at Inalende area, Ibadan; Mr Kazeem Ijadunola, a photographer at born foto (Gege area) Ibadan; Mrs Mistura Ibikule, a trader at Oja Oba area, Ibadan; Mr Olusola Anjorin, a motor spare part dealer, Adebisi Compound, Oje area, Ibadan. From 6/02/14
[101] Interview with Dr Godwin Ikwuyatum…..

industrial activities on the environment primarily stems from lack of master plan and the absence of zoning arrangement in post-colonial Ibadan (in which land will specifically allocated strictly for particular purposes and so demarcated). Personal visits and investigation reveal three main implications of this. In the first place, industrial and residential spaces are practically in one and the same place while factory production, especially of the large scale types is generally in buildings or premises separate from dwelling houses in the same vicinity.[102] Examples of this situation could be seen in Sanyo Nig. Ltd. along Ibadan -Lagos Express Road, Odo-Oba and Askar Paints Nig. Ltd. at Eleiyele. The traditional craft such as blacksmith industry is also organised on cottage or compound basis.[103] Second is the scattered nature of modern industries in Ibadan which is largely due to the location of the very few industrial estates namely: Oluyole, Old Lagos Road, Olubadan Industrial Estate, along New Ibadan/Ife express Road, Ajoda New Town and Eleiyele Light Industrial Estate. The Nigerian Breweries PLC has a modern brewery located next to Olubadan Estate along new Ife express way with some industries located round the place.[104] The third pattern (which shows a slight improvement from the previous ones) is what now constitutes the fringe of Ibadan. This seems like a response to the changing patterns of urban development in the city as some industries are now located in those areas. For examples we have: Gas Cylinders Ltd. at Ejioku; Leyland Nigeria Limited at Iyana – Church; the Nigeria Wire and Cable Ltd. along Ibadan-Abeokuta Road Owode; the Standard Breweries at Alegongo Village; Eagle Flower Mills, Toll-Gate, The British-American Tobacco Company on Lagos-Ibadan Express Road new Toll-Gate Ibadan and many more.

With the above stated haphazard arrangement, it has become obvious why industries have been pivotal to environmental degradation of Ibadan

---

[102] Outcome of personal visits and direct observation which was carried between January and March, 2015

[103] Outcome of personal visits and direct observation

[104] Interviews with Mr. Olusegun Johnson, student 30, University of Ibadan; Mrs Monsura Aremu, trader, Bodija market, Ibadan; Mr Alimi Okunade, a surveyor, Poly road, Ibadan; Mr Victor Bishop, a technician/interior decorator, Oremeji, Mokola, Ibadan; Mr Dare Olukayode, a teacher, at his residence New bodija area, Ibadan on 16/01/14

despite of extremely low industrial activities. Although, the few industries also impacted negatively on other aspects environmental problems but they are most notorious in aspects waste generation (solid, gaseous and effluent) and pollution. With their "generous" pollutants, these emissions are mostly dispersed either up their 'chimneys' or down their sewage plants. The industrial sector has been contributing to the scourge of pollution, as admitted by some of our informants, who divided these wastes into solid wastes, liquid wastes or effluent wastes and gaseous ones.[105] Though contribution of these industries to the economic growth of Ibadan may be relatively low; but the effects of their activities on the environment are very significant. Largely, industrial activities generate wastes as the industrial establishments struggle to make profit. Industrial pollution problems arise because of most manufacturing industries in Nigeria, only 18% usually attempt basic recycling of wastes before disposal.[106] The recent size and number of industries have in recent years increased due to incentives by the government to adopt Foreign Direct Investment (FDI) strategies to attract industries to the nation. Thus, the economic wealth they bring makes the government relax on industrialisation laws, especially on emissions. Ibadan as one of the centres of industrial activities (like other cities in Nigeria) is facing a lot of environmental challenges due largely to unregulated industrial businesses.

While some scholars attempt to differentiate contributions of each stakeholder to the environmental degradation of post-colonial Ibadan (as we have seen above); others argued that these negative contributions are hardly separable as they deeply connected poverty, cultural perspective and attitude of the people; uncooperative attitude of small and big

---

[105] Interviews with: Mr Isaiah Orelusi, a business man, Lebanon Street, Dugbe, Ibadan and Mr Odunayo Kilani, Accountant, Jericho, Ibadan; Mrs Adedeji Suliat, a teacher, Sango, area Ibadan; Mr Bankole A. a police officer 35, Police Headquarters, Eleyele, Ibadan; Mr Taye Adegun, a welder , Abebi area , Ibadan on 17/01/14. Interviews with Mr. Ndigi Emi, a teacher at Orogun area, Ibadan ; Mr. Ladipo Olabode ,a transporter at Ojoo area, Ibadan; Mr. Oduniyi Akeem, a businessman at Ojoo area, Ibadan; Mr. Jamiu Esuola, a farmer at Ijaye via Moniya, Ibadan; and Mr. Damilare Olusesan, Akinyele village, Ibadan from 03/02/14 to 04/02/14

[106] World Bank, www.worldbank.org/1992report

businesses; lack of capacity and absence of adequate planning philosophy on part of managers of waste. In the opinion of Ikwayatum:

> Indeed, there are networks of interwoven problems confronting Ibadan from lack of planning to lack of discipline among the people and governments. All output from human consumption, output from industrial production are bundled into environmental problems confronting Ibadan. Sadly, nobody is thinking ahead for Ibadan even for the next 5 years (unlike Lagos with plans for the next 20 to 50). Health is wealth as they say but for any city to health and wealth, the environment must be healthy and sustainable. It is all about planning, economic activities, population growth, production and consumption which increasing waste generation. As the population is growing and their productive activities expanding, so will the consumption increasing resulting in the explosion in waste generation with negative implications for the city due to lack of planning, absence of capacity and inadequate waste disposal strategies. Pollution is a major outcome.[107]

From all indications, all the stakeholders in Ibadan have made differing efforts in addressing the above mentioned environmental problems but the main issue here is that the attempts made to solve the problems were too minimal and negligible when compared to the ways those problems were created. This is why even in the post-colonial era critical environment-development issues in Ibadan largely remain: Inadequate provision and management of environment infrastructure, which consists of unrepaired water pipes, inability to generate sufficient funds, erratic power supply and poor management, all of which have prevented over half of Ibadan's residents from having access to potable water, with attendant health and economic problems; insufficient waste management services, that is a low solid waste collection rate which has resulted in illegal dumping, blocked drains, disruption of business in commercial areas, reduced road space and localised air pollution due to neighbourhood incineration; flooding exacerbated by uncontrolled urbanisation. Much of the city has no storm drains, sewers or gutters. This

---

[107] Interview with Dr Ikwuyatum....

lack of drainage combined with shallow valley floors, increasing impermeabilisation and poor solid waste management has resulted in at least ten devastating floods from 1902 to 1980 and many more between 1980 and year 2000.[108] These were worsened by settlement in flood plains and deforestation and hillsides. There are also problems of poor environmental health; inadequate water supply, water pollution, poor refuse disposal, crowded and sub-standard housing, contaminated food and disasters such as flooding have resulted in high health risks for Ibadan residents. As an extreme example a cholera epidemic claimed over, 10,000 in the early 1970s.[109]

This chapter concludes that a deeper and better participation of people in urban environmental management is essential. Improving people's sense of belonging of place can also lead to creation of better and liveable environments in the urban environment but building an enduring synergy between and among the most critical stakeholders (the government, people and industries/institutions) is the most fundamental way to regenerate and achieve sustainable development in Ibadan and other Nigerian cities. The next chapter will essentially evaluate the existing efforts at tackling environmental crises in the city.

---

[108] A. T. Busari and M.O. Olaleye, "Urban Flood and Remediating Strategies in Nigeria: The Ibadan Experience", *The International Journal of Environmental Issues*, Vol. 6, No. 1 and 2, (2004). 7; and Ebenezer Adurokiya, "The Search for Environmental Sanity in Oyo", *Nigerian Tribune*, 24 September, 2011. 36

[109] David Taiwo (SIP Project Manager), *Ibadan: Mobilising Resources through a Technical Coordinating Committee*, http://ww2.unhabitat.org/programmes/uef/cities/summary/ibadan.html

Chapter Seven

Evaluating the Existing Post-Colonial Environmental Management Strategies in Ibadan

Having drawn from overwhelming historical and contemporary accounts, this study contends that a city is a true reflection of level of development in any country. This primarily stems from the fact that as the peak and agglomeration of the best of human civilisation, city often embraces the world more than the village that merely encases a monolithic cosmology-world view. In the opinion of Sam Omatseye "that is why the village has collapsed as a barometer of a civilisation, especially since the city began its rise on the back of that great idea called capitalism."[1]Thus, it has become incontrovertible that, cities are the world's greatest assets to pursuing sustainable development. This is because cities are the areas with the highest ability to create economic growth and also continue to play an increasingly significant role in regional and spatial development policy.[2] To drive home this significant point, the United Nations once submitted that:

> How we plan, build and manage our cities now will determine the outcome of our efforts to achieve a sustainable and harmonious development tomorrow. We believe that the battle for a more sustainable future will be won or lost in cities.[3]

For any discerning mind, the above assertions clearly demonstrate that urbanisation could be a positive force to be harnessed in support of social equality, cultural vitality, economic prosperity and environmental

---

[1] Sam Omatseye, "Redeeming Ibadan", available online at http://thenationonlineng .net/new/sam-omatseye/redeeming-ibadan/,

[2] See the Executive summary of a 2013 special Report by the United Nations' Agency for Human Habitat : The State of European cities in transition: taking stock after 20 years of reform" available online at *www.unhabitat.org*

[3] the Executive summary of a 2013 special Report by The United Nations....'

sustainability. It also shows that urbanisation poses tremendous and complex opportunities for a shared, sustainable future and that city can be the driving force for sustainable development. Cities could also present solutions toward meeting global challenges with the potentials for: inclusive community building, support of diverse cultures; economies of scale and energy-efficient development. But cities have also become cynosure of disgrace, the repository of filth and psychic decay, where the criminal rumbles and the politician cons and the businessman profiteers and the child expires on the cheap.[4] Sadly, this is the ugly situation in most African cities in the post-colonial period. As noted in the previous chapter, the general consensus among experts and scholars is that most people living in most African cities today are still facing great challenges such as: the deterioration of basic services, housing and environment, mass unemployment and underemployment, the virtual absence of State welfare and many more. All of this culminated in what Ilda Lourenco called "urban crisis."[5] Although the impact of environmental problems in Africa's urban centres vary from place to place, no place in sub Saharan Africa seem to be immune against the tragedy.

As it has been demonstrated earlier, the current models of urbanisation in Africa are socially, economically and environmentally unsustainable. Post-colonial patterns of urbanisation in Africa present most cities with grave challenges. Indeed, most African cities today have lost values for humanity as most city planners only prioritise economic growth at the expense of human dignity. Therefore, sporadic and largely unmanaged urbanisation has resulted into the proliferation of slums and informality; chronic poverty and vulnerability; natural and health hazards and so on. These conditions threaten the safety, security and social cohesion of individuals, their neighborhoods, cities and nations. As one of the few countries in sub Saharan Africa, which had many large pre-industrial cities, Nigeria cannot possibly be an exception to the African urban environmental decay.[6] And Ibadan (which is the primary focus of this

---

[4] Sam Omatseye, 'Redeeming Ibadan,'....

[5] I. L. Lindell, *Walking the Tight Rope: Informal Livelihood and Social Networks in a West African City*, (Stockholm University Press, 2003) 9

[6] Laurent Fourchard, *Urban Slums Reports: The Case of Ibadan, Nigeria*, (Ibadan: Institut Francais de Recherche en Afrique (IFRA), 2003)

study) no doubt has been a great cosmopolitan city in Nigeria and the history of its environmental decay actually reflects the history of Nigeria's ecological conundrum.

But it will be an exercise in deep absurdity to assume that successive governments now or in the past did nothing to ameliorate the environmental decadence in Ibadan. So, at this juncture it is necessary to evaluate the existing environmental management strategies in Ibadan before coming up with some new paradigms towards ameliorating the nagging environmental issues.

Evaluating the Existing Post-colonial Environmental Management Strategies in Ibadan

There is no doubt that Ibadan during the period covered by this study witnessed tremendous physical and structural changes as a result of urbanisation process. The city's attempt to transit from a traditional metropolis to a modern city finds reflection in the socio-cultural, political and economic structures of the city. Ibadan which has been taunted to be the most urbanised in Nigeria after Lagos, grew from a population of 200,000 in 1898 to 625,000 in 1963 and to over 2 million in 1988 with an estimated annual growth rate of five percent.[7] According to the World Bank, Ibadan is the third largest metropolitan area in Nigeria after Lagos and Kano; with a population of 3.1 million and a land area of 3,850 square kilometers.[8] The city is also said to be the largest metropolitan geographical area in West Africa, housing almost half of Oyo state's population (45 percent); from around 60,000 in the early 1800s its

---

[7] T. Soji and A. Dayo, "Urban Governance and the Governance of Nigerian Cities", in T. Agbola et al ed., *Environmental Planning and Health in Nigeria*, (Ibadan: Department of Urban and Regional Planning, University of Ibadan, 2008) 234

[8] See the Report on the World Bank assisted Ibadan Urban Flood Management Project, available online at https://www.food-security.net/wp-content/uploads/2021/07/PAD 6780PAD0P13010Box385224B00OUO090.pdf

population grew to 200,000 in 1890, and to a million by 1930, the population is projected to reach 5.6 million by 2033.[9]

As such, it won't be wrong to assert that Ibadan has grown to become a large sprawling city of some millions of inhabitants and has been variously called the "World's largest indigenous city" or "the largest urban village in Africa."[10] Also, the city over the years has been characterised by an extremely complex socioeconomic and institutional structure with several layers of authority and influence (formal, informal, and traditional).

Since independence, several efforts have been made by governments at all levels to guide urban environmental management and development in Nigeria. For instances, the then Federal Military Government promulgated the land use decree in 1978 as a regulatory tool to control the use of land and ostensibly to ensure equitable access to it by all Nigerians; created the Infrastructure Development Fund (IDF) in 1985; adopted a National Housing Policy in 1991; promulgated Decree No. 3 of 1992 which established the National Housing Fund as a source of funds for Housing Finance; established an Urban Development Bank of Nigeria in 1992 to focus on urban infrastructure and public facilities; adopted the Nigerian Urban and Regional Decree No. 88 of 1992; and launched the National Urban Development Policy document in 1997.[11]

Certain measures were also taken to directly tackle environmental crises. Some of these include: Introduction of War Against Indiscipline (WAI) - the monthly national clean-up exercise in 1984; the establishment of environmental sanitation activities and Waste Disposal Boards; the creation of the Federal Environmental Protection Agency (FEPA) under Decree No. 50 of 1989; promulgation of the Environmental Impact

---

[9] See the Report on the World Bank assisted Ibadan Urban Flood Management Project, available online at

[10] Laurent Fourchard, *Urban Slums Reports: The Case of Ibadan, Nigeria*

[11] J. O. Okunfulire, "The Challenges of Managing The Nigerian Urban Environment", paper presented by the Directorate of Lands, Environment, Urban and Regional Development, Federal Ministry of Works and Housing, at the Conference Center, University of Ibadan, Ibadan, 5-17 December, 1997

Assessment Decree of 1992 and formulation of the National Guidelines and standards for Environmental pollution Control in 1992.[12]

Immediately after independence, three institutions were charged with the responsibility of ensuring proper management of the environment of Ibadan. These include: Ibadan Town Planning Authority (now Ministry of Urban and Physical Development); the Ibadan Solid Waste Board (currently known as the Oyo State Solid Waste Management Authority); and the Local Government Councils. The Ministry of Environment was later created in 2011 to complement the efforts of the existing environmental management agencies.

Ibadan Town Planning Authority

The main concern of physical planning (traditionally known as town and country planning and more recently as urban and regional planning) is the distribution and arrangement of different competing land uses and structures in spaces, in orderly, balanced and consistent forms.[13]As it has been established earlier, from inception, the rapid growth of Ibadan city has not followed any coordinated, long-term, strategic development plan. The attempt to prepare a master plan for the city some years back was abandoned part-way through, and as such, no local government within the Metropolitan area has a comprehensive plan or strategy.[14] The nearest thing to physical planning is probably industrial and residential layout plan. Most local governments have no comprehensive or strategic plans and planning is largely restricted to the annual estimates process. As a result, forward planning is weak in that most local governments do not engage in detailed long term financial planning and programming. As a consequence, the city of Ibadan experiences a range of problems including

---

[12] Ishaq Ishola Omoleke, "Management of Environmental Pollution in Ibadan, An African City: The Challenges of Health Hazard Facing Government and the People", available online at www.krepublishers.com/.../JHE-15-4-265-275-2004-omoleke.pdf

[13] F. Samuel, "Urban Planning and the Quality of Life of Nigeria", in T. Agbola et. al eds., *Environmental Planning and Health in Nigeria*, (Ibadan: Department of Urban and Regional Planning, University of Ibadan, 2008) 200

[14] T. Soji and A. Dayo, "Urban Governance and the Governance of Nigerian Cities", in T. Agboola et. al eds., *Environmental Planning and Health in Nigeria....*215

housing shortage, inadequate urban infrastructure and service, and deteriorating environment.[15]

The history of governance and town planning authority in postcolonial Ibadan has been that one of increasing centralisation. Prior to independence in 1960, Ibadan like other Nigerian cities was under British colonial rule and up to 1963, a British-type system remained with a provincial government which looked after both urban and rural areas. The Ibadan Native Authority during the Colonial Administration was responsible for planning and development control in the urbanised area.[16]By 1958, this had evolved into the Ibadan Town Planning Authority, still operating within the provincial government system. Before 1960, the provincial government in the West metamorphosed into the Western Regional government.

In 1967, Western State (consisting of areas now in the present Oyo, Ondo, Osun, Ekiti and Ogun States) was created along with 11 others in the new 12 states structure for the Nigerian Federation put in place by the Military Regime of that period.[17]In the early part of the 1970s, Ibadan was made up of two council areas namely: Ibadan city council Area; and Ibadan less city council area. The former referred to the urban settlements while the latter referred to the rural settlements of the region.

Similarly, the physical planning institution in Ibadan experienced a transformation such that by 1970 the Ibadan Town Planning Authority that has existed since 1959 was replaced by the Ibadan Area Planning Authority. It is important to know that planning at this stage was still a state responsibility. However, in 1976, local government reforms placed Town planning under the jurisdiction of the local governments and from this date onwards, the state's role in planning of Ibadan was much reduced, focusing only on policy development and advisory services.

Furthermore, the local government reforms of 1976 equally brought about changes to the existing governance structure at the local government level

---

[15] T. Soji and A. Dayo....
[16] T. Soji and A. Dayo....236
[17] T. Soji and A. Dayo....

in Ibadan. This led to the establishment of Ibadan Municipal Government (IMG) surrounded by three (3) rural local governments of Akinyele, Lagelu and Oluyole. This structure remained in existence until 1991 when the IMG was further divided into five (5) local governments namely: Ibadan North West; Ibadan North; Ibadan North-east; Ibadan South-East; Ibadan South-West. At the same time, the three rural local governments were also sub-divided into six (6) namely: Akinyele; Lagelu; Oluyole; Egbeda; Ona Ara; Oluyole and Ido. This structure has remained unchanged.

The problems inherent with a decentralised form of urban planning and management as in the case in Ibadan are many, and compounding these problems, is the lack of any holistic thinking on a metropolitan wide scale. The post-colonial governance structure in Ibadan is particularly problematic for the successful implementation of any city-wide development initiative. Decision making is fragmented and quite localised and the state government which presumably could provide the metropolitan cohesiveness which is lacking does not do so. For instance, the water corporation follows its metropolitan plan, the Waste Management Authority has its own plan, the Ministry of Works and Transport has its own, and also the Ministry of Environment. Similarly, the eleven (11) local governments in Ibadan follow their plans in their respective local governments without any attempt at coordinating these plans on metropolitan wide basis.[18]

Moreover, environmental matters are often quite broad in complexity and in spatial impact. They normally transcend political, administrative and socio-cultural boundaries and this require broad perspectives for dealing with them. Thus, there was a potential contradiction or a divergence between the need to work on broad environmental issues such as waste disposal, water supply, erosion and flooding among others which are metropolitan wide issues, (which will require the attention of the 11 local governments in Ibadan and other relevant state agencies and ministries) while at the same time nurturing community based local government aspirations.

---

[18] T. Soji and A. Dayo.... 240

Ibadan Solid Waste Management Board (ISWMB)

This Board came to existence through an Act of the Parliament in 1997 as Ibadan Waste Management Authority but was transformed to Oyo State Waste Management Authority in 2004.[19] The fundamental duty of the body is to regulate waste management in the Ibadan and the entire Oyo State. Prior to the enactment of Edict No. 8 of 1997 establishing the Ibadan Solid Waste Management Board, the management of the environment of Ibadan city was the responsibility of the defunct Ibadan city council. Consequently, Ibadan city and its environs were constitutionally broken into (11) eleven local government councils which shouldered the collection and disposal of solid wastes in Ibadan. However, with the commencement of the edict, the functions of the local government councils in Ibadan urban area under the 1979 Nigerian constitution, and the instrument establishing them to collect, transfer and dispose solid waste were delegated to the new Authority. The functions of the Authority include among others: collection, transfer and disposal of solid wastes for the Ibadan urban area directly or indirectly; collection and registration of private refuse contractors in the city. Each refuse collection firm will need to pay specified amounts to the authority annually. The Board has additional responsibilities which are; to hire/lease out and sell its equipment to refuse contractors at profitable rates; to maintain land fill sites around Ibadan and to charge economic rates; to make effective use of sanitary inspectors from the Local Government Service and to impose sanctions on any refuse contractor or citizen in form of fines for the contravention of any of the law in accordance with provision under offences and related matters.[20]

Local Government Areas

Today, there are eleven (11) local governments in Ibadan Metropolitan Area consisting of five urban local governments in the city and six semi-

---

[19] Interview with Mr T. G Safiu (the then Director of Enlightenment, Oyo State Waste Management Authority)
[20] Ishaq Ishola Omoleke, "Management of Environmental Pollution in Ibadan, An African City: The Challenges of Health Hazard Facing Government and the People...."

urban local governments in the city during our period. The five local government areas in Ibadan metropolitan city are: Ibadan South-West local government area; Ibadan South-East local government; Ibadan North-West local government area; and Ibadan North-east local government area. Other ones in the less-city are; Egbeda local government; Akinyele local government; Ido local government; Ona-Ara local Government; Oluyole local government; and Lagelu local government.

All the local government areas expected to liaise with the Oyo State Solid Waste Management Authority to make the objectives of OYSSWMA realisable in their respective areas. They also monitor and supervise the activities of the OYSSWMA in their respective areas and ensure compliance to environmental laws and policies. They try to achieve these by sending local government environmental officials to arrest any environmental offenders especially on sanitation exercise days.

Ministry of Environment

As part of efforts to take on environmental crises in Ibadan and other parts of the State, the House of Assembly, Oyo State passed a law to establish the Ministry of Environment and Water resources which commenced functioning on 1st January, 2011.[21]Its responsibility includes formulation and enforcements of policies, statutory rules and regulations on waste collection and disposal, general environmental protection, control and deregulation of the ecological system and all activities related hitherto; advice the government on the environmental policies and priorities and on scientific and technological activities affecting environment; coordinate the activities of the local governments and the government agencies on environmental and ecological matters; establish and take measure to ensure effective environmental structures in the state for flood control, solid and liquid collection and disposal, water and air pollution eradication, noise control and general sanitation; conduct public

---

[21] Ishaq Ishola Omoleke, "Management of Environmental Pollution in Ibadan, An African City: The Challenges of Health Hazard Facing Government and the People…."

enlightenment campaign and disseminate vital information on environmental and ecological matters among others.[22]

Regrettably, despite the presence of the above mentioned agencies, Ibadan contended with myriad of severe developmental and environmental crises during our period and many of these persisted till date. The most prominent among these problems include : lack of functioning sewage systems; erratic and unreliable water supply; inadequate waste disposal system; inadequate drainage facilities which cannot cope with seasonal rains; chaotic and congested traffic situation, and complete lack of comprehensive planning and inadequate control over land use. As emphasised in previous chapters, the persistence of these problems stems from the fact that the rapid growth of Ibadan city has not followed any coordinated, long-term, and strategic development plan. No local government within the metropolitan area has a comprehensive plan or strategy. Past planning efforts for Ibadan have been under-resourced and sporadic and have not systematically addressed the need for a dynamic, responsive, implementation-oriented and participatory development planning process (Environmental planning management) which would lead to preparation and implementation of a development plan.

Evidently, the Oyo state government, local governments and various other environmental agencies in post-colonial Ibadan were and are still largely incapable of dealing effectively with the day-to-day environmental issues facing them. As a result, apart from the fact that communities and neighbourhoods are increasingly turning inwards to themselves for solutions to their many problems; the city had experienced occasional intervention by international community. One of such (perhaps most critical) was the Sustainable Ibadan Project (SIP), which still exist till date.

---

[22] Ebenezer Adurokiya, "Oyo State and Battle for Environmental Cleanliness", *Nigerian Tribune*, 12 November, 2011. 36

Sustainable Ibadan Project (SIP)

As it has been established earlier, Cities are generally centres for ideas which are enhancing productivity, social, human and economic development. Unfortunately, in many cities (especially in sub-Saharan Africa), development is not usually environmental friendly and inclusive as a large percentage of the citizenry are often left behind in the developmental engagements. As noted by Olaniyan, "human beings can be generally selfish and may not consider what happens to other members of the society as long as he/she is included in the development process."[23] It is in order to ensure all-inclusive development (that no one is left behind in the development process) the concept of sustainable development was initiated.

In the opinions of Olaniyan and Ogujiuba Kanayo et-al Patrick, sustainable development is tripartite in nature; for sustainable development to be achieved, it is crucial to harmonise the three core elements of economic growth, social inclusion and environmental protection; these elements are interconnected; they are all crucial for the well–being of individuals and societies.[24] Guenther and Al Shawaf argue that sustainability needs cities as much as cities need sustainability -- not only because they are a linchpin for the survival of our people and planet, but also a lever for shared progress and prosperity -- and thus, that a greater share of sustainability effort should be expended within them.[25] This is why urban planning, transport systems, water, sanitation, waste management, disaster risk reduction, access to information, education and capacity-building are all relevant issues to sustainable urban development.

---

[23] Olanrewaju Olaniyan , *SUSTAINABLE DEVELOPMENT OF IBADAN: Past, Present and Future* ,Ibadan, CESDEV Issue Paper No. 2017/1, Centre for Sustainable Development ,University of Ibadan, Nigeria. 5-6

[24] Olanrewaju Olaniyan , *SUSTAINABLE DEVELOPMENT OF IBADAN: Past, Present and Future*....9 and Ogujiuba Kanayo et-al "The Challenges and Implications of Sustainable Development in Africa: Policy Options for Nigeria," in *Journal of Economic Cooperation and Development*, 34, 4 (2013). 82

[25] C. Guenther and M. Al-Shawaf, " *The 7 keys to sustainable cities*" https://www.green biz.com/blog/2012/03/13/*sustainability-and-renaissance-citystates*, 2012

According to the United Nations Centre for Human Settlements (UNCHS, Habitat),ensuring that our ever growing cities and towns are socially, economically and environmentally sustainable is possibly the greatest challenge for urban development policy-makers and practitioners.[26] For instance, it has been discovered that in most countries, rapid urban expansion has been accompanied by massive environmental crises. These problems not only have negative implications for well-being and health of the urban residents (especially of the poor) but also destabilizing the urban economy and threatening environmental sustainability. As we have contended earlier in consonance with the view of the UNCHS, it is not urban growth itself that "causes" environmental problems; instead, it is a series of policy and management weaknesses which mean that cities are generally not able to cope adequately with the physical and environmental consequences of growth and change. In order ameliorate the situation; UNCHS launched the global Sustainable Cities Programme (SCP) in 1991.

The programme was meant to help city governments and their partners in the public, private and community sectors to develop improved environmental planning and management capacities required by them to deal more effectively with the process of urban evolution and development. The United Nations Centre for Human Settlements (UNCHS, Habitat) launched the SCP as the operational arm of the global world Bank/UNCHS/UNDP Urban Management Programme (UMP).[27]

---

[26] *Sustainable Cities and Local Governance: The Sustainable Cities Programme,* written and published by the United Nations Centre for Human Settlements (Habitat) and the United Nations Environment programme (UNEP), Nairobi, Kenya, 1997; See also *the Urban Environment Agenda Volume 2: City Experiences and International Support Volume 3: The UEF Directory* Written and published by the United Nations Centre for Human Settlements (Habitat) and the United Nations Environment Programme (UNEP), Nairobi, Kenya, 1997

[27] K. Idem, "The Institutionalisation of the Environmental Planning and Management Process", in T. Agboola et. al eds., *Environmental Planning and Management: Concepts and Application to Nigeria,* (Ibadan: Counstellation Book, 2006), 48

The 1992 United Nations Conference on Environment and Development (UNCED) in Rio de Janeiro –otherwise known as "Earth Summit" gave the much needed impetus to the SCP on many fronts. At one level, in what seemed to be the first of such attempt at a global scale, the "Earth Summit" focused the world's attention on the great implications of environment for social and economic development.[28] At another level, it resulted in widespread adoption of the famous 'Agenda 21,' which "articulated a range of desirable policies and concepts, including an emphasis on cross-sectoral coordination, decentralisation of decision-making and broad-based participatory approaches to development management."[29] More importantly, the summit also recognised and supported the potential of the SCP "as a vehicle for implementing Agenda 21 at the city level."[30] This role was further strengthened at the Second United Nations Conference on Human Settlements - Habitat II tagged- "City Summit" in Istanbul in 1996, which also inaugurated- the Habitat Agenda.[31] Overall, in order to ensure efficiency of the SCP as a new global development initiative which centred on partnership, mutual learning and mutual assistance, sharing of experience, with primary reliance on local resources supported by international programmes in the role of facilitator.2; in 1996 the UNCHS and of UNEP (United Nations Environment Programme) jointly took over the operations of the programme.[32]

In line with the fundamental argument of this study, the fact must be reiterated here that, the SCP did not view environmental deterioration as

[28] *The SCP Process Activities: A Snapshot of what they are and how they are implemented* Written and published by the United Nations Centre for Human settlements (Habitat) and the United Nations Environment Programme (UNEP), Nairobi, Kenya, 1998

[29] *Sustainable Human Settlements Development: Implementing Agenda* 21,prepared by the United Nations Centre for Human Settlements, for the United Nations Commission on Sustainable Development, Nairobi, Kenya, 1994

[30] *Sustainable Human Settlements Development: Implementing Agenda 21....*

[31] *The Habitat Agenda: Goals and Principles, Commitments, and Global Plan of Action* Agreed at the United Nations Conference on Human Settlements (Habitat II), Istanbul, Turkey, June 1996

[32] *UNCHS (Habitat) and UNEP Join Forces on Urban Environment* Briefing Note prepared for the United Nations Commission on Human Settlements (CHS15) and the Governing Council of UNEP (GC18) Prepared by the United Nations Centre for Human Settlements (UNCHS), Nairobi, Kenya, 1995

a necessary or inevitable consequence of rapid urban growth. The SCP rather considered: inappropriate urban development policies and policy implementation; poorly planned and managed urban growth (without adequate regard for the natural environment); inadequate and inappropriate urban infrastructure; and lack of coordination and cooperation among key institutions and groups as the major causes of environmental conundrum bedeviling cities across the globe (especially African cities).[33] To address the issues more creatively and constructively, the SCP explicitly adopted urban environmental planning and management (EPM).[34] The scheme worked directly with local governments and their partners to develop and nurture local capacities, system-wide, for more effective and responsive local governance, highlighting: more relevant and more appropriately utilised environmental information and technical expertise; better identification and understanding of priority environmental issues, leading to more soundly-based decision-making about urban development and environment; improved processes and mechanisms for formulating coordinated environmental strategies and for implementing them effectively; enhanced and institutionalised managerial capacities in the public, private and community sector partners; more effective mobilisation and use of available technical and financial resources.[35]

Another key component of the SCP (which makes more relevant to this study) was its emphasis on understanding the two way relationship between environment and development. According to its proponents-the urban development affects the environment (air pollution, exhaustion of ground water supplies, draining of wetlands, etc); but the environment in turn affects urban development (water supply shortages, flooding, land subsidence, etc). It is therefore not surprising that, the SCP also

---

[33] *The SCP Process Activities: A Snapshot of what they are and how they are implemented*....28

[34] *The Environmental Planning and Management (EPM) Source Book. Volume 1: Implementing the Urban* Written and published by the United Nations Centre for Human Settlements (Habitat) and the United Nations Environment Programme (UNEP), Nairobi, Kenya, 1997

[35] *The Sustainable Cities Programme: Approach and Implementation* Written and published by the United Nations Centre for Human Settlements (Habitat), Nairobi, Kenya, 2nd edition 1998

encouraged proper understanding the *long-term implications* of the environment-development relationships.[36] Hence, the development and strengthening of cross-sectoral and inter- institutional connectivity was central feature of every SCP city project.

In order to benefit from the phenomenon of Sustainable Cities Programme, the Oyo State Government requested to join the selected cities of the world in experimenting with the SCP using Ibadan as the experimental site. It would be recalled that the first SCP city demonstration project began in January 1992 in Dar es Salaam (Tanzania).Other cities that followed almost immediately include: Accra (Ghana), Cagayan de Oro, Tagbilaran, and Lipa Philippines), Concepción (Chile), Dakar (Senegal), La Habana (Cuba), Ibadan (Nigeria), Ismailia (Egypt), Katowice (Poland), Lusaka (Zambia), Madras (India), Maputo and Nampula (Mozambique), Moscow & St Petersburg (Russia), Shenyang & Wuhan (China), and Tunis (Tunisia), Amman (Jordan), Asuncion (Paraguay), Belo Horizonte (Brazil), Gaza (Palestine), Harare (Zimbabwe), and Kampala (Uganda).[37] In 1994, Oyo State government signed an agreement with the UNCHS (HABITAT) which made Ibadan to become the first city in Nigeria to join the SCP worldwide through the Sustainable Ibadan Project (SIP).[38]

The SIP was designed as a broad-based process which involves stakeholders in the discussion of key environmental issues. Consequences on how to deal with priority problems was to be achieved from the bottom up, than via the classic top-down master planning approach. The project document itself was based on the participation of state government's officials, parastatals, eleven local governments, and local administrators, the chamber of commerce, major community-based

---

[36] *The Sustainable Cities Programme: Approach and Implementation.....*

[37] A. G. Onibokun, "The EPM Process in Sustainable Development and Management of Nigerian Cities", in Agbola, T. ed., *Environmental Planning and Management: Concepts and Application to Nigeria,* (Ibadan: Constellation Book, 2006). 5-6.

[38] Bisi Adesola, "Application of Environmental Planning and Management (EPM) process to sustainable Urban Planning and development in Nigeria", (an opening address by the special guest of honour and Secretary to the Oyo State Government, at a conference organised by the Institute for Human Settlement and Environment IHSE, 5 – 17 December, 1997) .

organisations (CBOs), NGOs, and regional research institutions.[39]The main module of action in the sustainable Ibadan project is the EPM process.[40] The process is a succession of steps to carry out environmental management plans. Its goal is to make the management of Ibadan urban city more responsive to environmental considerations. The EPM process in Ibadan unleashed considerable enthusiasm among community groups and informal sector operators for taking part in implementation of investment demonstration projects, and revived the strong tradition of communities working in partnership with local governments. The resources mobilised through the Sustainable Ibadan Project from various sources such as Oyo State, local governments, Ministries, Parastatals, organisations, the SIP Trust Fund, and UNICEF, provided the catalytic push necessary to translate this surge of interest into tangible community contributions and involvement.[41] As a result, during the project period, a number of community based small-scale investment projects were accomplished such as: spring water development in Odo-Akeu community, establishment of a compost plant, drilling of two bore-holes in two underserved areas; and development of a number of project proposals for improving the situation in the major markets in the city.[42]

In order to initiate the EPM process, two key activities were initiated, namely: building public awareness and creating a sustainable institutional structure. Sensitisation began with the public sector where local government officials attended a one-day workshop to discuss SIP goals, procedures, and outputs. Subsequent awareness programmes were conducted with CBOs, community development groups and specific

---

[39] David Taiwo (SIP Project Manager), *Ibadan: Mobilising Resources through a Technical Coordinating Committee*, http://ww2.unhabitat.org/programmes/uef/cities/summary/ibadan.html,

[40] A.G. Onibokun, "The EPM Process in Sustainable Development and Management of Nigerian Cities"....6

[41] The SCP Source Book Series, Volume 5 Institutionalising the Environmental Planning and Management Process, Ibadan, Nigeria: Prepared and written by staff and consultants of the Sustainable Cities Programme The United Nations Centre for Human Settlements (UNCHS) and the United Nations Environment Programme (UNEP), 1999. 48.

[42] The SCP Source Book Series, Volume 5....

sectors (business, trade and transport, education and health, traditional authorities and the media).

To facilitate the SIP, a participatory institutional structure was created. A steering committee, representing a broad mix of the city's stakeholders was created to provide general guidance. Also, a technical coordinating committee was established to serve as a think-tank and its first task was the organisation of the city consultation.[43] Since many of the working groups in Ibadan were community based and area specific, with limited capacity of expertise of their own, the role of the professionals from the academic institutions, NGOs and the private sector became very important. To this end, the academics from the Ibadan University and other relevant local institutions were actively involved in the different SIP Working Groups, and played very commendable supportive roles.[44]

In order to creatively tackle urban environmental issues in Ibadan at that period, a city consultation was prepared in 1995 with the main objectives of: identifying and prioritising environmental issues; finding solutions to the identified crises; selecting those who would be involved in implementing solutions and determining how the projects would be funded.[45]The week-long consultation brought together three hundred selected stakeholders. The consultation was organised around three major cross-cutting issues namely: sanitation, health and solid waste management, health and water supply, and institutional arrangements for improved solid waste and water supply management. It was after the city consultation that various working groups were formed to develop options, design interventions, mobilise funds, and pursue implementations.

At the heartbeat of the implementation of the EPM process were the working groups were seen as mechanisms representing the key stakeholders to address commonly agreed and prioritised cross-cutting environmental issues in an operationally feasible manner. The groups also ensured the participation of stakeholders in resolving specific

---

[43] The SCP Source Book Series, Volume 5....
[44] The SCP Source Book Series, Volume 5....56
[45] The SCP Source Book Series, Volume 5....

environmental problems such as water supply, waste management, flood among others. As broad based participatory groups, there were not meant to supplant but rather complement the efforts of exiting agencies in addressing environmental issues in post-colonial Ibadan. Most of the active working groups were housed by Parastatals in charge of the various environmental issues.[46] For examples: The State waste management authority housed the Waste Working Group, while the water one was housed by the State water Corporation. Since the end of the City Consultation in October 1995, the Sustainable Ibadan Project has successfully established six Working Groups and eleven Sub-Working Groups to address the various identified and prioritised environmental issues of concern in the city.[47] Some areas were later selected for the practical demonstration of what SIP was meant to achieve in Ibadan.

Some of the demonstration projects included: Odo-Osun (formerly Odo-Akeu) natural spring development (at Oke-Ofa Babasale in Ibadan North LGA), pace setter organic fertilizer plant at Bodija Market (Ibadan North LGA), bore-hole and toilets improvement projects at Bodija Market (Ibadan North LGA); community-based Waste sorting Centre at Ayeye-Agbeni area (Ibadan North-West LGA).[48]Similarly, the state government over the years has invested in a composting project in the Bodija Market among others through the use of the EPM process and with the assistance of the waste recycling working group. The state government also provided financial and logistical support for the SIP through certain budgetary provisions since 1988 ostensibly to enhance its operations. [49]

To take care of challenges of funding, the Sustainable Ibadan Trust Fund, which was created through the SCP process, was founded as an innovative funding mechanism. The Trust Fund was established by

[46] L. Busari, "Operationalising the EPM Process: Working Group Concepts and Dynamics", Agbola, T. ed. *Environmental Planning and Management: Concepts and Application to Nigeria,* Ibadan (Constellation book, 2006), 39

[47] L. Busari, "Operationalising the EPM Process....

[48] Bisi Adesola, "Application of Environmental Planning and Management (EPM) Process to Sustainable Urban Planning and Development in Nigeria...."

[49] B. Wahab, "Some Aspects of Physical Planning and Development in Nigeria", in T. Agbola et. al eds., *Environmental Planning and Health in Nigeria,* (Ibadan: Department of Urban and Regional Planning, University of Ibadan, 2008). 298

wealthy private sector operators in the city and designed to finance SCP process-led investment projects through soft loans and grants; it also very helpfully cemented the participation of the private sector in the sustainable development process.[50]

Although, the Sustainable Ibadan Project has been adjudged successful with a number of demonstration projects show-cased above; in the long run it also became unsustainable due to a number of related issues. Failure of successive governments in Ibadan to act on or institutionalised lessons of participatory, collaborative, holistic and sustainable urban environmental planning and management which SCP offered through SIP in addressing developmental problems of Ibadan was perhaps the fundamental foundation of the environmental crises that are confronting the city up till today.[51]

Paucity of fund which has always bedeviling urbanisation processes in Ibadan since the colonial period was also a great challenge to SIP. As is the case generally in Nigeria, the lion share of finance for local governments in Ibadan comes from the Federal and State governments. During the period 1960-90, the internally generated revenue of Local Governments in Ibadan has been a mere 3 percent of their total revenue, while the Federal and State Government contributions were 86 and 11 percent respectively. The limited capacity to generate internal finance, and the dwindling Federal and State contributions have left little scope for the cash-strapped local governments to undertake big capital investments. In addition, lack of mechanisms that would have coordinated investments in the eleven local governments of Ibadan, made financial consolidation and joint actions difficult, further reducing the capacity of local governments to effectively absorb funds and utilise them most effectively. This situation underlined the need for more innovative municipal financing instruments, as well as the need to tap the resources of the private and community sectors. The Sustainable Ibadan Trust Fund, which was created through the SCP process, was one such innovative funding mechanism. The Trust Fund was established by wealthy private

---

[50] B. Wahab....49
[51] L. Busari, "Operationalising the EPM Process: Working Group Concepts and Dynamics", 41

sector operators in the city and designed to finance SCP process-led investment projects through soft loans and grants. Although it was also very helpfully in cementing the participation of the private sector in the sustainable development process but it was never enough in ensuring overall sustainability of development in Ibadan.[52]

As we have demonstrated in previous chapter, absence of city governance in Ibadan was a major hindrance to sustainable development in Ibadan. This also posed serious impediment towards the success of SIP. The fact that metropolitan Ibadan comprises eleven separate local governments has made the institutionalisation process much more complex. Not only did this fragmentation impede operational coordination, but it also undermined the possibilities for consolidated large investments which may be necessary to address certain city wide problems at a more efficient and economic scale. Certainly, the meager resources each local government dispose can do very little in addressing issues and areas where effective and optimal solutions are sensitive to economic scale and technology. With inability of SCP through SIP (just like existing indigenous efforts) to entrench sustainability in urban environmental governance in Ibadan, studies such as the current effort will continue to be relevant as Ibadan continues to search for solutions to her multidimensional environmental challenges. The issues incessant flood appears to tower above other environmental crises confronting the post-colonial Ibadan; the Ibadan Urban Flood Management is a major effort towards tackling the issue.

The Ibadan Urban Flood Management Project

About half of Nigeria's total population now lives in cities compared to 35.3 percent in 1990, generating 60 percent of the country's GDP and the urban population is projected to increase to about 65 percent by 2025.[53] Sadly, Nigeria is not reaping expected benefits from this rapid

---

[52] L. Busari, "Operationalising the EPM Process: Working Group Concepts and Dynamics...." 46

[53] See the Report on the World Bank assisted Ibadan Urban Flood Management Project, available online at https://www.food-security.net/wp-content/uploads/2021/07/PAD 6780PAD0P13010Box385224B00OUO090.pdf. 15

urbanisation because it is largely unplanned and uncoordinated. Rather it has deepened the tremendous challenges in Nigerian cities. Some of the prominent crises include: convoluted urban planning, poor solid waste and waste water management, lack of access to basic infrastructure, weak local governance and many more. Due to those challenges, there is always inadequate preparedness to deal with natural disasters and climate change induced vulnerabilities such as frequent flooding in most of Nigerian cities (including Ibadan).

As it has been demonstrated earlier, Ibadan is highly exposed to frequent flooding. For examples, from 1963-2013, the city is said to have experienced 16 major flood events (with those of 1980 and 2011 as the most devastating within the period). Although the most recent flooding has become a recurring decimal in postcolonial Ibadan (with the major one as recent as July, 2023) but that of August 26, 2011 was the most tragic in recent memory as it caused significant human and economic losses in the city, primarily in the housing, education, agriculture and transport sectors. Settlements located in unstable and risky locations such as along Ogunpa, Kudeti, Ogbere and Orogun floodplains and the hillsides of Oke-Are, Oke-Aremo, Sapati and Mokola were seriously affected with over 120 fatalities reported.[54] Land use within the city is primarily residential and a majority of Ibadan's urban poor live in crowded slums within the core residential areas of Ayeye, Agbeni, and Beere to mention but a few are almost permanently at an increased risk from flood events due to their location in low lying areas. Ibadan city setting is characterized by rugged terrain with wide valley plains.

Recognising the need for an integrated and long term solution to flooding in Ibadan, the Oyo State Government requested the World Bank's support to finance a flood management project. This culminated in to The Ibadan Urban Flood Management Project with an objective to improve the capacity of Oyo State to effectively manage flood risk in the city of Ibadan. The project consists of three main components described in subsequent paragraphs: (i) Flood Risk Identification, Prevention and Preparedness Measures; (ii) Flood Risk Reduction; and (ii) Project Administration and

---

[54] See the Report on the World Bank assisted Ibadan Urban Flood Management Project....18

Management Support. The first component assess flood risk in the city of Ibadan, plan risk reduction measures, and finance preventive structural and non-structural measures to enhance flood preparedness. The second component is to ensure flood risk mitigation through structural measures by financing public infrastructure investments for flood mitigation and drainage improvements. The third component will finance incremental operational costs related to the implementation of the project for goods, equipment, staff, travel, and Project Management Units consultant services.[55]

The Ibadan Urban Flood Management Project was partly funded by a World Bank IDA credit of $200 million which was secured by the former Governor of Oyo State, Abiola Ajimobi in 2014. The total amount for the project is $220 million; the Oyo State Government would contribute $20 million. The IUFMP which was a response to the devastating flood of 2011 and aimed at creating a robust flood risk management system in Ibadan commenced in February 2015. According to a report by the Oyo State Government, on June 7, 2019, nine days after Governor Seyi Makinde assumed office, the World Bank released an assessment of the Ibadan Urban Flood Management Project in which it stated that "the Project has shown moderately satisfactory signs of progress: 4 of the 18 ongoing works have been completed. Eleyele Dam rehabilitation (20%), and the 13 Priority Sites (5%) remain ongoing...."[56] In addition, the Early Warning and Response System inception report has now been completed. The contractor for the First Pool of Long Term Investments (Lots 1 and 2) has been selected. Lot 3 and 4 of long-term investments detailed designs and bidding documents are close to completion."So, from February 2015, the effective commencement date of the Ibadan Urban Flood Management Project to June 2019, only 4 of the 18 projects earmarked had been completed while the remaining 14 were at 5-20% completion.

The true status of the project was contained in the summary of a six page document released by the World Bank on June 28th 2021 that:

---

[55] See the online Report at https://projects.worldbank.org/en/projects-operations/project-detail/P130840. 1
[56] "An Update on the Ongoing Ibadan Urban Flood Management Projects (IUFMP)," June, 9, 2022,https://feedbackoysg.com/ongoing-ibadan-urban-flood-management-projects/

The project has made achievement in the following: (i) All 18 priority rehabilitation sites have now been completed, including the Eleyele Dam rehabilitation; (ii) As of date, the number of direct project beneficiaries has reached 118,014 well exceeding its end-of-project (EOP) target of36,600; (iii) The completion of the restoration works of the Eleyele Dam has provided protection to a population of 7,925 against the EOP target of 5,065; (iv) 9,000ha of land has been protected from the 25 year return period of flood event against the EOP target of 3,500ha; (v) 17 flood prone sites have made flood resilient in Ibadan; (vi) Flood Risk Management and drainage master plan, solid waste master plan have been completed. Though the project has made some progress in some areas the following challenges are being experienced: (i) The implementation of First Pool ofLong-Term Investment (Lot 1 and 2) has experienced significant delays and will not delivered within current contract period; (ii) Though No Objection was issued on Second Pool of Long-Term Investment (PLTI2) (Lot 3A, 3B, 4A, and 4B) awarding of contract has delayed due to safeguards compliance requirements posing a big challenge to deliver PLTI2 within the remaining project period; (iii) The Flood Risk Management & Drainage Master plan and Solid Waste Management Master Plan have been finalized but there is no legal framework to facilitate operationalisation; and (iv) Asset Management Plan, Early Warning and Response System and Dredging strategy are still to be delivered....[57]

A cross-examination of the above reports reveals that the IUMP could be regarded as an 'uncompleted project' (quite a familiar phrase in government circles in Nigeria). Indeed by the end of August, 2021, the World Bank has decided not to fund the second pool of Long Term Investment (PLTI2) which ultimately led to the withdrawal of the loan facility and untimely death of the project. The failure of the Oyo State Government's to meet safeguard compliant requirements was given as

---

[57] For the details of the World bank on June 28th 2021 report please visit https://documents1.worldbank.org/curated/en/387131624912950665/pdf/Disclosable-Version-of-the-ISR-Ibadan-Urban-Flood-Management-Project-P130840-Sequence-No-14.pdf

the main reason cancellation of the contract. Further probe also reveal that other critical aspects of the project such as the Flood Risk Management & Drainage Masterplan; Solid Waste Management Master Plan, Asset Management Plan, Early Warning and Response System and Dredging strategy have been finalized but they are yet to be is operationalised or delivered."[58]

Expectedly, after the expiry date of the contract (June,30,2022); when confronted with the question about what will happen to the bulk of the remaining uncompleted parts of the IUMP; the Oyo State Government simply state that they will seek alternative funding to ensure that all proposed projects under the Ibadan Urban Flood Management Project are duly completed.[59] These remain to be seen as Ibadan continue to groan under severe environmental challenges till date. Establishment of the Rural Water Sanitation Agency by the Oyo State Government is a further step towards addressing these challenges.

Rural Water Supply Sanitation Agency (RUWASSA)

As I have argued elsewhere "before we blame all the environmental crises in cities in sub- Saharan Africa on climate change; we must remind the current African city dwellers that most of their cities are still like what late Tai Solarin called 'civilizations without toilets,'"[60] due to their very poor attitude towards the environment. Olaniyan also agrees with this, when he observed that "…. some of the recent flooding witnessed in the city were not entirely the results of climate changes but were due to the

---

[58] https://documents1.worldbank.org/curated/en/387131624912950665/pdf/Disclosable-Version-of-the-ISR-Ibadan-Urban-Flood-Management-Project-P130840-Sequence-No-14.pdf ;See also See also See also "Oyo loses out on World Bank $70m flood intervention funds" available online at https://tribuneonlineng.com/oyo-loses-out-on-world-bank-70m-flood-intervention-funds/ August 10, 2021

[59] See "An Update on the Ongoing Ibadan Urban Flood Management Projects (IUFMP)…."

[60] Soji Oyeranmi, "How Fourth Industrial Revolution Can Aid Sustainable Development in African Cities: Lessons from China," January 21, 2023 https://ascir.org/2023/01/21/how-fourth-industrial-revolution-can-aid-sustainable-development-in-african-cities-lessons-from-china/

actions of the residents of Ibadan...."[61] This explains why 21st century Ibadan still battles open defecation and annual floodings without a single sea or ocean!

As a national effort towards improvement of environment, the Federal Government of Nigeria selected some States for special intervention by the UNICEF on Programme on Education, Health, Nutrition and Water Supply. Luckily, Oyo State was one of the States selected for this UNICEF Programme labeled as Assisted Water and Sanitation (WATSAN) Project. WATSAN as a project existed until May, 2011 when it fully assumed the status of Agency when the name was changed to Rural Water Supply and Sanitation Agency.[62] Although the hardware components of the Agency's mandate include: Geophysical Survey, Drilling Operation, Platform construction, Pump Installation, Latrine construction; other objectives are:

- To increase access to potable water supply by provision of boreholes with durable hand-pumps and eventually eradicate faecal and water related diseases in our rural communities.
- To improve standard of living and increase productivity in rural areas.
- To increase access to proper waste disposal through promotion, construction, use and maintenance of low cost "sanplat" (sanitation platform) latrines.
- To promote behavioral changes with respect to water use, personal hygiene and efficient means of excreta disposal.
- To enable participating communities to operate and maintain their own installations through specially designed training and cost recovery schemes.

To really demonstrate the menace of open defecation in the 21st century Ibadan, in a 2020 publication, the Oyo State Government acknowledged the problem, charged RUWASSA to make total stoppage of open

---

[61] Olanrewaju Olaniyan , SUSTAINABLE DEVELOPMENT OF IBADAN: Past, Present and Future ,Ibadan, CESDEV Issue Paper No. 2017/1, Centre for Sustainable Development ,University of Ibadan, Nigeria

[62] https://old.oyostate.gov.ng/oyo-state-rural-water-supply-and-sanitation-agency-ruwassa/#1496768094884-2b68d237-9abc

defecation in Oyo State (especially in Ibadan) its topmost priority and set a target of 2025 to achieve this. According to the publication, Oyo State Government has vowed to put all efforts into ending the menace of open defecation in the State before the target date of 2025, promising that the State would be the first to achieve the goal in the country. The then Secretary to the State Government, Mrs Olubamiwo Adeosun proudly said the State has always been known for taking the lead in good governance and human development and declared that the State Government would continue to support Oyo RUWASSA to ensure that the aims and targets of the agency were achieved. On open defecation, (during a five day training organised by RUWASSA) she specifically stressed that:

> We have set our own target to end this menace of open defecation [2025] and with what we have today and the next four days in training and drive of the desk officers from the 33 Local Government areas, we assure the people that once again, Oyo State will set the pace among other States in stamping out the menace of open defecation.[63]

Without being pessimistic, one can only say that only time will tell whether the 2025 target will be achievable or not. But as a current resident of Ibadan (for more than four decades) and as an eye-witness to the age-long/enduring unsustainable environmental management strategies in Ibadan; this target may likely go the way of the several failed promises of the past.

In concluding this chapter, there is no need for any fruitless repetition. Today, despite all the identified attempts at ameliorating the multi-dimensional environmental crises confronting post-colonial Ibadan; the city is still witnessing catastrophic crises in the critical areas of urban planning, sanitation and waste management. Even a glimpse of the "Oyo

---

[63] "Open Defecation: Oyo Vows To Set Pace in Ending Menace Before 2025," available online at https://oyostate.gov.ng/open-defecation-oyo-vows-to-set-pace-in-ending-menace-before-2025/ , 10th December, 2020

State Roadmap to Sustainable Development, 2023-2027,"[64] which seems to be the latest codified proposal by the current government leave so much to desired in terms of achieving sustainable development in cities in Oyo State (including Ibadan). Sadly, the so-called roadmap to sustainable development primarily focused on Economy, Education, Healthcare and Security, while the environment which is the centre piece of sustainable development was treated as a mere appendage attracting sparse attention from the writers of that policy document. As latest as July, 2023, Ibadan still witnessed the annual severe flood incidences with usual tales of severe sorrow and tears. With this, the struggle to achieve sustainable development continues in Ibadan (like most other cities in Oyo State/Nigeria and sub-Saharan Africa). Expectedly, the next chapter (the last one for this study) will focus on some possible new paradigms in regenerating, making the city resilient and achieve sustainable development in post-colonial Ibadan; that are applicable in other cities in the Global South and particularly in sub-Saharan Africa and Nigeria.

---

[64] Seyi Makinde Campaign Organisation ,*Oyo State Roadmap to Sustainable Development, 2023-2027*https://seyimakinde.com/wp-content/uploads/2023/02/Roadmap-To-Sustainable-Development-2023-2027.pdf

Chapter Eight
Examining Some New Paradigms Towards Achieving Sustainable Development and Environmental Regeneration in Ibadan

As noted in the previous chapter, today, despite all the identified attempts at ameliorating the multi-faceted environmental crises confronting post-colonial Ibadan; the city is still witnessing catastrophic crises in the critical areas of urban planning, sanitation and waste management. Therefore, this chapter (which concludes the study) will focus on some possible new paradigms: in regenerating the city, making it resilient and globally competitive towards achieving sustainable development. Interestingly, these suggestions are also applicable to other cities in Nigeria, sub-Saharan Africa and the Global South.

As reflected in the thought of Anna K. Tibaijuka, Africa (like most undeveloped parts such as Asia and Latin America), high rates of urbanisation have been accompanied by high levels of poverty and inequality, due mostly to near-absence of good governance and urban leadership in most of the cities. She lamented further that:

> Urban inequality has a direct impact on all aspects of human development, including health, nutrition, gender equality and education. In cities where spatial and social divisions are deep, lack of social mobility tends to reduce people's participation in the formal sector of the economy and their integration in society. This exacerbates insecurity and social unrest which, in turn, divert public and private resources from social services and productive investments to expenditures for safety and security.[1]

---

[1] Foreword by Anna K. Tibaijuka Under-Secretary-General and Executive Director United Nations Human Settlements Programme (UN-HABITAT) "São Paulo: A Tale of Two Cities: Bridging the Urban Divide" in UN-HABITAT *Cities & Citizens Series* ht ©

The above assertion is in no way suggesting that cities in the most developed parts of the world are totally immune against urban malaise. Indeed according to Sam Omatseye, "no city in the world, whether it is London, New York, Tokyo or Paris that did not pass through the foul rhythm of grime and crime before surging to a place of envy.[2] However, the difference is that most of the cities in advanced parts of the world, if not all of them were fortunate enough to have leaders who are innovative, progressive and competent. These leaders were able to inspire them out of doom and transform them to become glorious human habitats and choice destinations for people around the world. For instance, the city of London today is celebrated for orderliness and enviable beauty, but it once groveled under Hitler's bombardments. However, it was that same tragedy that threw up the genius of the great Winston Churchill with his speeches of inspirational growl. Sadly, it is a gory story for post-colonial Ibadan (as in most other sub-Saharan African cities).

Ibadan contrary to early post–independence optimism remains a city-village characterised by near- absolute inequality. This can still be partly attributed to the heinous colonial legacy of 'duality' that led to existence of 'two cities' in one-the old core traditional centre and the new or emerging areas, which were (and still are) diametrically opposed to each other. With this contradiction, the city could not sustain the promises of an emerging 'global city' but rather became a city of massive physical distortions and vast disparities. To worsen the situation, most post colonial leaders in Ibadan (with possible exception of Chief Obafemi Awolowo) did more damage to the urban environmental management and development of the city. Indeed, the beleaguered city fell into the hands of pillagers and bumblers who preferred the vanity of their sartorial splendour to the environment. The leaders loved themselves so much that there was not much love left for the city they pretend to govern. Consequently, rather than inspiring the city, 'they inspired filth.' At a point in time within the scope of this chapter and beyond, Ibadan literally became overwhelmed with filth so much that, at every turn you

---

United Nations Human Settlements Programme (UN-HABITAT), 2010, www.unhabitat.org

[2] Sam Omatseye "Redeeming Ibadan" http://thenationonlineng.net/new/sam-omatseye/redeeming-ibadan/

are confronted with heaps of refuse intimidating traffic into paralysis. Things got so grave that the city was declared as one of the dirtiest cities in the world at the dawn of the new Millennium.[3]

Despite these enormous difficulties; this study shared the optimism from the UN that 'our cities are up to the challenge to deliver a sustainable future.'[4] More so, since these problems were created by man, it is also within the people's capacity to proffer possible panacea. As a study of 'cautious optimism,' it suggests that due to presence of great disequilibrium and disparities in the evolution and growth of Ibadan, there is need for real evidence before 'we start dancing to a new rhythm of hope of urban rejuvenation.' Therefore, this study suggests a paradigm shift which must be anchored on attitudinal change in people and constructive changes in policies on part of the decision makers. The main suggestions are discussed below:

As it has been firmly established earlier, this study sees cities as drivers of sustainable development. The concept of sustainable development is a people-centred innovation that has initiated a positive drastic shift away from the existing traditional norms towards proffering enduring solutions to multi-faceted and dire difficulties confronting the global community (in the cities or rural areas) today. Apart from the fact that sustainable development is rooted in systems thinking; it also helps people around the world to understand themselves and the world. This is why, we strongly argue here that: while global community may not be able to address all the enormous problems that the planet earth is facing the same way the people have created them; our world can address the environmental crises more effectively if only people across the globe genuinely allow core principles of sustainable development to always guide their conducts.[5]

---

[3] Sam Omatseye "Redeeming Ibadan"....
[4] UN-HABITAT (United Nations Human Settlements Programme), "I'm a City Changer", available online at www.worldurbancampaign.org and www.imacity changer.org
[5] Darshini Mahadevia, "Sustainable Urban Development in India: An Inclusive Perspective", in David Westerndorff ed., *From Unstainable to Inclusive Cities*, (Geneva:

Furthermore, as a research that envisions cities as drivers of sustainable development, the current work is also guided by a new paradigm such as the urban ecosystem approach (which calls for concerted human efforts at tackling environmental crises). The urban ecosystem approach provides a framework for understanding the interactions between economic, social and ecological factors in the urban environment.[6] This is because to solve nagging environmental challenges facing humanity successfully, it is imperative for policy makers at the local, national, regional and international levels to recognise the crucial role of cities in catalysing shift towards sustainable development. The dawn of new Millennium in the year 2000 saw the concerted global effort at achieving this for cities around the world increased tremendously.

In addition to the urban ecosystem approach, the study also adopts the 'whole-of-society approach' to creating healthy, resilient and sustainable cities as being enunciated and propagated by the United Nations.[7] This post Covid-19 virtual training was jointly organized by: United Nations Office for Disaster Risk Reduction (UNDRR); Global Education and Training Institute (GETI); United Nations Office for South-South Cooperation (UNOSSC); World Health Organization (WHO) Health Emergencies Programme; Pan American Health Organization (PAHO/WHO) 7, 14, 21, 27 June 2023. This training serves as an introductory training for urban leaders, planners, and practitioners, aiming to:

- Increase the understanding and capacities on the long-term whole-of-society engagement approach to managing complex and

---

An UNRISD Publication in Collaboration with Swiss Agency for Development Cooperation, October 2002). 3.

[6] G. E. Machlis et. al, "The Human Ecosystem Part I: The Human Ecosystem as an Organizing Concept in Ecosystem Management", *Society & Natural Resources*, 10, (1997). 347-367; C.H. Nilon et. al "Editorial: Understanding Urban Ecosystems: A New Frontier for Science and Education", *Urban Ecosystems*, 3, (1999). 3-4.

[7] For details see UNOSSC, UNDRR GETI, PAHO/WHO, WHO Joint Certificate Training Program 2023 "Whole-of-society approach to creating healthy, resilient and sustainable cities: Harnessing South-South Cooperation for a post-COVID era" available online at https://www.undrr.org/event/undrr-unossc-who-paho2023

systematic urban disaster risks and enhancing public health emergency response preparedness, and South-South cooperation.

- Better prepare city stakeholders and engage them in making cities resilient for future crises, health and non-health emergencies and uncertainties.
- Facilitate learning through South-South and triangular cooperation (SSTC) towards sustainable development.
- Inspire and motivate whole-of-society (especially young people) to play key and active role in securing a resilient and sustainable future.

As it has been established earlier, battle for sustainable development will be lost or won in cities. It is therefore not surprising this latest global effort from the UN is focusing on promoting and propagating multi-sectoral efforts at 'making cities resilient' around the world with 2030 as the target year of achieving this. According to their concept note, the programme is primarily targeting:

Local and national government officials in charge of disaster risk reduction and management, urban development and planning, and public health emergency preparedness, national associations of municipalities, urban resilience and development practitioners, as well as civil society, private sector, and academia. Among the focused participants, efforts will be made to mobilize youth including young experts, local government officials and other young stakeholders to raise their awareness on SSC, DRR and health resilience issues facing the world globally. An outreach and engagement efforts will be made with universities in the global south, youth networks and organizations, including with UN agencies such as the Office of the Secretary General's Envoy on Youth, UNFPA, among others.[8]

This study frantically catalyses a shift away from conceptualising cities as parasites, to seeing cities as key agents in driving sustainable development. Main issues discussed include how policy makers can

---

[8] 'Concept Note', available online at https://www.undrr.org/event/undrr-unossc-who-paho2023.3

balance different information relevant at different spatial and temporal scale and how to balance between conflicting priorities. The study also offers practical tools that could enable both people and governments to identify those obstacles which have been preventing positive response to different types of environmental challenges. It equally provides examples of the kind of policies that can increase synergies between environmental, social and economic cycles and promote the creation of positive environmental feedback flows within cities. All these are mainly to generate a communal commitment and general interest in recreating Ibadan as driver of sustainable development and suggest more scientific and people oriented urban environmental management strategies.

First and foremost, in order to achieve sustainable development and effectively tackle various environmental challenges confronting post colonial Ibadan, the leaders must be ready "to provide leadership and encourage partnership in caring for the environment by inspiring, informing, and enabling communities and peoples to improve their quality of life without compromising that of future generations."[9] The people must also be prepared to make necessary sacrifice. To achieve this, city managers and policy makers in Ibadan (and cities in Nigeria/Africa) must shift away from the old paradigm of mere keying or begging to benefit of existing global city development programmes to key players and partners in any universal efforts at remedying the existing global environmental crises or forestalling further damages to the environment.

While it is absolutely impossible change the past of Ibadan, it is certainly possible to take advantage of the new opportunities (especially from international arena) that today is presenting to tackle the grinding environmental challenges of today in order a better future for the city and its people. This is why it is so important for Ibadan and other Nigerian/African cities to embrace and key in to the latest global effort from the UN is focusing on promoting and propagating 'whole-of-society

---

[9] United Nations Environmental Programme (UNEP) *Global Environment Outlook 3: Past, Present and the Future Perspectives*, [On-line], 2002.Available at: http://www.unep.org/geo/geo3/

approach' at 'making cities resilient' around the world.[10] As a participant in the inaugural edition, I am convinced beyond any reasonable doubt, cities around Global South (including Ibadan) have a lot to gain in making their cities resilient and achieve sustainable development if they key into the programme and apply lessons learnt constructively.

Some of the existing relevant paradigms should also be reviewed to meet the current challenges. For example, the Sustainable Ibadan Project (SIP) must be revisited and creatively revised. The Sustainable Cities Programme (under which SIP was founded) essentially gave credence to the Urban Ecosystem and Whole Society Approaches as it has a central commitment to the widest possible range of participation a wide range of "stakeholders" in urban environmental planning and management. This commitment according to SCP's proponents was not based on an abstract theory but on the practical reality that effective and sustainable environmental management requires the active and meaningful involvement of all those different groups, organisations and interests whose cooperation is necessary for successful action and implementation.[11]

Ultimately, as we have established earlier, success of environment management depends on proper involvement and synergy between the public sector and private sector. At the Public sector level, SIP focused particularly on the local or metropolitan level and including all the relevant agencies, public tertiary institutions, departments, authorities, politicians, officials etc. On the private sector focus was on: the economic sectors (trade, business, industry etc), both large-scale modern participants; those in the "informal" sector; the Community Sector; Non-Government Organisations (both at local and national levels); Community Based Organisations; Private Voluntary Organisations, special environ-

---

[10] Concept Note', available online at https://www.undrr.org/event/undrr-unossc-who-paho2023.3
[11] *The Sustainable Cities Programme: Approach and Implementation,* written and published by the United Nations Centre for Human Settlements (Habitat), Nairobi, Kenya, 2nd edition 1998.

mental interest groups and many more.[12] This is line with the suggestions of the advocates of the Whole Society Approach.[13]

Obviously, creative environmental management is significant to cities everywhere. Not only can well managed environmentally cities become sustainable, contribute to health, welfare and productive capacity of their own citizens, but they can also make a major contribution to the world environment. No doubt, the SCP initiative with its responsiveness, flexibility, and a practical demonstration of Urban Ecosystem approach offered one best opportunities to achieve that. This critical initiative also sees the environment is a critical ingredient for the success or failure of urban development, and that participatory management is the most effective response to environmental concerns. Most importantly, the SCP (as an embodiment of epitome of sustainable development) establishes all-inclusive management strategies. As a true paradigm shift away from the existing traditional norms, it seeks to synergies numerous stakeholders in urban development. It is based on the above stated premise that this study strongly advocates for an absolute revaluation of SIP documents (which are currently lying fallow at the SIP building in Ibadan) by both State and the 11 LGAs of Ibadan. If this could be done with utmost diligence and sincerity of purpose; its wide application and institutionalisation in Ibadan will certainly inaugurates the much desired urban environmental revolution in Ibadan, which will have multiplier effects on other Nigerian cities.

Closely related to the need to revive SIP is the critical need to revolutionise urban governance in Ibadan. As it has been established earlier, one of the most constant features of environmental history of Ibadan was the near zero presence of properly coordinated or efficient urban governance (which became more apparent during colonial and post-colonial eras). This has been hampering positive urbanism and making it impossible to achieve sustainable development in the city. While commenting on the need for engendering of 'good urban

---

[12] The SCP Source Book Series, Volume 5 Institutionalising the Environmental Planning and Management Process Ibadan, Nigeria....

[13] Concept Note', available online at https://www.undrr.org/event/undrr-unossc-who-paho2023.3

governance in cities across the world UN-HABITAT's Global Campaign on Urban Governance once argued that:

There has never been a more important time than now, to focus on the quality of governance at the local level. The new social contract arising out of the emerging democratic dispensation, the strong reemergence of the civil society and the expansion of the public space, foster the need for taking responsibility and accounting for outcomes and impacts. In addition, the forces of globalisation and the movement towards decentralisation are putting cities and local governments under tremendous pressure to deliver an ever-expanding range of benefits. The realisation of these expectations, however, is affected by several important realities related to urban governance. The emerging partnership between local government, private sector and civil society in sustainable development requires not only a reconfiguration of public space, which was earlier dominated by government only, but also new mechanisms for creating operational linkages among these spheres. This is all the more important due to the enlarged area of competence of local authorities, the increased Volume of resources under its custody, and the expanded scope of transactions being handled by local governments.[14]

The above submission clearly demonstrates the importance any discerning city planners/managers should place on the quality of urban governance. According to the United Nations Development Programme (UNDP) governance is: "the exercise of economic, political and administrative authority to manage a country's affairs at all levels. It comprises the mechanisms, processes and institutions through which citizens and groups articulate their interests, exercise their legal rights, meet their obligations and mediate their differences."[15]

---

[14] UN-HABITAT (United Nations Human Settlements Programme) and Transparency International Tools to Support Transparency in Local Governance Urban Governance Tool Kit Series, Nairobi, UN-HABITAT, Secretariat.
[15] UNDP *Governance and Sustainable Development*, (New York, 1997). 2-3.

At the level of city administration, concept of governance centres on the responsibilities in the management of cities, which is a very complex phenomenon. This involved a wide variety of stakeholders including local governments, States, national and regional governments; the private sector; non-governmental and community-based organisations (NGOs/CBOs), the media, professional associations and generality of the people. Achieving sustainability in our cities with such complexities certainly require technicalities of extreme proportions in propositions and implementation. Unfortunately, this had been largely lacking in Ibadan throughout her turbulent environmental history till today. As the quest for sustainable development in Ibadan continues, this study strongly advocates the inculcation and integration of the twin core principles of 'good urban governance –transparency and inclusiveness in to the city's urban environmental management strategies.

Transparency is widely recognised as a core principle of good governance.[16] As the UNDP has rightly posited, transparency is very central to urban governance because it promotes: "information sharing and acting in an open manner; allows stakeholders to gather information that may be critical to uncovering abuses and defending their interests; presents clear procedures for public decision-making and open channels of communication between stakeholders and officials; and makes a wide range of information available.[17]Access to free information usually plays a significant role in promoting transparency. It must, however, be mentioned here that for this information to effective for good urban governance, it must be relevant, accurate and timely. This is why who produces what information, and for what purpose have become important issues in urban governance.

It has also been argued that improving transparency can contribute to inclusion in cities by promoting public participation in development

---

[16] See World Bank, *Cities in Transition: World Bank Urban and Local Government Strategy.* (Washington, DC 2000); UNDP *Governance for Sustainable Human Development,* (New York, 2000); Asian Development Bank, "Elements of Governance", http://www.adb.org/Governance/gov_elements.asp; UN-HABITAT Global Campaign on Urban Governance: Concept Paper, (Nairobi, 2002)

[17] UNDP *Governance for Sustainable Development...* 1997. 36.

decision-making, enhancing accountability of various actors and stakeholders, and improving the quality of urban governance in general.[18] According to Tibaijuka an enlightened and committed political leadership combined with effective urban planning, governance and management that promote equity and sustainability are critical components to the building of cities for all (inclusive cities).[19] An "Inclusive City" is defined by UN-HABITAT as "a place where everyone, regardless of wealth, gender, age, race or religion, is enabled to participate productively and positively in opportunities that cities have to offer."[20]Without transparency at the local level the marginalised stakeholders (especially the poor) will be negatively affected in many ways, as this often enhance exclusion and limit their access to urban resources and opportunities. At the macro level, lack of transparency can reduce the number of jobs and livelihood opportunities available for the urban poor.

The most apparent historical fact here is that, poor urban governance inherited from the colonial era by post-colonial Ibadan is mainly responsible for many maladies ravaging the city today. Lack of transparency and absence of free flow of information have led to non-inclusive administrative styles. This became manifest in the non-responsive allocation of resources leading to a disproportionate spending on the priorities of the few elite and the wealthy (especially large infrastructure projects), at the expense of the vast majority poor (who usually lack access to basic benefits of urbanisation such as employment, housing, water supply, proper sanitation etc).[21] Worse still, the poor residents are forcefully pushed to the urban periphery and hazard prone areas through extremely exorbitant cost of land at the city centres and non-transparent land allocation practices by governments at all levels. In order to regenerate and turn Ibadan into 'an inclusive city' most of these anti-people's environmental policies must be reversed in favour of the people (particularly the poor).

---

[18] UN-HABITAT (United Nations Human Settlements Programme) and Transparency International....7
[19] Foreword by Anna K. Tibaijuka Under-Secretary-General and Executive Director United Nations Human Settlements Programme (UN-HABITAT)....
[20] UN-HABITAT, *Global Campaign on Urban Governance: Concept Paper, 2nd Edition,* (Nairobi, 2002). 5.
[21] UN-HABITAT....5

Certainly, efforts at ensuring inclusiveness and regeneration of Ibadan without a master plan and land reform would be in complete nullity. This is why most experts and scholars consulted agreed that a master plan and land reform are compulsory towards regenerating Ibadan.[22] This will inaugurate decentralised and well coordinated city governance by empowering eleven Local Government Areas to discharge their duties more effectively in the areas of sanitation and waste management. The land reform will also put an end to the gruesome 'land grabbing' phenomenon that is currently ravaging the city and ensure adequate management and redistribution of land for developmental purposes. The Local Government Areas must be resuscitated from their current moribund status in order to be alive to their responsibilities. They must also be harmonised in terms of environmental management philosophy, policies and implementation. All of this must supported by elaborate legislation to back environmental management techniques expected in the master plan for continuity and sustainable development.

While everybody can not benefit from infrastructural provisions at the same level, the expected Ibadan master plan must be an all inclusive document containing environmental strategies that will embrace and take care of all relevant stakeholders like: the poor; the wealthy; politicians from diverse political affiliations and orientations; landlord associations; trade unions; market unions; NGOs; CBOs and relevant government agencies. The environmental challenges should also be comprehensively identified, studied and cleverly prioritised to ensure justice for all. For examples, the proposed master plan must be able to initiate drainage revolution and provide basic infrastructure such as feeder roads linking the communities to major roads. While the drainage revolution will eradicate or minimise incessant floods; feeder roads will equally greatly ease waste evacuation (especially in traditional core areas which have battled with congestion over the years). These apart from tackling issues of floods, waste management, sanitation and planning which are the 'deadliest' environmental crises confronting Ibadan, today; the actions will be serving the interest of the majority of the people.

---

[22] Interviews Dr Femi Afolayan and Mr. S. B. Taiwo....

Since poverty has been identified as one of the major factors debarring sustainable development in Ibadan; any effort at regenerating the city must include achieving economic development and prosperity for the people. This study rejects the common notion that urbanisation and economic development are antithetical to environmental sustainability. Rather, it argues that drives for urban and economic development could complement environmental sustainability if the decisions are also committed to strengthening the role of cities in urban planning, waste management, sanitation and general protection of the environment. At their best, cities can be the principal drivers of economic, social development and even transformation. Business and industry can take advantage of shared access to labour, services, infrastructure, and information to lower costs and increase productivity. If these are in place, then urban consumers will continue to benefit from increased choice and the improved quality of goods and services. With such opportunities, cities (especially those in the developed part of the world) have become dynamic centres of economic growth and development: providing jobs, education, and markets and many more. Most of these advanced cities have been able to achieve sustainable growth because they are able to capitalise on the same trends and resources that led them to economic growth and invested heavily in environmental infrastructure. However, cities in most parts of the undeveloped world (especially in sub-Saharan Africa) are still very far away from sustainable development due to the prevalent urban poverty. Ibadan is no exception.

In 2022, the World Bank announced that we will fail to meet the goal of ending extreme poverty by 2030, and that 'global progress in reducing extreme poverty has come to a halt,' amid what it said was likely to be the largest increase in global inequality and the largest setback in addressing global poverty since World War II.[23] Sadly, Africa remains the world's last frontier in the fight against extreme poverty. Today, one in three Africans—422 million people—live below the global poverty line. They represent more than 70 percent of the world's poorest people, making

---

[23] Quoted in Oxfam, UK 2023 Report, *Survival of the Richest*, available online at https://oxfamilibrary.openrepository.com/bitstream/handle/10546/621477/mn-survival-of-the-richest-methodology-160123-en.pdf ,4.

their cities most susceptible to grievous environmental crises.[24] To reveal the level of extreme poverty in Nigeria, a 2022 report by the National Bureau of Statistics also disclosed that 133 million (out of 200 million) Nigerians are multi-dimensionally poor.[25] With this all pervasive and grinding poverty, it is impossible to achieve sustainable development in Ibadan and other parts of Nigeria without waging a successful war against this monster. It is very apt here to refer to the immortal thought of late Obafemi Awolowo that the best way to achieve genuine peace and development in Nigeria/Africa is by successfully waging "….the war against grinding poverty, hunger, preventable diseases, squalor, and ignorance among the masses of our people…."[26] with purposeful leadership, patriotism, and grim determination.

The poor in post-colonial Ibadan were (still are) often victims of urbanisation processes in many respects. Apart from the fact that most of them live in areas without access to basic amenities such as water, roads, toilets; they also work in most inhuman conditions without dignity of labour. Due to their level of poverty, thousands of them that engage in petty trading usually sell on the roads and sidewalks due to lack financial capacity to occupy legal lands or shops. Most of them are informal workers who operate on a day-to-day economic cycle. They work today in order to eat today. On the days they can't work—because they are ill, or because they are chased away from their workplace—they don't earn, and therefore don't eat. Expectedly, planners in Ibadan often had rough and hazardous entanglement with these unfortunate Nigerians for embarking on so called illegal activities for bare daily survival. But in order to have all-inclusive Ibadan that we are advocating for, authorities in the city must also take care of the socio-economic wellbeing of the poor (in their millions) as they plan and implement urban renewal programmes.

---

[24] See Soji Oyeranmi, "How Fourth Industrial Revolution Can Aid Sustainable Development in African Cities: Lessons from China," available online at https://ascir.org/2023/01/21/how-fourth-industrial-revolution-can-aid-sustainable-development-in-african-cities-lessons-from-china/ , January 21, 2023.1.

[25] Sami Tunji, (2022,18 November) " Nigeria's poverty exceeds World Bank projection, five states lead," *The Punch*, https://punchng.com/nigerias-poverty-exceeds-world-bank-projection-five-states-lead/

[26] Quoted in Tekena N. Tamuno, 2012, *Stakeholders at War in Nigeria, from Lord Lugard to Goodluck Jonathan, Vol.2*, Ibadan, Stirling-Holden Publishers Ltd.65

This must start from the recognition of the sacred fact by those in authority that the poor in Ibadan also have 'the right to the City' as spelt out by Habitat III's Policy Unit 1.[27] This policy framework for sustainable cities is to ensure that cities prioritise people over profit.[28] The Right to the City is a set of principles meant to guide decision-making processes in the best way possible to ensure good quality of life for all citizens irrespective of their socio-economic status by promoting and propagating: sustainability, equity and socio-economic justice. Such equitable relationship won't be so easy to accomplish but certainly not impossible. In the case of Ibadan and her millions of impoverished workers/residents, these principles can only be realised through organisation, mobilisation and meaningful engagement between city authorities and the people. The authorities must find ways to implement participatory consultation processes to make the voices of all the inhabitants heard.[29] If Ibadan city authorities can achieve this, they will be in a better position to find ways to support existing livelihoods and to recognize the social production of the city. This in other words will mean that the processes are being carried out with the initiative and contribution of ordinary people.

Most importantly in order to ensure economic survival of the average citizens in Ibadan, the public and private authorities must evolve what Roever calls "the people's economy"- with the target of lifting the poor out of their misery."[30] This will allow them to grow their enterprises over time by providing stability and certainty in their work environment: whether through licenses, uniforms or identification cards that help them fend off those harassing them. Stability in the workplace will definitely bring a long-term perspective and inspires further investments in homes and businesses. It will also make a lot economic sense if State and Local governments can make fees payable to the government/private agencies in charge of environmental management/planning (environmental sanitation, health, waste management, land survey, building plans, C of

[27] Available online at www.unhabitat.org/issue-papers-and-policy-units
[28] Sally Roever "The People's Economy and the Future of Cities", available online at www.wiego.org/.../ the people's economy and the future of cities....
[29] This view according to the United Cities and Local Governments' Global Charter-Agenda for Human Rights in the City is available online at www.cities-localgovernments.org .
[30] Roever Sally....

O) affordable. This will not only reduce the economic burden of the poor, reduce deviance of the people and combat corruption among the enforcement agents but also lead to desirable increase in revenues accruable the city authorities; thereby making more money available for proper environmental management in Ibadan.[31]

Arising from the above analysis, it has become obvious that development in Ibadan will only become sustainable when there is steady and sustained economic growth; when there is entrenchment of good governance which allows markets to flourish and makes the state/local government institutions and private sector work together as key promoters of development; when the city can raise more funds to fight poverty and manage wastes and sanitation in favour of the poor. The fact that there are many indications that there is high correlation between citizens' poverty and environmental management problems, should make governments at all levels pursue vigorously the program of poverty reduction as a way of addressing urban environmental management problems and to ensure its sustainability.[32] Furthermore, environmental managers in Ibadan as a matter of urgency must recognize the fact that in order to achieve environmental sustainability and economic development, they have to make deliberate choice of solid and pragmatic environmental management strategies. Ultimately, they must know that 'to manage the life of a city is ultimately to choose a future: to identify priority objectives and the risks that threaten them and then to mobilise resources effectively with which to meet those threats.'[33]

Beyond utilising socio-economic transformation as tool as urban environmental revolution is the need for attitudinal rejuvenation of the

---

[31] Interview with Mr Kehinde Ibrahim.
[32] B.O. Uwadiegwu and K. E. Chukwu, "Strategies for Effective Urban Solid Waste Management in Nigeria", *European Scientific Journal*, Vol. No. 8, (March 2013 Edition). 296-307.
[33] Soeren Jeppesen, Joergen Eskemose Andersen and Peter Vangsbo Madsen "Urban Environmental Management in Developing Countries – Land Use, Environmental Health and Pollution Management – A Review", (ReNED – Research Network on Environment and Development, www.ReNED.dk) and A. M. Kjaer, "Central Government Intervention as Obstacle to Local Participatory Governance: The Case of Uganda." Paper prepared for the ILO Conference on Governance, 2005. 9-10.

people. While more than half of our informants were honest enough to express their disdain and lack of awareness about the environment; few that showed keen interest in environmental related issues are doing so either superficially or due to professional callings.[34] It was in few instances that one comes across those fighting to protect the environment as a personal calling or conviction. Many of course have reasons for the care-free attitude towards the environment (genuine or otherwise) such as poverty, unemployment, failure of government policies to cater for their needs, cultural inhibitions etc. But sincerely most of them are with their present state of mind due to ignorance and lack of environmental consciousness. To this end, this study recommends raising environmental consciousness of the citizens of Ibadan as one of the most critical measure towards regenerating the city. This should be done through entrenchment of environmental ethics, environmental education, public enlightenment/ mobilisation and application of technology.

Environmental ethics is the part of environmental philosophy which considers extending the traditional boundaries of ethics from solely including humans to including the non-human world. It exerts influence on a large range of disciplines including environmental law, environmental sociology, eco-theology, ecological economics, ecology and environmental geography.[35] It addresses many ethical decisions that human beings make with respect to the environment. For examples it attempts to answer some of these questions: should we continue to clear cut forests for the sake of human consumption? Why should we continue to propagate our species, and life itself? Should we continue to make

---

[34] Interviews already cited earlier

[35] For details on Environmental ethics see: Peter Vardy and Paul Grosch, *The Puzzle of Ethics*, (New York: Harper Collins, 1999); Alan Marshall, "Ethics and the Extraterrestrial Environment", *Journal of Applied Philosophy*, Vol. 10, No. 2, (1993). 1468-5930; Peter Singer, *Practical Ethics*, (Cambridge: Cambridge University Press, 2011); Peter Scott, "Ecology: Religious or Secular?" *Heythrop Journal*, Vol. 38, No. 1, (January 1997); William Cronon, ed., *Uncommon Ground: Rethinking the Human Place in Nature*, (New York: W.W. Norton & Co., 1995). 69-90; Peter Singer, "Environmental Values", in Ian Marsh ed., *The Oxford Book of Travel Stories*, (Melbourne, Australia: Longman Chesire, 1991). 12-16; Lynn White, "The Historical Roots of our Ecologic Crisis", *Science*, Vol. 155, No. 3767, (1967). 1203–1207; Garrett Hardin, "The Tragedy of the Commons", *Science*, Vol. 162, No. 3859, (1968). 1243–8. Aldo Leopold, "The Land Ethic". *A Sand County Almanac*.1949.

gasoline powered vehicles? What environmental obligations do we need to keep for future generations? Is it right for humans to knowingly cause the extinction of some species for the convenience of humanity? How should we best use and conserve the space environment to secure and expand life? The academic field of environmental ethics grew up in response to the work of scientists such as: Pre-20th century thinkers include Saint Francis of Assisi and the American transcendentalists Ralph Waldo Emerson and Henry David Thoreau; Aldo Leopold (1887-1948); Rachel Carson (1907-1964) and events such as the first Earth Day in 1970, when environmentalists started urging philosophers to consider the philosophical aspects of environmental problems. A major theme in ecological science and environmental ethics is interdependence.

Just like every other branch of philosophy, environmental ethics is embedded with many theories and approaches in trying to justify the need for environmental codes of conduct. For examples we have: anthropocentricism and eco-centricism. Anthropocentricism sees humans as the most important or critical element in any given situation and concern with human-centered values such as social justice and economic equity. Peter Vardy distinguished between strong and weak anthropocentrism. A strong anthropocentric ethic argues that humans are at the center of reality and it is right for them to be so. Weak anthropocentrism, however, argues that reality can only be interpreted from a human point of view, thus humans have to be at the centre of reality as they see it.[36] Eco-centricism on the other hand focuses on non-human components of the environment. There are also: social ecology, deep ecology, environmentalism, environmental pragmatism, agrarianism, eco-humanism, eco-feminism, bio-regionalism, religious environmentalism, holism and so on.[37]

In all, this study opts for holism because it truly reflects centrality of interdependence in environmental ethics. On one hand, it teaches the world our dependence on the environment for our collective survival. On

---

[36] Peter Vardy and Paul Grosch, *The Puzzle of Ethics*, (New York: Harper Collins, 1999). 5-7

[37] Alan Marshall, "Ethics and the Extraterrestrial Environment", *Journal of Applied Philosophy*, Vol. 10, No. 2, (1993). 1468-5930.

the other, the ideology shows the people clearly that if we are to survive the current and impending environmental catastrophe, we must take care of the environment with absolute connectivity and interdependence. When views and handling of environment is holistic, it takes of care all (human, non human, biotic, abiotic, biological, abiological etc). In this model, the ecosystem is primary and individuals are of less concern. Cities everywhere must take the lead in this race to save the human race by being holistic in the management of the environment.

For effective environmental management in Ibadan, the concerned authorities must incorporate principles of holistic environmental ethics as enunciated above into their systems. The governments at all levels must ensure constant training and retraining of officials in charge of environmental management in the city. The citizens must also be mobilised through sensitisation and environmental education. Sensitisation will bring about adequate citizen participation and private sector partnership. The public agencies such as Ibadan Waste Management Authority (now Oyo State Waste Management Authority); Ibadan Town Planning Authority (now Ministry of Urban and Physical Development); the eleven Local Government Areas that make up metropolitan Ibadan and Water Corporation cannot alone achieve success in managing the environment without clearly understood working ethics or philosophy which will enhance constructive/positive partnership with the citizens and the private sector. Citizens and public sensitisation will bring about the much needed development of desirable environmental behavior and initiation of neighborhood spirit. [38]

Through Public enlightenment and mobilisation and environmental education, the attitude of the people towards improving and maintaining the neighborhood quality will be achieved. The public enlightenment may take several forms and shapes such as: occasional workshops/ seminars/trainings for environmental personnel and residents; production of relevant handbills and leaflets; street to street campaigns; regular meetings between authorities and trade unions, market associations,

---

[38] Linda Steg and Charles Vlek, "Encouraging Pro-Environmental Behaviour: An Integrative Review and Research Agenda", *Journal of Environmental Psychology*, Vol. xxx, (2008). 1–9.

CBOs, NGOs, Landlord/Tenant associations and other concerned groups etc ate. With this the residents will develop a sense of belonging, some emotional attachment to their neighborhood and thus display a sense of commitment to the cleanliness of the neighborhood and Ibadan as a whole. If residents of Ibadan (from neighbourhood to neighbourhood) are properly organised, sensitised and educated, they can be mobilised to team up and on their own become committed to their neighborhood to the extent that they can on their own take decisions to clean up their neighborhoods and thus improve the processes of environmental management.

If the above stated environmental camaraderie could be entrenched among neighbourhoods in Ibadan, then it won't be difficult to introduce 'the Civic Exnora Concept' to the people of Ibadan as a strategy to tackle waste management and sanitation crises in the city. According to Anand, EXNORA International is a non-governmental organisation based in Madras, India, which promoted the idea of forming neighbourhood associations for managing primary waste collection.[39] Civic Exnora units are formed of households from one or a set of streets, and a small number of office bearers (either elected or, more commonly, filled by Volunteers) form the committee that manages the Civic Exnora. One person, responsible for collecting the wastes (called "street beautifier"), is appointed and trained; often a tricycle waste collection cart is purchased with a bank loan or funds from private sponsors. Wastes are collected from each household once daily and are taken to a municipal bin or (increasingly) to a municipal corporation transfer station. Each household contributes a monthly fee to the Civic Exnora. Based on the contributions, a monthly salary is paid to the street beautifier and the remainder is used for repaying the loan for the purchase of the tricycle and undertaking any other programmes.

In evaluating the success of the programme, Furedy mentioned that more than 60,000 people were receiving waste services from some 500 roads in

---

[39] P.B. Anand, "Waste Management in Madras Revisited", *Environment & Urbanization*, Vol. 11, No. 2, (October 1999).

about 80 neighbourhoods, organised by 150 Civic Exnora units.[40]Another scholar also estimated that there are about 1,500 Civic Exnoras covering approximately 0.45 million people.[41] The concept of mobilising people's cooperation for tackling waste collection and sanitation is said to be fairly simple and not necessarily unique to Exnora. For instances, Anand opined that residents' associations in Calcutta are using the same concept[42] and Chatterjee also cited a similar programme in Hyderabad, India, where the municipal corporation is also contributing to some of the costs.[43] With these examples and their relative successes, this study is also optimistic that neighbourhood environments in Ibadan will benefit greatly from Exnora concept in combination with other panaceas earlier recommended.

Environmental education through the entrenchment of environmental ethics into the school curriculum from primary to tertiary levels in Ibadan is another potent measure to institutionalise the proposed solutions. This must go beyond aspects of teaching of personal hygiene which has been incorporated into subjects like civic education, social studies, healh science etc. Environmental ethics and sustainable development should be floated as separate subjects (at basic level at both primary and secondary) at advanced level at tertiary level which may start as part of General Studies in all our tertiary institutions in Oyo State (Ibadan) at the first instance. In order to achieve maximum results, traditional/ indigenous environmental ethics must be taught at both formal and informal levels.[44] As part of mass sustainability education effort, the basic explanation of sustainable development and Sustainable Development Goals (SDGs) should be translated to most popular local languages. In case of Ibadan,

---

[40] C. Furedy, "Garbage: Exploring Non-Conventional Options in Asian Cities", *Environment and Urbanization,* Vol.4, No.2, (1992). 42-61.

[41] T.K. Ramkumar, "Community Initiative for Environmental Management: The Work of Exnora International, An NGO", paper presented at the International Conference on Environmental Strategies for Asian Cities, Madras, India, February 14-17,1996.

[42] P.B. Anand, "Waste Management in Madras Revisited....."45

[43] R. Chatterjee, "Response", in I. Serageldin, M.A. Cohen and K.C. Sivaramakrishnan eds., *The Human Face of the Urban Environment,* (Washington DC: The World Bank, 1995). 283-286.

[44] P. A. Ojomo, "Environmental ethics: An African Understanding", *Journal of Environmental Science and Technology,* Vol. 5, No. 8. (August, 2011). 572-578. Available online at http://www.academicjournals.org/AJEST, 1-7

the Yoruba language may be chosen in the first instance for this experiment.

At formal level, indigenous and foreign environmental ethics may be incorporated into the Yoruba Language and Literature curricular from primary to tertiary level. Special trainings in environmental ethics and management must be organised for the teachers and lecturers at regular intervals. In order to really localise the concept of sustainable development for effective urban governance, I will suggest the translation of the concept and Sustainable Development Goals (SDGs) into major local languages in Ibadan. This may be produced in leaflets/ pamphlets for onward distributions at regular intervals among the local populace as part of mobilization and sensitisation to raise environmental consciousness among the people. Both Oyo State and the affected Local Governments may also seek financial assistance and partnership from corporate bodies, academic and research institutes, Federal Government of Nigeria and international development partners such as UNDP, UNEP, UN-Habitat, City Alliance etc.

Application of technology will also go a long way in improving environmental standard in Ibadan.[45]For instance, there is the urgent need for authorities in Ibadan to evolve a new technology that will pave way for integrated waste management (IWM) which is best defined as a set of management alternatives including reuse, source reduction, recycling, composting, landfill and incineration.[46] With this technology, waste will no longer exist because it will not be produced and if produced, would be a resource to be used again. At best, this will lead to what is referred to as 'waste to wealth' concept or industrial ecology in which Ibadan would be functioning more like an ecological system where waste from one part of

---

[45] B.O. Uwadiegwu and K.E. Chukwu, "Strategies for Effective Urban Solid Waste Management in Nigeria", *European Scientific Journal,* Vol. 9, No. 8, (March 2013 Edition); G.E.D. Omuta, "Urban Solid Waste Generation and Management; Towards an Environmental Sanitation Policy", in P.O Sada and F.O. Odemerho eds., *Environmental Issues and Management in Nigerian Development,* (Ibadan: Evans Brothers Ltd. (Pub), 1988).

[46] P. Relis and A. Dominiski, "Beyond the Crisis, Integrated Waste Management", (Santa, Barbara: Community Environmental Council Publication, 1987).

the system would be a resource for another part.[47] These measures will also encourage maximum protection against possible health risks that commonly emanate from careless refuse disposal practices and landfill sites. Geographic Information Management Systems is equally crucial to Ibadan city planning and management. Unfortunately, previous and existing developmental initiatives in Ibadan are often based on fragmented information; hence, the constant failures. Therefore, it has become most important for environmental managers in Ibadan to take advantage of Geographic Information technology to gather, analyse and manage information in order to identify, prioritise and handle problems that emerge as a result of developmental efforts.[48]

Will the earlier stark revelation about grinding poverty in urban centres across Africa (including Ibadan), any effort at regenerating their cities must also include poverty alleviation strategies. This why the entrenchment of bio-technology/bio-economy in to management of Ibadan and other Nigerian/African cities is so crucial. This will not only ensure both urbanisation and economic development plans are complimentary to environmental sustainability but also that urbanisation becomes real tool for lifting the people out of their economic misery /extreme poverty without compromising the environment. According to the European Union Commission, the main goal of bio-economy is a more innovative and low-emissions economy, reconciling demands for sustainable agriculture and fisheries, food security, and the sustainable use of renewable biological resources for industrial purposes, while ensuring biodiversity and environmental protection. Wastes to wealth/energy components are particularly useful for waste management in African cities.

Since sustainable development is about solving today's problems and securing better future for the generations to come; the people of Ibadan must take possession of solutions to the myriad of environmental problems facing the city. The best way to achieve this is for the

---

[47] B. O. Uwadiegwu and K. E. Chukwu, "Strategies for Effective Urban Solid Waste Management in Nigeria...."
[48] David T. Afolayan, Chief Executive Officer, GISKonsult Limited, Internet: www.gisk nigeria.com

environmental management authorities to incorporate indigenous knowledge into the modern environmental practices and also entrench traditional/informal ideas about protection of the environment into formal environmental education highlighted earlier. These traditional ideas are generally referred to as 'Taboos'. Taboos were introduced to regulate the moral order of the society. They took their origin from the fact that people discerned that there were certain things which were morally approved or disapproved by the deity.[49] These are not contained in any written law but are preserved in the tradition. Mabawonku and Agbola sometimes ago submitted that taboos and superstitions were often regarded as integral part of traditional education.[50] Every society in the world cares for its tradition because in a society without schools, a type of education known as traditional/informal education goes on. In Yoruba societies, traditional education is supported and encouraged because of its contribution to the growth, renewal and development of the society.[51]

Taboo was derived from a Polynesian term -'*tapu*' which means forbidden.[52] It is similar to the sacer in the Greek, Kadesh in Hebrew and Nso in Igbo language of Nigeria and mmusu in Akan language.[53] Contrary to the common view that taboos are essentially guide only the pre-industrial societies; the ideas are universal. In the opinion of Holden, taboos are not a feature of 'primitive' societies' as it was assumed sometime ago by some anthropologists but it is a characteristic of any society.[54] Indeed as contended by Durkheim, taboos a "phenomenon that

[49] Cecilia Omobola Odejobi, "An Overview of Taboo and Superstition among the Yoruba of Southwest of Nigeria", *Mediterranean Journal of Social Sciences*, Vol. 4, No. 2, (May 2013). 221-223.

[50] Tunde Agbola, and A. O. Mabawonku, "Indigenous Knowledge, Environmental Education and Sanitation; Application to an Indigenous African City", in D.M. Warren, Layi Egunjobi and Bolanle Wahab eds., *Indigenous Knowledge in Education*, (Ibadan: Indigenous Knowledge Study Group, University of Ibadan, 1996).78-94.

[51] E. B. Idowu, *Olodumare: God in Yoruba Belief*, (Ibadan: Longman, 1962); O. Oduyale, "Traditional Education in Nigeria", in O. Y Oyeneye and O. M. Shoremi, eds. *Nigerian Life and Culture*, (Ogun state University, 1985), 230-244.

[52] C. Blakemore, and J. Shelia, *Taboos*, (Oxford: Oxford University Press, 2001).

[53] Cecilia Omobola Odejobi, "An Overview of Taboo and Superstition among the Yoruba of Southwest of Nigeria...."222.

[54] L. Holden, *Encyclopedia of Taboos*, (Oxford: ABC CLIO Ltd., 2000).

is universal."[55] While commenting on the importance of taboos, Steiner argued that this lies in the maintaining of harmony between God and spirits (invisible world) and human beings and the rest of creation (visible world).[56] This harmony would be ruled "by moral order which is preserved by tradition and, if followed, have the power or force to sustain the existence and operation of the universe, ensuring a bountiful life for humanity.[57] Taboos clarify which attitudes and behaviours are not socially acceptable because they do not assure the continuation of life in its fullness, do not enhance the quality of life of the community and do not preserve the social code of behaviour. Hence, breaking of a taboo endangers life and is seen as bad and wrong because it interrupts peace and harmony.[58]

In the traditional African society especially in Yoruba society, taboos also played significant and positive roles.[59] Thorpe highlighted seven fundamental functions of taboo namely: to: (a) avoid accident; (b) have respect for religion; (c) respect elders; (d) obey rules of cleanliness; (e) teach moral values; (f) guide against being wasteful; (g) explain things that are difficult to understand.[60] Therefore, if people kept taboos, some of the difficult situations such as serious environmental crises in post-colonial Ibadan and in other Yoruba/Nigerian cities would have been avoided or drastically reduced. While It is true that the impact taboo exercising in the contemporary Ibadan had significantly diminished when compared with pre-colonial period (due to influence of modernisation); this study strongly argue for revitalization of these original codes of conduct in combination with the modern solutions as the surest way to ensure holistic /sustainable development in Ibadan.

[55] E. Durkheim, *Incest: The Nature and the Origin of the Taboo*, (New York: Lyle Stuart, 1963).

[56] F. Steiner, *Taboo*, (London, Cohen &West Ltd., 1956).

[57] L. Magesa, *African Religion: The Moral Traditions of Abundant Life*, (Maryknoll, New York: Orbis Books, 1997).

[58] M. Andemariam, "Place of Taboos in Gikuyu Morality", in L. Magesa, ed., MIASMU Research Integration Papers to Moral Teaching and Practices of African Religion, August Session, 2001.

[59] P. A. Ojomo, "Environmental Ethics: An African Understanding", *Journal of Environmental Science and Technology* ....

[60] C. O. Thorpe, *Àwon Èèwò Ilè Yorùbá*, (Ibadan: Onibon-Oje Press, 1972).

These traditional environmental codes of conduct are as important to the regeneration of Ibadan as all other earlier suggested solutions towards ameliorating the ugly trend. Post colonial Ibadan still largely remains pre-industrial and a 'city-village': where traditionalism and modernism frenetically mixed and co-exist side by side; where there are incessant and chaotic political situation, mass illiteracy, endemic poverty, very low/ unregulated industrialisation, inadequate town planning; where sprawling agglomeration of buildings, spread out in numerous directions without any coherent order; where majority of the people are still very poor and mostly occupying greater percentage of the landscape(usually congested without the basic amenities). Yet, governments at all levels are seemingly growing more apathetic about environmental issues. With this negligence from the concerned authorities, the people (especially those in the core traditional areas) often resort to self-help utilising some of these traditional codes to address their environmental issues with little or no success due to lack of coordination and cooperation. Governments and private enterprises in charge of environmental management in Ibadan could therefore build on this by educating and sensitising the people about the usefulness of environmental taboos in tackling the prevalent ecological problems facing the city today.

From all said and done, this study concludes that while it impossible to rebuild Ibadan from the scratch; transformation is a possibility. Therefore, resilience, regenerating and making development in Ibadan sustainable is herculean but not impossible. If Ibadan will become sustainable and resilient, the city must rely more on critical human infrastructure which will be the bedrock for physical infrastructure. With this, Ibadan will become all-inclusive where all residents irrespective of their socio-economic status will be captured by developmental programmes and environmental policies of government at all levels. All agencies and personnel in charge of the environment must be empowered for proper public enlightenment and enforcement of environmental laws and regulations.

Furthermore, the government agencies and people of the city must heed the admonition of the UN-Habitat, that all major stakeholders must recognise the ever-changing dynamics and the diversity of the city that requires a combination of interdependent skills, investment, technology and well informed decision making to harness the development potential of urbanisation. The recent 'Whole Society Approach' to make cities resilient propagated by the United Nations Office for Disaster Risk Reduction (UNDRR) is equally very important. They must also realise that in order to harness Ibadan as asset, all stakeholders must be committed to humanize the city in order to be able generate justice, knowledge and happiness. To achieve all these, the city must rely on the power of intelligence, audacity, wise decision-makers, men and women, from youth to elders based on a better understanding of the landscape, history, culture and eco-systems of the city. The policy makers and the people must always remember that decisions taken today in Ibadan and other cities across the world will shape not only their own destinies alone but the social and environmental future of humankind.[61] If these recommendations are followed sincerely, one can hope that in no distant future, Ibadan will successfully transit from 'a city of thousands environmental crises'[62] to a hub of ecological transformation and sustainable development.

---

[61] C. O. Thorpe....
[62] Tai Solarin, "Civilization without Toilets", *Nigerian Tribune*, August 16, 1972. 2.

Bibliography

Primary Sources

Archives

Annual Report for the Colony of Southern Nigeria, 1907.

NAI CMS C/A2/049/103, D. Hinderer, Journal entry, June 2, 1856.

Donald Cameron, *The Principles of Native Administration and their Application,* Lagos, 1934.

NAI IBADIV 1/895 Vol. II, Europeans Resident in Ibadan, 1947.

Maxwell-Fry-Farm Report on Ibadan Town Planning in File 1400, National Archives, Ibadan, 1945.

Memo No. 713/92/1924 of 14 March 1924, D.O. Ibadan to Resident Oyo province, Handing over Notes by H.L. Ward Price.

NAI AD 110/28 Ibadan Health Committee Minutes, 1952-1958.

NAI CSO 15965, "Infectious Diseases treatment, Ibadan: Memorandum from Senior Resident, Oyo to the Secretary, Southern Provinces," 8.

NAI CSO 26/2 (para. 11).

NAI CSO 26/2. File 12723 Vol. V, Oyo Province Annual Report 1927, (para 138).

NAI File No PX/F2 "Reports on Colonial Nigeria Federal Development, 1949-1952.

NAI Iba Div 1 File No 1400 Vol. I-VI, "Town Planning: Ibadan".

NAI Iba Div 1/1 File No 355-306, Acquisition of Land for Ibadan Waterworks, 1938-53.

NAI IBA DIV Vol.1-4, Ibadan Divisional Office, 1893-1957.

NAI Iba Div. 1/1, 1978 Vol. Health Committee, Ibadan: Annual Sanitation Report — Oyo Province, 1941.

NAI Iba Div. 1/1, 1978 Vol. 1, Health Committee, Ibadan: Annual Sanitation Report — Oyo Province, 1941, I.

NAI Iba Prof. 1/3 File No IB.7/1951-1954, Materials for Ibadan Waterworks Major Scheme.

NAI Iba Prof. 1\1 File No 843 "Western Region Production Development, 1952-1954". NAI File No PR/E5 "Western Government on Survey-Mother of all Developments, 1953".

NAI Iba Prof. File No 32 "Land Rules etc Acquisition of Land for by Nature Authorities" 1923-1938.

NAI IBA Prof. Vol. 1-4: Ibadan Provincial Office, 1897-1960.

NAI Iba, Div. 1/1 2642/S.3, 2642/S.3 Vol.2, and 2994 Vol. 1.

NAI Iba. Div. 1/1 File No. 1400 Vol. IV.

NAI Iba. Div. 1/1 File No. 2997 'Waterworks Undertaking'.

NAI Iba. Div. 1/1 File No. 937 Vol. 3, "Combined Waterworks and Electric Lights Scheme for Ibadan.

NAI Iba. Prof. 253/1/G. NAI. Iba, Div. 1/1 2642/S.3, 2642/S.3 Vol.2, and 2994 Vol. 1.

NAI IBADIV 1/489/Vol.XIX, Ibadan Division Annual Report, 1959.

NAI Iba Div. 1/1 File No 1978 Vol. 1, Health Committee, Ibadan: Minutes of Meeting of Ibadan Health Committee.

NAI Iba Div. 1/1 File, 1978 Vol. 1, Health Committee, Ibadan: Annual Sanitation Report, Oyo Province, 1941.

NAI Iba Div. 1/1 File, 1978 Vol. 1, Health Committee, Ibadan: Annual Sanitation Report, Oyo Province, 1948.

NAI Iba Div. 1/1 File, 1978 Vol. 1, Health Committee, Ibadan: Annual Sanitation Report, Oyo Province, 1941.

NAI Oyo Prof 21/1 File No 394, Vol I, Water Works Schemes Oyo Province, 1927-1937.

NAI Oyo Prof. 21/1 File No 394, Vol II-IV, "Water Works II Urban and Rural Water Supply: Development, 1937-1957".

NAI Oyo Prof. 21/1 File No 4730 Colonial Economic Research, 1941.

NAI Oyo Prof. 811 File No 2919, 'European Reservation Areas'.

NAI Oyo Prof. 811 File No 2919, 'European Reservation Areas', 1949.

NAI Oyo Prof. 1 895 Vol. IV Sanitation — Oyo Province: Ibadan Sanitary Committee, p. 137.

NAI Oyo Prof. 2/1 File No 773 Vol. III, Ogunpa Water Supply Ibadan, 1827-1932.

NAI Oyo Prof. 2/1 File No 772 Ibadan Water Right Schemes, 1921-1929.

NAI Oyo Prof. 2/2 File No 556 Vol. I, "Nigerian Ordinances".

NAI Oyo Prof. 2/2 File No 556 Vol. I, "Nigerian Ordinances", (Section 13, 14, 30, 34, 44, 45, 45, 47, 49 of the 1917 Township Ordinances).

NAI Oyo Prof. 2/2 File No 556 Vol. I-IV, "Nigerian Ordinances".

NAI Oyo Prof. 2/2 File No 556 Vol. I-VI, "Nigerian Ordinances".

NAI Oyo Prof. 2/2 File No 676 Vol. I &II, "System of Land Tenure in Yorubaland".

NAI Oyo Prof. 2/2 File No 799 Vol. II, "Slaughter House in Ibadan.

NAI Oyo Prof. 2/2 File No 968 Vol. I & II, "Land in Colonial Nigeria."

NAI Oyo Prof. 21/1 File No 1340 Vol. II, "Medical & Sanitary Department Oyo Province".

NAI Oyo Prof. 21/1 File No 1340 Vol. II….a.

NAI Oyo Prof. 21/1 File No 1566 Colonial Development Fund.

NAI Oyo Prof. 21/1 File No 1566, 'Colonial Development Fund'.

NAI Oyo Prof. 21/1 File No 1918 Vol II, "Electricity Schemes, 1937-1950".

NAI Oyo Prof. 21/1 File No 1918 Vol II, "Electricity Schemes, 1937-1950".

NAI Oyo Prof. 21/1 File No 885 Vol. III, "Sanitation Oyo Province Matters, 1921-1922".

NAI Oyo Prof. 21/1, File No 4504 Vol. I and II, 'General Progress of Development' and Community Development Progress Reports.

NAI Oyo Prof. 4/12, File No 59/1923.

NAI Oyo Prof. 811 File No 2080 "European Reservation Ibadan, 1937-1950.

NAI Oyo Prof. 811 File No 2080 "European Reservation Ibadan".

NAI Oyo Prof. 811 File No 2663 Colonial Development Scheme, Vol. I.

NAI Oyo Prof. 811, File No 2663 Vol. I and II, "Colonial Development Schemes".

NAI Oyo Prof. 811, File No 4104 Provincial Geography, 1944.

NAI Oyo Prof. 811, File No 811 Vol. I and II, 'Colonial Development Scheme, 1940-1941'

NAI Oyo Prof. File No 1187 Vol. I and II, "Ibadan Business Area Allocation Intelligence Report, 1933-1937.

NAI Oyo Prof. File No 1323 Vol. I-III, "Town Planning Committee for the Southern provinces, 1930-1956".

NAI Oyo Prof. File No 808 Residential Layout at Ibadan, 1932-1935.

NAI Oyo Prof. I., 895 Vol. IV, Sanitation-Oyo Province.

NAI Oyo Prof. I 895 Vol. IV, Sanitation-Oyo Province: Sanitation Committee, Appointment of.

NAI Oyo Prof.1895, Vol. IV, Extract from the Minutes of the I.N.A. Inner Council Meeting held 13 August, 1945.

NAI Oyo Province Annual Report, 1920, (para. 176).

NAI Oyo Province Annual Report, 1930, (para. 98)

Pollitt, H.W.W., *Colonial Road Problems: Impressions from Visits to Nigeria*, (London: His Majesty Stationery Office, 1950).

Report on the Ibadan Water Supply Expansion Project, Western State Ministry of Home Affairs and Information, Ibadan, 1972.

The Nigerian Handbook, Lagos 1919-98.

Secondary Sources

Books and Monographs

Abumere, S. I., 'Urbanisation and Urban Decay in Nigeria', in: Onibokun, A.G., Olokesusi, F. and Egunjobi, L. eds. *Urban Renewal in Nigeria*, (Ibadan: NISER, 1987).

Adams, Robert M., *Heartland of Cities*, (Chicago: University of Chicago Press, 1981).

Adelugba, Dapo ed., *Ibadan Mesiogo: A Celebration of a City, its History and People*, (Ibadan: Bookcraft Ltd., 2002).

Adeniyi, E.O., *Environmental Management and Protection in Nigeria*, (Ibadan: NISER 1986).

Adewoye, O., *The Judicial System in Southern Nigeria, 1854-1954: Law and Justice in Dependency*, (London: Longman, 1977).

Adigun, O., "The Problem of Housing in Nigeria", in Obilade, A.O. ed., *A Blueprint for Nigerian Law*, (Faculty of Law, University of Lagos, 1995).

Aduwo, A., "Historical Preview of Town Planning in Lagos before 1929", in
    Olaseni, A. M. ed., *Urban and Regional planning in Nigeria.* (Lagos: Nigerian
    Institute of Town Planners, Lagos State Chapter, 1999).

Agbaje-Williams, Babatunde, "Yoruba Urbanism: the Archaeology and Historical
    Ethnography of Ile-Ife and Old Oyo", in Akinwumi Adediran ed., *Pre-Colonial*
    Nigeria Essays in Honour of Toyin Falola, (New Jersey: Africa Press World
    Incorporated, 2005).

Agbola, Tunde and Mabawonku, A.O., "Indigenous Knowledge, Environmental
    Education and Sanitation; Application to an Indigenous African City", in
    Warren, D. M., Egunjobi, Layi and Wahab, Bolanle eds., *Indigenous Knowledge*
    in Education, (Ibadan: Indigenous Knowledge Study Group, University of
    Ibadan, 1996).78-94.

Aidan, Campbell, *Western Primitivism: African Ethnicity - A Study in Cultural*
    Relations, (London: Cassell Press, 1997).

Ake, C., "The New World Order: The View from Africa", in Hans-Henrik and
    Sovensen, G. (eds.), *Whose World Order: Uneven Globalisation and End of Cold*
    War, (London: West View Press, 1995).

Akinboye, S.O., Globalisation *and the Challenge for Nigeria Development in the 21st*
    Century, (Lagos: University of Lagos Press, 2008).

Akintola, F.O., *Flooding Phenomenon,* (Ibadan Region: Rex Charles Publication,
    1994).

Akinyele, I. B., *Outlines of Ibadan History,* (Lagos, Ist Edition, 1946). *Iwe Itan Ibadan*
    (England, 2nd Edition, 1950).

Andemariam, M., "Place of Taboos in Gikuyu Morality", in Magesa, L. ed.,
    MIASMU Research Integration Papers to Moral Teaching and Practices of
    African Religion, August session, 2001.

Anderson, David and Rathbone, Richard eds., *Africa's Urban Past,* (Oxford: James
    Currey, 2000).

Andre, Gunder Frank, *Reorient: Global Economy in the Asian Age,* (Los Angelis:
    University of California Press, 1998).

Awe, Bolanle, "Ibadan,Its Early Beginnings", in Lloyd, P.C., Awe, Bolanle and
    Mabogunje, Akin eds., *The City of Ibadan,* (Ibadan: Institute of African Studies,
    University of Ibadan, 1967).

Ayeni, Bola, "The Metropolitan Area of Ibadan: Its Growth and Structure", in
    Filani, M.O. et. al eds., (Ibadan Region, Ibadan: Rex Charles Publication, 1994).

Ayorinde, F., *Solid Waste Management in Commercial Area,* (Lagos: Bambee Press, 2001).

Bamikole, A. "Human Impact on the Environment", in Ismaila, B.R. ed., *Problems in Nigeria,* (Oyo: Odumalt Publishers, 2008).

Barkley, Paul and Seckler, David, *Economic Growth and Environmental Decay: The Solution Becomes the Problem,* (New York: Harcourt Brace Jovanovich, 1972).

Barnhart, Robert K. ed., *The World Book Dictionary,* Vol. One, (Chicago: World Book, Inc.,1995).

Bascom, William, "Urbanisation among the Yoruba", in Burke, Gerald, *Towns in the Making,* (London: Edward Arnold, 1955).

Beder, Sharon, *The Nature of Sustainable Development,* (Melbourne: Scribe Publications, 1997).

Beinard, William & Coates, Peter, *Environment and History: The Taming of Nature in the USA and South Africa,* (London, 1995).

Bell-Gam, W. I. et. al (eds.), *Perspectives on the Human Environment,* (Port-Harcourt: Amethyst & Co Publishers, 2004).

Blackwell, Jonathan M., et. al, *Environment and Development in African Selected Case Studies,* (Washington D.C: The World Bank, 1992).

Blakemore, C. and Shelia J., *Taboos.* (Oxford: Oxford University Press, 2001).

Bohannan, C. P. and Bohannan, L., *The Tiv of Central Nigeria,* (London, 1962).

Bolkin, D.B. and Keller, F.A., *Environmental science, Earth as a Living Planet,* (Boston: Von. Hoffmann Press, 1995).

Boroffice, R. A., "Environment and Development", in Otokiti, S. and Odewunmi, S. G. eds., *Issues in Management and Development,* (Ibadan: Rex Charles Publication, 2001).

Braudel, Fernand, "Pre-modern Towns", in Clark, Peter ed., *The Early Modern Town: A Reader,* (London: Longman, 1976).

Bruce, Russett et. al, *World Politics, The Menu for Choice,* Sixth edition, (Boston: Bedford/ St. Martins, 2000).

Burke, Gerald, *Towns in the Making of London,* (London: Edward Arnold Publishers Ltd., 1975).

Busari, L. "Operationalising the EPM process: Working Group Concepts and Dynamics", Agbola, T. ed., *Environmental Planning and Management: Concepts and Application to Nigeria,* (Ibadan: Counstellation Book, 2006).

Chatterjee, R., "Response" , in Serageldin I., Cohen, M.A. and Sivaramakrishnan, K.C. eds., *The Human Face of the Urban Environment*, (The World Bank: Washington DC, 1995).

Chickering, Roger, *A World at Total War: Global Conflict and the Politics of Destruction*, (UK: Cambridge University Press, 2006).

Clapperton, H., *Journal of Second Expedition into the Interior of Africa*, (Philadelphia: Carey, 1829).

Clapperton, Hugh, *Journal of a Second Expedition into the Interior of Africa*, (London: John Murray, 1829).

Clark, I., *Globalisation and International Relations Theory*, (Oxford: Oxford University Press, 1999).

Climate Change 2007: The Physical Science Basis, Contribution of Working Group 1to the fourth Assessment Report of the Intergovernmental Panel on Climate Change(IPCC), (Cambridge: Cambridge University Press, 2007).

Commonwealth Government, *Ecologically Sustainable Development: A Commonwealth Discussion Paper*, (Canberra: AGPS, 1990).

Connah, Graham, *African Civilisation, Pre-colonial Cities and States in Tropical Africa: An Archaeological Perspective*, (Cambridge: Cambridge University Press, 1995).

Connor, A.M.O. *The African City*, (London: Hutchinson University Library for Africa, 1983).

Coquery-Vidrovitch, C., *Histoire des villesd'Afrique noire. Des origines à la colonisation*, (Paris: Albin Michel, 1993).

Cougler, Josef and Flanagun, William F., *Urbanisation and Social Change in West Africa*, (Cambridge: Cambridge University Press, 1981).

Cronon, William, ed., *Uncommon Ground: Rethinking the Human Place in Nature*, (New York: W. W. Norton & Co., 1995).

Daly, Herman E., and Cobb, John B, Jr, For the Common Good: Redirecting the Economy Towards Community, the Environment, and a Sustainable Future, (Boston: Beacon Press, 1989).

Davidson, Basil, The Lost Cities of Africa, (Boston: Little Brown and Co., 1970).

Dodman, D., et-al, "Cities, Settlements and Key Infrastructure Climate Change 2022: Impacts, Adaptation and Vulnerability," in H.-O. Pörtner (eds. Contribution of Working Group II to the Sixth Assessment Report of the Intergovernmental Panel on Climate Change, Cambridge, UK and New York,

NY, USA Cambridge University Press.907–1040, doi:10.1017/9781009325844.008. Also available online at https://www.ipcc.ch/report/ar6/wg2/chapter/chapter-6/.

Douglas, Mary, Natural *Symbols*, (London: Penguin, 1973).

Droste, B. von and Dogse, P., "Sustainable development: The Role of Investment", in Goodland, R., Daly, H., El Seraty, S., and Droste, B. von eds., *Environmentally Sustainable Economic Development: Building on Brundtland*, (Paris: UNESCO, 1991).

Durkheim, E., *Incest: The Nature and the Origin of the Taboo*, (New York: Lyle Stuart, 1963).

Egunjobi, J. K., *Solid Waste Management in an Increasingly Urbanised Nigeria*, (Ado Ekiti: Proceedings of the National Practical Training Workshop, 2004).

Ehrenfeld, David, "Why put a value on biodiversity?" in Wilson, E.O. ed. *Biodiversity*, (Washington DC: National Academy Press, 1988).

Ekeh, Peter P., *Colonialism and Social Structure: An Inaugural Lecture*. (Ibadan: University of Ibadan Press, 1983).

Ekpo, U., *The Niger Delta and Oil Politics,* (Lagos: Orit. Egwa Press, 2004).

Emeribe, A.C. *Policy and Contending Issues in Nigeria, National Developmental Strategy* (Enugu: John Jacob Classic Publishers Ltd.)

Enger, E.D., Smith, B.F. and Bockarie, A.Y., *Environmental Science: A Study of Interrelationship*, (New York: McGraw Hill Companies, 2006).

Environmental Performance Index (EPI), (Yale University and University of Columbia, 2008).

Eyoh, Dickson and Stren, Richard eds., *Decentralization and the Politics of Urban Development in West Africa*, (Washington D.C.: Woodrow Wilson International Centre for Scholars).

Fabiyi, Seyi, "Colonial and Postcolonial Architecture and Urbanism", Tijani, in Hakeem Ibikunle ed., *Nigeria's Urban History, Past and Present*, (Maryland: University of America Press, 1984).

Falola, Toyin and Oguntomisin, G.O., *Yoruba Warlords of the 19th Century*, (Trenton, NJ: Africa World Press, 2001).

_____ and Salm, Steven J. eds., *Nigerian Cities*, (New Jersey: African World Press Inc., 2004).

_____, *Ibadan: Foundation, Growth and Change, 1830-1960*, (Ibadan: Bookcraft, 2012).

_____, *The Political Economy of a Pre-colonial African State, Ibadan, 1830-1900*, (Ile- Ife: University of Ife, 1984)

_____, *The Political Economy of a Pre-Colonial African State, Ibadan, 1830-1990*, (Lagos: Modelor Design Aids Ltd., 1989).

Faniran, A. *Solid Waste Management*, (Ibadan Region, Ibadan: Rex Charles Publication, 1994).

Fanon, Frantz, *The Wretched of the Earth*, (Great Britain: Penguins Books, 1983).

Filani, M.O., Akintola, F.O. and lkporukpo, C.O. eds., *Ibadan Region*, (lbadan: Rex Charles Publication, 1994).

Foster, L. M., Osunwole, M., Samuel A., and Wahab, W., "Imototo: Indigenous Yoruba Sanitation Knowledge Systems and their Implications for Nigerian Health Policy", in (Frank Fairfax III, 2005)

Fourchard, Laurent, *Urban Slums Reports: The Case of Ibadan, Nigeria*, (Ibadan: Institut Francais de Recherche en Afrique (IFRA), University of Ibadan, 2003).

Freire, Mila, and Polèse, Mario, *Connecting Cities with Macro-economic Concerns: The Missing Link*, (Washington, DC: The World Bank, 2003).

Giddens, Anthony, *The Consequences of Modernity*, (Cambridge: Polity Press, 1999).

Glasson, J.A., Lawrence, W.B. and Biddinger, G.B. eds., *Introduction to Environmental Impact Assessment*, (U.S.A: SETAC Publication, 1998).

Glencoe, G.L., and McGraw-Hill, M.C., *Merrill Earth Science*, (USA: Macmillan Press, 1993).

Gray, Colm, *Another Bloody Century: Future Warfare*, (United States: Phoenix Press, 2007).

Harrison, Paul, *The Third World Revolution: Population, Environment and a Sustainable World*, (New York: Penguin, 2003).

Haslam, Jonathan, Russia's Cold War: From the October Revolution to the Fall of the Wall, (United States: Yale University Press, 2011).

Hinderer, David, Provided these figures in Anna Hinderer, Seventeen Years in Yoruba Country, (London, 1873).

Holden, L., Encyclopedia of Taboos, (Oxford: ABC CLIO Ltd. 2000).

Hull, R.W., African Cities and Towns before European Conquest, (New York: W.W. Norton and Co., 1976).

Idem, K., "The Institutionalisation of the Environmental Planning and Management Process", in Agbola, T. ed., Environmental Planning and Management: Concepts and Application to Nigeria, (Ibadan: Constellation Book, 2006).

Idowu, E. B., Olodumare: God in Yoruba Belief, (Ibadan: Longman,1962).

International, Tools to Support Transparency in Local Governance, (Nairobi: UN-HABITAT, Secretariat, Urban Governance Tool Kit Series, 2004).

IPCC, Climate Change, (Cambridge: Cambridge University Press, 1996).

Jamaica, Economic Issues for Environmental Management, (Washington DC: World Bank, 1993).

Johnson, Samuel, The History of the Yorubas, (Lagos: CSS Press, 1969).

Kamack, Andrew, The Tropics and Economic Development a Provocative Inquiry in Poverty of Nations, (Baltimore: Penguin, 1976).

Kimble, H. T. Tropical Africa, Land and Livelihood, (New York: Twentieth Century Fund Press, 1998).

Labinjoh, Justin, "Ibadan and the Phenomenon of Urbanism", in Ogunremi, G.O. ed., A Historical and Socio-Cultural Study of an African City, (Ibadan: Oluyole Club, 1999).

_____, Modernity and Tradition in the Politics of Ibadan, 1900-1975, (Ibadan: Fountain Publications, 1991).

Lander, Richard and Lander, John, Journal of an Expedition to Explore the Course and Termination of the Niger, (London: Thomas Tegg and Son, 1832).

Lander, Richard, Records of Captain Clapperton's Last Expedition to Africa, (London: Henry Colburn and Richard Bentley, 1830).

Larson, Thomas, The Race to the Top: the Real story of Globalisation, (Washington DC: Cato Institute, 2001).

Leopold, Aldo, "The Land Ethic", A Sand County Almanac, (1949).

Lindell, I.L., Walking the Tight Rope: Informal Livelihood and Social Networks in a West African City, (Stockholm, Stockholm University Press, 2003).

Lloyd, P.C., Africa in Social Change, (Baltimore: Penguin, 1967).

_____, Awe, Bolanle and Mabogunje, Akin eds., The City of Ibadan, (Ibadan: Institute of African Studies, University of Ibadan, 1967).

Mabogunje, A.L., "The problems of a Metropolis", in Lloyd, Mabogunje and Awe eds., The City of Ibadan, (London: Cambridge University Press, 1967).

_____, Cities and African Development, (Ibadan: Oxford University Press, 1976).

_____, Urbanisation in Nigeria, (London: University of London Press, 1968).

Magesa, L., African Religion: The Moral Traditions of Abundant Life, (Maryknoll, New York: Orbis Books, 1997).

Mahadevia, Darshini, "Sustainable Urban Development in India: An inclusive Perspective", in Westerndorff, David ed., From Unsustainable to Inclusive Cities, (Geneva: An UNRISD Publication in Collaboration with Swiss Agency for Development Cooperation, October 2002).

Mamadou, Diouf and Rosalind, Fredricks eds., The Art of Citizenship in African Cities: Infrastructures and Spaces of Belonging, (New York: Palgrave Macmillan, 2014).

Martins, W.F. et. al, Hazardous Waste Handbook for Health and Safety, (London: Butler and Tanner Ltd., 1987).

Mbiti, J., African Religions and Philosophy, (London: Heinemann, 1969).

McGranahan, G., Mitlin, D., Satterthwaite, D., Tacoli, C. and Turok, I., "Africa's Urban Transition and the Role of Regional Collaboration", Human Settlements Working Paper Series, Theme: Urban Change-5, (London: IIED, 2009).

McNeill, John R., Something New Under the Sun: An Environmental History of the Twentieth Century, (2000).

Merchant, Carolyn, The Columbia Guide to American Environmental History, (New York: Columbia University Press, 2002).

Metz, H.C. ed., Nigeria: Country Studies, (USA: Library of Congress, 1991).

Millson, A.W., 'The Yoruba Country, West Africa', Proceedings of the Royal Geographical Society, 13, (1891).

Mink, Stephen D., "Poverty, Population and the Environment", World Bank Discussion Paper, No 189.

Mintzer, I. M. ed, Confronting Climate Change: Risks Implications and Responses, (Cambridge: Cambridge University Press, 1992).

Mumford, Lewis, The City in History, (Penguin Book, 1961).

Nigerian Environmental Study/Action Team, Nigeria's Threatened Environment: A National Profile, (Ibadan: Intec Printers Limited, 1991).

NITP, Twenty-Five Years of Physical Panning in Nigeria, (Lagos: Nigeria Institute of Town Planners, 1991).

Obateru, O.I., The Yoruba City in History, 11th Century to the Present, (Ibadan: Penthouse Publications (NIG), 2003).

Obateru, Oluremi I., The Yoruba City in History, 11th Century to the Present, (Ibadan: Penthouse Publications (Nig), 2006).

Obialo, D.C., Town and Country Planning in Nigeria, (Owerri, Nigeria: Asumpta Printing and Publication, 1999).

Oduyale, O., "Traditional education in Nigeria", in Oyeneye, O. Y. and Shoremi, O. M. eds., Nigerian Life and Culture, (Ogun State University, 1985).

Ogungbemi, S., "An African Perspective on the Environmental Crisis", in Pojman, Louis J. ed., Environmental Ethics: Readings in Theory and Application, (Belmont, CA: Wadsworth Publishing House, 1997).

Ojo, A.O. The Climatic Dilemma, Lagos State University 32nd Inaugural Lecture, August 19, 2007.

Okebukola, Peter "Global Warming and Ozone Layer Depletion", in Otokiti, S.O. and Odewumi, G. eds., Issues in Management and Development, (Ibadan: Rex Publications, 2001).

Ola, C.S., Town and Country Planning and Environmental Laws in Nigeria, (Ibadan: University Press, 1993).

Olajire, Olaniran, "The Geographical Setting of Ibadan", in Ogunremi, G.O., Ibadan: A Historical, Cultural and Socio-Economic Study of an African City, (Ibadan: Oluyole Club, 1999).

Olanipekun, A. O. "Environmental Hazards in Nigeria", in B. R. Ismaila ed., Problems in Nigeria, (Oyo: Odumalt Publishers, August, 2005).

Olaniyan,Olanrewaju SUSTAINABLE DEVELOPMENT OF IBADAN: Past, Present and Future ,Ibadan, CESDEV Issue Paper No. 2017/1, Centre for Sustainable Development ,University of Ibadan, Nigeria.5-6.

Olson, James and Roberts, Randy, Where the Domino Fell: America and Vietnam 1945 – 1990, (New York: St. Martin's Press, 1991).

Olukoju, Ayodeji, "Historical Background of Nigerian Cities", in Toyin Falola and Salm, Steven J., (eds.) Nigerian Cities, (New Jersey: African World Press Inc. 2004).

Oluya, S.I. and Olu-Buraimoh, H. "Environmental Protection", in Oluya, S.I. et al. eds., Compendium of Issues in Citizenship Education in Nigeria, (Ibadan: Remco Press, 1998).

Omoregie, E. and Onwuliri, C.O. "Interactions between Crude Oil and Biological Productivity in Aquatic Environment: A Case Study of the Niger Delta", in Osuntokun, A. ed., Democracy and Sustainable Development in Nigeria, (Lagos: Frankad, 2002).

Omuta, G.E.D, "Urban Solid Waste Generation and Management; Towards an Environmental Sanitation Policy". in Sada, P.O. and Odemerho, F.O. eds., Environmental Issues and Management in Nigerian Development, (Ibadan: Evans Brothers Ltd (Pub), 1988).

Onibokun, A.G. ed., Housing in Nigeria: A Book of Reading, (Ibadan: Nigerian Institute of Social and Economic Research, NISER).

Onibokun, A.G. et. al, Affordable Technology and Strategies for Waste Management in Africa: Lessons From Experiences, (Ibadan: Centre for African settlement – Studies and Development (CASSAD), 2000).

Onibokun, A.G., "The EPM Process in Sustainable Development and Management of Nigerian Cities", Agbola, T., ed. in Environmental Planning and Management: Concepts and Application to Nigeria, (Ibadan: Constellation Book, 2006).

Onibokun, A.G., *Public Utilities and Social Services in Nigerian Urban Centres: Problems and Guides for Action,* (Canada: IDRC and Ibadan: NISER, 1987).

Onokerhoraye, Andrew G., Urbanisation and Environment in Nigeria: Implications for Sustainable Development, (Benin: The Benin Social Sciences Series for Africa, 1995).

Owolabi, K. A., Because of our Future: The Imperative for an Environmental Ethic for Africa, (Ibadan: IFRA, 1996).

Oyedeji, Babatunde ed., Readings in Political Economy and Governance in Nigeria; Selected Speeches and Articles of Chief T.A. Akinyele. (Lagos: CSS Ltd., 2002).

Oyeranmi, Soji, "Climate Change and Sustainable Development in Nigeria", in Falola, Toyin and Amutabi, Maurice eds., Perspectives on African

Environment, Science and Technology, (New Jersey: African World Press, 2012).

_____, "May, 29th 1999 Democracy Returns to Nigeria" in The Great Events in History, 1971-2000, (New York: Salem Press).

Pearce, David, Markandya, Anil & Barbier, Edward, Blueprint for a Green Economy, Earthscan, (London, 1989).

Pelling, M., "Toward a Political Ecology of Environmental Risk: The Case of Guyana", in Zimmerer, Karl and Bassett, Thomas eds., Political Ecology: An Integrative Approach to Geography and Environment-Development Studies, (The Guildford Press, 2003).

Perham, Margery, Native Administration in Nigeria, (London, 1962).

Pickering, K.T. and Owen, H.A., An Introduction is Global Environmental Issues, (London: Butler and Tanner Ltd.)

Pielke, Roger"Tracking progress on the economic costs of disasters under the indicators of the sustainable development goals,"in Environmental Hazards 18(1):1-6, October 2018,DOI: 10.1080/17477891.2018.1540343

Pourtier R., "Migrations et dynamiques de l'environnement", in Pontier, G. and Gaud, M. eds., Afrique contemporaine. L'environnement en Afrique, (Paris: La Documentation française, 1992).

Redfield, Robert, A Mexican Village, (Chicago: Chicago University Press, 1930).

Relis, P. and Dominiski, A. "Beyond the Crisis, Integrated Waste Management", Community Environmental Council Publications, (Santa: Barbara, 1987).

Report of the South Commission, The Challenge to the South, (Oxford: Oxford University Press, 1990).

Rodney, Walter, How Europe Underdeveloped Africa, (London: Bogle L Ouverture Publishers, 1986).

Ronald, Fletcher, "African Urbanism: Scale, Mobility and Transformation", in Graham, Connah ed., Transformations in Africa: Essays on Africa's Later Past, (London: Leicester University Press, 1998).

Rueben K. Udo "Environments and Peoples of Nigeria, A Geographical Introduction to the History of Nigeria", in Obaro, Ikime ed., Groundwork of Nigerian History, (Ibadan: Heinmann, 1980).

Sachs, J., The End of Poverty: Economic Possibilities for Our Time, (New York: Penguin).

Samuel, F. "Urban Planning and the Quality of Life of Nigeria", in Agbola, T. et. al (eds.) Environmental Planning and Health in Nigeria, (Ibadan: Department of Urban and Regional Planning, University of Ibadan, 2008).

Sandburg, Carl, Abraham Lincoln; The Prairie Years and The War Years, (New York: Dell Publishing Co. Incorporation, 1939).

Schaltegger, Stefan et. al, An Introduction to Corporate Environmental Management. Striving for Sustainability, (Sheffield: Greenleaf, 2003).

Schwela, D., "Review of Urban Air Quality in Sub-Saharan Africa", Clean Air Initiative in Sub-Saharan African Cities, (Washington DC: The World Bank, 2007).

Seneca, J. and Taussig, M. 1984, Environmental Economics, (New Jersey: Prentice-Hall, 1984).

Simmons, Ian G., Environmental History: A Concise Introduction, (Oxford: Oxford University Press, 1993).

Singer, Peter, "Environmental Values", in Ian Marsh ed., The Oxford Book of Travel Stories, Melbourne, (Australia: Longman Chesire, 1991).

Singer, Peter, Practical Ethics, (Cambridge: Cambridge University Press, 2011).

Sjoberg, Gideon, The Preindustrial City, Past and Present, (Glencoe, IL: The Free Press, 1960)

Soji, T. and Dayo, A. "Urban Governance and the Governance of Nigerian Cities", in Agbola, T. et al eds., Environmental Planning and Health in Nigeria, (Department of Urban and Regional Planning, University of Ibadan, 2008).

Sonthamier, Sally, ed., Women and the Environment: A Reader: Crisis and Development in the Third World, (London: Earthscan Publication, 1991).

Steiner, F. Taboo, (London: Cohen & West Ltd, 1956).

Stren, Richard, Coping with Rapid Urban Growth in Africa: An Annotated Bibliography in English and French on Policy and Management on Urban Affairs in the 1980s, (Montreal: Centre For Developing Areas Studies, Mcgrill University, 1986).

Sustainable Cities and Local Governance: The Sustainable Cities Programme, written and published by the United Nations Centre for Human Settlements (Habitat) and the United Nations Environment programme (UNEP), (Nairobi, Kenya, 1997).

Thampapillai, D. J., Environmental Economics, (Melbourne: Oxford University Press, 1991).

The Cities Alliance, Foundations for Urban Development in Africa: The Legacy of Akin Mabogunje, (Washington, DC: The Cities Alliance Secretariat, 2006).

The Environmental Planning and Management (EPM) Source Book, Volume 1: Implementing the Urban Written and published by the United Nations Centre for Human Settlements (Habitat) and the United Nations Environment Programme (UNEP), (Nairobi, Kenya, 1997).

The Habitat Agenda: Goals and Principles, Commitments, and Global Plan of Action Agreed at the United Nations Conference on Human Settlements (Habitat II), (Istanbul, Turkey, June 1996).

The SCP Process Activities: A Snapshot of what they are and how they are implemented Written and published by the United Nations Centre for Human settlements (Habitat) and the United Nations Environment Programme (UNEP), (Nairobi, Kenya, 1998).

The SCP Source Book Series, Volume 5 Institutionalising the Environmental Planning and Management Process Ibadan, Nigeria: Prepared and written by staff and consultants of the Sustainable Cities Programme The United Nations Centre for Human Settlements (UNCHS) and the United Nations Environment Programme (UNEP), 1999.

The Sustainable Cities Programme: Approach and Implementation Written and published by the United Nations Centre for Human Settlements (Habitat), (Nairobi, Kenya, 2nd Edition, 1998).

The United Nations Centre for Human Settlements, (for the United Nations Commission on Sustainable Development) Sustainable Human Settlements Development: Implementing Agenda 21, (Nairobi, Kenya, 1994).

The United Nations Centre for Human Settlements, UNCHS (Habitat) and UNEP Join Forces on Urban Environment prepared for the United Nations Commission on Human Settlements (CHS15) and the Governing Council of UNEP (GC18) , (Nairobi, Kenya, 1995).

The Urban Environment Agenda Volume 2: City Experiences and International Support Volume 3: The UEF Directory, written and published by the United Nations Centre for Human Settlements (Habitat) and the United Nations Environment Programme (UNEP), (Nairobi, Kenya, 1997).

Thijs De La Court, Beyond Brundtland: Green Development in the 1990s, (London: Zed Books Ltd., 1990).

Thomas, Larsson, The Race to the Top: The Real Story of Globalisation, (Washington D.C.: Cato Institute, 2001).

Thorpe, C. O., Àwon Èèwò Ilè Yorùbá. (Ibadan: Onibon-Oje Press, 1972).

Tibaijuka, Anna K., Under-Secretary-General and Executive Director United Nations Human Settlements Programme, "São Paulo: A Tale of Two Cities: Bridging the Urban Divide" in UN-HABITAT Cities & Citizens Series ht © United Nations Human Settlements Programme (UN-HABITAT), 2010.

Tijani, Hakeem I. "The New Lagos Town Council and Urban Administration, 1950-1953", in Falola, T. and Salm, S. eds., Nigerian Cities, (Trenton: Africa World Press, 2003).

Tijani, Hakeem I. ed., Nigerian Urban History Past and Present, (Maryland: University Press of America, 2006).

Tijani, Hakeem I., "Reflection on Nigeria's Urban History", in Tijani, Hakeem I. ed. Nigeria's Urban History: Past and Present, (Oxford: University Press of America, 2006).

Ukiwo, Ukoha, "Indigenous and Received knowledge in the conservation of the Niger-Delta Environment", in Fayemi, Kayode et. al eds., Toward an Integrated Development of the Niger Delta, (Ikeja: Centre for Democracy and Development, 2005).

UN-HABITAT, Global Campaign on Urban Governance: Concept Paper, (Nairobi, 2002).

Vardy, Peter and Grosch, Paul, The Puzzle of Ethics, (New York: Harper Collins, 1999).

Vaidya,Hitesh and Tathagata Chatterji, "SDG 11 Sustainable Cities and Communities: SDG 11 and the New Urban Agenda: Global Sustainability Frameworks for Local Action" in I. B. Franco et-al (eds.) Actioning the Global Goals for Local Impact, Science for Sustainable Societies, Singapore, Springer Nature .January 2020 DOI: 10.1007/978-981-32-9927-6_12

Wahab, B., "Some Aspects of Physical Planning and Development in Nigeria", in Agboola, T. ed., Environmental Planning and Health in Nigeria, (Ibadan: Department of Urban and Regional Planning, University of Ibadan, 2008).

_____, "The Institutionalisation of the Environmental Planning and Management Process", in Agboola, T. ed., Environmental Planning and Management: Concepts and Application to Nigeria, (Ibadan: Constellation Book, 2006).

Wahab, Bolanle, Egunjobi, Layi and Warren, Michael D. eds., Alafia and Well-Being in Nigeria Studies, (IOWA: Technology and Social Change, No. 25, Ames Centre for Indigenous Knowledge for Agriculture and Rural Development, IOWA University, 1996).

Watson, Ruth, Civil Disorder is the Disease of Ibadan - Ija Igboro Larun Ibadan: Chieftaincy and Civic Culture in a Yoruba City, (Ibadan: Heinemann Educational Books Nigeria Plc, 2005). 5

Werlin, Herbert, Governing an African City: A Study of Nairobi, (New York: Africana Publishing Company, 1974).

Whyte, Anne V., "Women, Environmental Perception and Participatory Research", in Rathgeber, Eva M. ed., Women's Role in Natural Resource Management in Africa, (Ottawa: International Research Development Centre (IDRC), 2004).

World Bank, "Climate Change, Disaster Risk, and the Urban Poor", (Washington: World Bank, 2012).

World Bank, Cities in Transition: World Bank Urban and Local Government Strategy. (Washington, DC 2000)

World Bank, Jamaica, Economic Issues for Environmental Management, (Washington DC: World Bank Press, 1993).

World Bank, Sustainable Development in a Dynamic World: Transforming Institutions, Growth and Quality of Life, (World Bank and Oxford University Press, 2003).

World Commission on Environment and Development (WCED), Our Common Future, (Oxford: Oxford University Press,1987)

Worster, Donald ed., The Ends of the Earth: Perspectives on Modern Environmental History, (Cambridge: Cambridge University Press, 1988).

Worster, Donald, The Wealth of Nature: Environmental History and the Ecological Imagination, (Oxford: Oxford University Press, 1993).

Wraith, R., Local Government in West Africa, (London: George Allen and Unwin, 1964).

Yi-Fu, Tuan, Landscapes of Fear, (New York: Pantheon Books, 2000).

Journal Articles and Occasional Papers

Abiodun Areola, Environmental Justice and Green Spaces in Ibadan Metropolis, Nigeria: Implications on Sustainable Development in Urban Construction *Environ. Sci. Proc.* 2022, *15*(1), 57; https://doi.org/10.3390/environsciproc2022015057, 24 May 2022.

Adebayo, Y. R., "An Analysis of Environmental Emergencies, the African Crises and Aspects of Energy Issues", *African Journal of International Affairs and Development,* Vol. 4, No. 2, (1999). 116 – 143.

Adebisi, Adedayo, "Environment Sanitation and Waste Management at the Local Level in Nigeria," *Geo-Studies Forum,* Vol, 1 and 2, (2000).

Adeyemi, A.S., Olorunfemi, J.F. and Adewoye, "Waste Scavenging in Third World Cities: A Case Study in Ilorin, Nigeria", *The Environmentalist,* 21, (2001). 93–96.

Africa Renewal (Formerly Africa Recovery), United Nations Department of Public Information, Vol. 21, No. 2, (July 2007), 14 – 16.

Aguwanba, J. C, "Solid Waste Management in Nigeria: Problems and Issues", *Environmental Management,* Vol. 22, No. 6. 849-856.

Akindele, S.T. and Olaopo, O.R., "Globalisation: Its Implications and Consequence for Africa", *Nigeria,* (2002). 1.

Alex Asakitikpi, "Environmental and Behavioural Risk Factors Associated with Childhood Diarrhoea in Ibadan Metropolis, Oyo State", *The Journal of Environment and Culture,* Vol. 2, No. 1, (2005). 1-13.

Anand, P.B., "Waste Management in Madras Revisited", *Environment& Urbanisation,* Vol. 11, No. 2, (October 1999).

Atanda, J. A., "Government of Yorubaland in the Pre-colonial Period", in *Tarikh,* Vol. 2, No. 1, (1974).

Bartlett, S., "Climate Change and Children: Impacts and Implications for Adaptation in low to middle income countries", *Environment and Urbanisation,* Vol. 20, No. 2, (2008). 501–519.

Bascom, William, "Urbanisation among the Yoruba", *The American Journal of Sociology,* Vol. 60, No. 5, (1995).

Beder, Sharon, "Economy and Environment: Competitors or Partners?" *Pacific Ecologist* 3, (Spring 2002), 50-56.

Bess, Michael et al., "Anniversary Forum: What Next for Environmental History?" *Environmental History,* Vol. 10, No. 1, (2005). 30–109.

Bess, Michael, "Artificialization and its Discontents", *Environmental History*, Vol. 10, No. 1, (2005).

Bess, Michael, Cioc, Mark and Sievert, James, "Environmental History Writing in Southern Europe", *Environmental History*, 5, (2000). 545-556.

Busari, A.T. and Olaleye, M.O., "Urban Flood and Remediating Strategies in Nigeria: The Ibadan experience", *The International Journal of Environmental Issues*, Vol. 6, No. 1 and 2, (2004).

Callaway, A., "Nigeria's Indigenous Education: The Apprentice System", *Odu*, 1, (1964). 62-79.

Catherine Vidrovitch Conquery, "The Process of Urbanisation in Africa (from Origins to the Beginning of Independence)", *African Studies Review*, Vol. 34, No. 1, (April, 1991).

Chinedu ,E., and Chukwuemeka, C.K., "Oil Spillage and Heavy Metals Toxicity Risk in the Niger Delta, Nigeria," *J Health Pollut*. 2018 Aug 21;8 (19):180905. doi: 10.5696/2156-9614-8.19.180905. PMID: 30524864; PMCID: PMC6257162. https://www.ncbi.nlm.nih.gov/pmc/articles/PMC6257162/pdf/i2156-9614-8-19-180905.pdf

Chukwuemeka, Emma E. O., Osisioma, B. C. and Ugwu, Joy, "The Challenges of Waste Management to Nigeria Sustainable Development: A Study of Enugu State", *International Journal of Research Studies in Management*, Vol. 1, No. 2, (October, 2012). 79-87.

Cox, Oliver C. "The Preindustrial City Reconsidered", *The Sociological Quarterly*, 5, (1964), 133-147.

Donald Worster, "The Two Cultures: Environmental History and the Environmental Sciences", *Environment and History*, 2, (1996). 3-14.

Donald, J. Hughes, "Ecology and Development as Narrative Themes of World History", *Environmental History Review*, 19, (Spring 1995). 1-16.

Egunjobi, L and Oladoja A., 'Administrative Constraints in Urban Planning and Development". *The Environmental Management*, Vol. 17, No.1, (1993). 15-30.

Egunjobi, L., "Human Elements in Urban Planning and Development: Ibadan", *Habitat International*, Vol. 10, No. 4, (1986).147-153.

Falola, Toyin, "Lebanese traders in Southwestern Nigeria, 1900-1960", *African Affairs*, 89, 357, 523-553.

Furedy, C., "Garbage: Exploring Non-Conventional Options in Asian cities", *Environment and Urbanisation*, Vol. 4, No. 2, (1992). 42-61.

Gene, R. and Krueger, Alan, "Economic Growth and the Environment", *The Quarterly Journal of Economics*, MIT Press, Vol. 110, No. 2, (May 1995).

Gordon, Childe, "The Urban Revolution," *Town Planning Review*, 21, (1950). 3-17.

Gordon, O. and Breach, C. *The International Journal on Environmental Studies*, 8, (1976). 227-233.

Gordy, Slack, "Africa's Environment in Crisis", *African Journal of Environmental Studies*, Vol. 24, No. 74, (December 2002). 1.

Hardin, Garrett, "The Tragedy of the Commons", *Science*, Vol. 162, No. 3859, (1968). 1243–8.

Hart, Jennifer "Fruity" Smells, City Streets, and the Politics of Sanitation in Colonial Accra," *Urban Forum* (2022) 33:107–127, 111 https://doi.org/10.1007/s12132-021-09446-4

Hoppit, Julian, "The Nation, the State, and the First Industrial Revolution", *Journal of British Studies*, (April 2011).

Idris, S. Mohammed, "Going Green – A Third World Perspective", *Chain Reaction*, 62, (October 1990). 16-17.

Lawoyin, T.O., et. al, "Outbreak of Cholera in Ibadan, Nigeria", *European Journal of Epidemology*, Vol. 15, No. 4, (1999).

Lyun, F., "Hospital Service Areas in Ibadan City", *Social Science and Medicine*, Vol. 17 No. 9, (1983).

Machlis, G.E. et. al "The Human Ecosystem Part I: The Human Ecosystem as an Organizing Concept in Ecosystem Management", *Society & Natural Resources*,10, (1997). 347-367

MacIntosh, S.K. and MacIntosh, R.J., "The Early City in West Africa: Towards an Understanding", *The African Archaeological Review*, 2, (1984). 73-98.

MacNeill, Jim, "Strategies for Sustainable Economic Development", *Scientific American*, (September 1989).

Marchand, Anne, "Multinationals Immunity and the African Environment", *African Journal of Environmental Studies*, Vol. 24, No. 74, (December 2002), 18 – 21.

Marshall, Alan, "Ethics and the Extraterrestrial Environment", *Journal of Applied Philosophy*, Vol. 10, No. 2, (1993). 1468-5930.

McIntosh, Roderick, "Early Urban Clusters in China and Africa: The Arbitration of Social Ambiguity", *Journal of Field Archaeology*, 18, (1991).

Mitchell, N. C., "Some Comments on the Growth and Character of Ibadan Population", *Research Notes*, (Ibadan: Department of Geography, University College, 1953). 4, 9-10

Nilon, C. H., Berkowitz, A. R and Hollweg, K. S, "Editorial: Understanding Urban Ecosystems: A New Frontier for Science and Education", *Urban Ecosystems*, 3, (1999), 3-4.

Odejobi Cecilia Omobola, "An Overview of Taboo and Superstition among the Yoruba of Southwest of Nigeria", *Mediterranean Journal of Social Sciences,* Vol. 4, No. 2, (May 2013). 221-223.

Ogundele, F. O., Ayo, O., Odewumi, S. G. and Aigbe, G. O., "Challenges and Pospects of Physical Development Control: A Case Study of Festac Town, Lagos, Nigeria", in *African Journal of Political Science and International Relations*, Vol. 5, No. 4, (April 2011). 174–178.

Oguntala, A. B. and Oguntoyinbo, J.S., "Ibadan Urban Nature", *Urban Ecology*, Vol. 7, No. 1, (September 1982). 39-46.

Ojo, G.J.A., "The Journey to Agricultural Work in Yorubaland", *Annals of the Association of American Geographers*.

Olabode, B.O. and Siyanbola, S.O., "Proverbs and Taboos as Panacea to Environmental Problems in Nigeria, a Case of Selected Yoruba Proverbs", in *Journal of Arts and Contemporary Society*, Vol. 5, No. 2, (2013). 56 – 66.

Omar Nagat *et-al* "Localising the SDGs in African Cities: A Grounded Methodology" in *Africa Development*, Volume XLVII, 4, 157-184 ,Council for the Development of Social Science Research in Africa, 2022, (https://doi.org/10.57054/ad.v47i4.2981) .

(https://au.int/en/pressreleases/20190526/making-african-cities-more-habitable-ministerial-meeting-kicks)

Onibokun, Adepoju, "Forces Shaping the Physical Environment of Cities in the Developing Countries: The Ibadan Case", *Land Economics*, Vol. 49, No. 4, (Nov., 1973). 424-431.

Ordinioha B., and Brisibe,S., "The human health implications of crude oil spills in the Niger delta, Nigeria: An interpretation of published studies," *Niger Med J.* 2013 Jan;54(1):10-6. doi: 10.4103/0300-1652.108887. PMID: 23661893; PMCID: PMC3644738. https://www.ncbi.nlm.nih.gov/pmc/articles/PMC3644738/

Oyesiku, O. K., Asiyanbola, R. A. and Sokefun, J. A. "Review of the Nigerian Urban and Regional Planning Law". *Journal of Public Law and Practice*, Vol. 1, No. 1, (June, 1999). 180 - 191.

Oyeranmi, Soji "A Civilization without Toilets? Ibadan and her Environment in the Postcolonial Era" *Sociology and Anthropology*, 6(2): 187-202. 2018 available online at https://www.hrpub.org/journals/article_info.php?aid=6771

Parnell, Susan and Robinson, Jenny, "Development and Urban Policy: Johannesburg's City Development Strategy", *Urban Stud*, Vol. 43, No. 337, (2006).

Perry, Smith and Espinosa, J. Andres, "Environment and Trade Policies, some Methodological Lessons", *Working Papers*, (Duke University Press, Department of Economics, 1995).

Petra, Christman and Glen, Taylor "Globalisation and Environment: Determinant of Firm Self Regulation in China", *Journal of International Business Studies*, (2001). 32.

Polycarp A. Ikuenobe, "Traditional African Environmental Ethics and Colonial Legacy", *International Journal of Philosophy and Theology*, Vol. 2, No. 4, (December 2014). 1-21.

Richards, John, "Documenting Environmental History: Global Patterns of Land Conversion", *Environment*, Vol. 26, No. 9, (1984).

Robert, Redfield and Sigler, Milton, "The Cultural Role of Cities", *Economic Development and Cultural Change*, Vol. 1, No. 3, (1955). 53-73.

Robert, S. Jordan and John Renninger, "The New Environment of Nation Building", *Journal of Modern African Studies*, Vol. xii, No. 2, (1975).187-207.

Safari, N. Guarzan and Trinedy, R.K. "Global and Regional Availability and the Future of Renewable Freshwater Supplies, Demand and Human Health", *Journal of World Water Resources*, Vol. 41.

Sanni, Lekan, "Forty Years of Urban and Regional Planning Profession and National Development in Nigeria", *Journal of the Nigerian Institute of Town Planners*, Vol. xix, No.1, (November, 2006).1-6.

Satoshi, Ishii, "Urban Air Pollution and Urban Management: Applicability of Ecosystem Approach and the Way Forward", UNU-IAS Working Paper No. 120, (March 2003).

Schwab, William, "Urbanism, Corporate Groups and Culture Change in Africa below the Sahara", *Anthropological Quarterly*, Vol. 43, No. 3, (1970).

Scott, Peter, "Ecology: Religious or Secular?" *Heythrop Journal*, Vol. 38, No. 1, (January 1997).

Self, Peter, "Market Ideology and Good Government", *Current Affairs Bulletin*, (September 1990). 4-10.

Shaw, Timothy M. and Grieve, Malcolm J., "The Political Economy of Resources: Future in the Global Environment", *The Journal of Modern African Studies*, Vol. 16, No. 11. 1-32.

Steg, Linda and Vlek, Charles "Encouraging Pro-Environmental Behaviour: An Integrative Review and Research Agenda", *Journal of Environmental Psychology*, Vol. xxx, (2008). 1–9.

United Nations Conference on Trade and Development (UNCTAD) , *Promoting Investment For Sustainable Development in Cities. The IPA Observer*, Issue.7, 2019.

UN-Habitat and UNCTAD, (Urban Economy Branch Discussion Paper #7. 2017.

Uwadiegwu, B.O. and Chukwu, K.E., "Strategies for Effective Urban Solid Waste Management in Nigeria", *European Scientific Journal*, Vol. 9, No. 8, (March 2013 Edition). 296-307.

Wheatley, Paul, "The Significance of Traditional Urbanism, Comparative Studies", *Society and History*, 12, (1970).

Wheatley, Paul, "The Significance of Traditional Yoruba Urbanism", *Comparative Studies of Society and History*, Vol. 12, No. 4, (1972). 393-423

White, Lynn, "The Historical Roots of our Ecologic Crisis", *Science*, 155, (1967). 1203–1207.

Winters, Christopher, "Traditional Urbanism in the North Central Sudan", *Annals of The American Geographers Association*, VOl. 67, No. 4, (1983). 500-520.

Wirth, Louis, "Urbanism as a Way of Life", *American Journal of Sociology*, 44, (1938).

Newspapers and Magazines

Awake, February 8, 2000, 3-7.

Awake, August 2008, 3-7.

Awake, June 22, 2003, 4.

Time International Magazine, Special Edition, An Earth Day, 2000.

Mabogunje, A.L., "Ibadan-Black Metropolis", *Nigeria Magazine*, No. 68, March, 1961.

Nigerian Tribune, August 16, 1975, 4.

Nigerian Tribune, January 10, 2006, 40.

International Herald Tribune, June 9, 2007

The Environment Magazine, Vol. 8, No. 2, February 2008, 54.

Tell Magazine, August 11, 2008, Special Edition, 20-31.

Tell Magazine, Special Edition, No. 29, July 2011.

Daily Times, September 2, 2011.

Nigerian Tribune, September 24, 2011, 36

Nigerian Tribune, November 12, 2011, 36.

Online Materials

Rural Water Supply and Sanitation Agency,https://old.oyostate.gov.ng/oyo-state-rural-water-supply-and-sanitation-agency-ruwassa/#1496768094884-2b68d237-9abc

Oyo State Government, "Open Defecation: Oyo Vows To Set Pace in Ending Menace Before 2025," available online at https://oyostate.gov.ng/open-defecation-oyo-vows-to-set-pace-in-ending-menace-before-2025/ , 10th December,2020.

Seyi Makinde Campaign Organisation ,*Oyo State Roadmap to Sustainable Development,* 2023-2027https://seyimakinde.com/wp-content/uploads/2023/02/Roadmap-To-Sustainable-Development-2023-2027.pdf

The World bank report on Ibadan, https://documents1.worldbank.org/curated/en/387131624912950665/pdf/Disclosable-Version-of-the-ISR-Ibadan-Urban-Flood-Management-Project-P130840-Sequence-No-14.pdf , June 28th 2021.

Nigerian Tribune,"Oyo loses out on World Bank $70m flood intervention funds," https://tribuneonlineng.com/oyo-loses-out-on-world-bank-70m-flood-intervention-funds/ August 10, 2021

"An Update on the Ongoing Ibadan Urban Flood Management Projects (IUFMP)," https://feedbackoysg.com/ongoing-ibadan-urban-flood-management-projects/, June, 9, 2022,

The Report on the World Bank assisted Ibadan Urban Flood Management Project, available online at https://www.food-security.net/wp-content/uploads/2021/07/PAD6780PAD0P13010Box385224B00OUO090.pdf

Sustainable Development Goals (SDGs), https://sustainabledevelopment.un.org/content/documents/18785E_HLPF_201 8_2_Add.4_ECAadvanceduneditedversion.pdf

https://www.undp.org/sdg-accelerator/background-goals and https://www.undp.org/content/undp/en/home/sustainable-development-goals/background/

The United Nations Organisation, *High Level Political Forum on Sustainable Development SDG 11 Sustainable Cities and Communities: SDG 11 and the New Urban Agenda: Global Sustainability Frameworks for Local Action*, 2018, Available online at https://www.un.org/sustainabledevelopment/wp-content/uploads/2018/09/Goal-11.pdf

António Guterres, UN Secretary-General's "The State of the Planet" an address at Columbia University,New York,02 December 2020,available online at https://www.un.org/sg/en/content/sg/speeches/2020-12-02/address-columbia-university-the-state-of-the-planet

Olasunkanmi Habeeb Okunolaa and Saskia Werners "Nigerian cities are underprepared for flooding caused by climate change," March 16th, 2023,https://blogs.lse.ac.uk/africaatlse/2023/03/16/nigerian-cities-are-not-prepared-for-flooding-caused-by-climate-change/

Aljezeera (English) "Death toll climbs above 50,000 after Turkey, Syria earthquakes," https://www.aljazeera.com/news/2023/2/25/death-toll-climbs-above-50000-after-turkey-syria-earthquakes, February 2nd, 2023

World Meteorological Organization, "Weather-related disasters increase over past 50 years, causing more damage but fewer deaths," 31 August 2021, https://public.wmo.int/en/media/press-release/weather-related-disasters-increase-over-past-50-years-causing-more-damage-fewer ,WMO Atlas of Mortality and Economic Losses from Weather, Climate and Water Extremes (1970 – 2019) and https://public.wmo.int/en/media/press-release/weather-related-disasters-increase-over-past-50-years-causing-more-damage-fewer, August 31, 2021.

Sylvia Croese and Susan Parnell (eds.) *Localizing the SDGs in African Cities*, Springer's Sustainable Development Goals Series (electronic/e-book), 2022, available at https://doi.org/10.1007/978-3-030-95979-1 an https://link.springer.com/bookseries/15486

United Nations Environment Programme (UNEP), "About Disasters and
    Conflicts" https://www.unep.org/explore-topics/disasters-conflicts/about-
    disasters-conflicts

Hannah Ritchie, Pablo Rosado and Max Roser "Natural Disasters"
    https://ourworldindata.org/natural-disasters, 2022

Global Volcanism Program, Volcanoes of the World, v. 4.7.3. Venzke, E (ed.).
    Smithsonian Institution. https://doi.org/10.5479/si.GVP.VOTW4-2013 2013;
    National Geophysical Data Center / World Data Service (NGDC/WDS):
    Significant Earthquake Database, available at:
    https://www.ngdc.noaa.gov/nndc/struts/form?t=101650&s=1&d=1;

African Development Bank (AfDB), *African Economic Outlook, 2022 Highlights*,
    https://www.afdb.org/en/documents/african-economic-outlook-2022-
    highlights

Susanne E. Bauer *et-al* "Desert Dust, Industrialization, and Agricultural Fires:
    Health Impacts of Outdoor Air Pollution in Africa," 17 February 2019,
    https://doi.org/10.1029/2018JD029336 and
    https://agupubs.onlinelibrary.wiley.com/doi/full/10.1029/2018JD029336

Lei Nguyen "5 Biggest Environmental Issues In Africa In 2023," Mar 20th 2023,
    https://earth.org/environmental-issues-in-africa/

United Nations Children Fund (UNICEF) "Silent suffocation in Africa: Air
    pollution is a growing menace: Hitting the poorest children hardest," 2019,
    https://www.unicef.org/reports/silent-suffocation-in-africa-air-pollution-2019

United Nations Development Programme (UNDP), "What are the Sustainable
    Development Goals?" https://www.undp.org/sustainable-development-goals

Fermin Koop, "Report: Half a billion people in Africa don't have safe access to
    water" March 21,2022,https://www.zmescience.com/science/water-insecurity-
    concern-africa-21032022/

Vicky Hallett, "Millions Of Women Take A Long Walk With A 40-Pound Water
    Can"July 7, 2016,
    https://www.npr.org/sections/goatsandsoda/2016/07/07/484793736/millions-of-
    women-take-a-long-walk-with-a-40-pound-water-can

"What Causes Water Pollution In Africa?", https://thelastwell.org/2019/06/what-
    causes-water-pollution-in-africa/, June 28, 2019.

Updates about global population, https://www.worldometers.info/world-
    population/

Oghenekevwe Uchechukwu,"OIL THEFT: Nigeria now Africa's fourth largest oil producer," September,2022, https://thenewsguru.com/news/oil-theft-nigeria-now-africas-fourth-largest-oil-producer/

Chris Stein, "Shell Accused of Failing to Clean Up Nigeria Oil Spills," November 03, 2015 https://www.voanews.com/a/amnesty-says-shell-did-not-clean-up-spills-in-niger-delta/3034447.html

Amnesty International, "Nigeria: Amnesty activists uncover serious negligence by oil giants Shell and Eni," March 16,2018, https://www.amnesty.org/en/latest/news/2018/03/nigeria-amnesty-activists-uncover-serious-negligence-by-oil-giants-shell-and-eni/

United Nations Environmental Programme (UNEP), *Environmental Assessment of Ogoniland,* 2011, http://www.unep.org

Neil Munshi and William Clowes, "One of World's Most Polluted Spots Gets Worse as $1 Billion Cleanup Drags On," August 31, 2022,https://www.bloomberg.com/news/features/2022-08-31/shell-s-1b-oil-cleanup-left-one-of-world-s-most-polluted-spots-dirtier-for-now#xj4y7vzkg

Amnesty International, "Nigeria: Amnesty activists uncover serious negligence by oil giants Shell and Eni," March 16 ,2018, https://www.amnesty.org/en/latest/news/2018/03/nigeria-amnesty-activists-uncover-serious-negligence-by-oil-giants-shell-and-eni/

Philip Andrew Churm ,"Shell's bid to clean-up polluted Ogoniland labeled "incompetent""03/09/2022, https://www.africanews.com/2022/09/03/shells-bid-to-clean-up-polluted-ogoniland-labelled-incompetent//

Oyeranmi Soji, "How Fourth Industrial Revolution Can Aid Sustainable Development in African Cities: Lessons from China," available online at https://ascir.org/2023/01/21/how-fourth-industrial-revolution-can-aid-sustainable-development-in-african-cities-lessons-from-china/ , January 21, 2023

2022 IPCC Report, https://www.ipcc.ch/report/ar6/wg2/chapter/chapter-9/

Olasunkanmi Habeeb Okunola and Prof. Mulala Danny Simatele "Climate change in urban Nigeria - 4 factors that affect how residents adapt" February 26, 2023, https://theconversation.com/climate-change-in-urban-nigeria-4-factors-that-affect-how-residents-adapt-198802

Nigeria: Urbanization from 2011 to 2021, https://www.statista.com/statistics/455904/urbanization-in-nigeria/

Terkula Igidi, "Major events that shaped environment sector in 2022," https://dailytrust.com/major-events-that-shaped-environment-sector-in-2022/, 29 Dec 2022

Abiola Durodola et-al "Nigeria's cities are at severe risk from climate change. Time to build resilience, and fast," November 10, 2022,https://climatechampions.unfccc.int/nigerias-cities-are-at-severe-risk-from-climate-change-time-to-build-resilience-and-fast/

United Nations Climate Change , "Mobilizing more financial support for developing countries," https://unfccc.int/process-and-meetings/conferences/sharm-el-sheikh-climate-change-conference-november-2022/five-key-takeaways-from-cop27/mobilizing-more-financial-support-for-developing-countries, November 2022.

Sustainable Development Goals Center for Africa, "Africa 2030: Sustainable Development Goals Three-Year Reality Check,", June 2019, https://sdgcafrica.org/wp-content/uploads/2019/06/AFRICA-2030-SDGs-THREE-YEAR-REALITY-CHECK-REPORT.pdf

Belay Begashaw, "Africa and the Sustainable Development Goals: A long way to go,"Monday, July 29, 2019, https://www.brookings.edu/blog/africa-in-focus/2019/07/29/africa-and-the-sustainable-development-goals-a-long-way-to-go/ and SDGs Implementation in Africa, https://www.youtube.com/watch?v=96Obg0Quk2Y

UNOSSC, UNDRR GETI, PAHO/WHO, WHO Joint Certificate Training Program 2023 "Whole-of-society approach to creating healthy, resilient and sustainable cities: Harnessing South-South Cooperation for a post-COVID era" available online at https://www.undrr.org/event/undrr-unossc-who-paho2023

'Concept Note', available online at https://www.undrr.org/event/undrr-unossc-who-paho2023.3

"Constraints on Sustainable Development in Sub-Saharan Africa", http://unu.edu

"Developments Magazine of the DFID", www.development.org.uk

"Impunity in the Niger Delta", Essential Action, 2000, www.essentialaction.org.

Victor Ukaogo, Environmental Security and the role of foreign interest in the Niger Delta, http://www.google.com/Nigeriapetroleumindustry

Afolayan, David T., "Chief Executive Officer, GISKonsult Limited", www.gisknigeria.com

Akinboye, S.O., Globalisation and the challenge for Nigeria's Development in the 21st Century available at http://globalisation.icaap.org/content/v7.1/Akinboye.html,

Church, Dennis, 1992, "The Economy Vs. The Environment: Is There A Conflict?" Available online at http://www.ecoiq.com/dc-products/prod_conflict.html,

Diouf, Mamadou, "Social crises and Political Restructuring in West African Cities", in Dickson Eyoh and Richard Stren eds. *Decentralization and the Politics of Urban Development in West Africa*, Washington DC, 2007 Woodrow Wilson International Center for Scholars, 2007. 95, available online at www.wilsoncenter.org/sites/default/files/Stren.pdf.

Jeppesen, Soeren, Andersen, Joergen Eskemose and Madsen, Peter Vangsbo, "Urban Environmental Management in Developing Countries – Land Use, Environmental Health and Pollution Management, Research Network on Environment and Development," available online at www.ReNED.dk,

Kavanagh, Jim, "2011, Year of Billion-Dollar Disasters", 20th August, 2011, http://edition.cnn.com/2011/US/08/20/weather.disasters/index.html,

Kitto, Elizabeth, "Before European colonialism was Africa Essentially Rural?" *Identity Academic Winter 2012* (Online Edition) http://retrospectjournal.co.uk/portfolio/before-european-colonialism-was-africa-essentially-rural/).

McLuhan, Marshall, "Globalisation" Available online at http://en.wikipedia.ord/wiki/globalisation

Montgomery, Carle, "Environmental Geology," www.google.com/montgomery

Nagam, A. D. and Runnalts, Halle M, "Environment and Globalisation Understanding the Linkages" available online at www.google/environmental-2bisuses.band.com,

Nagam, A. D. and Runnalts, Halle M, "Environment and Globalisation Understanding the Linkages,"www.google/environmental-2bisuses.band.com

Nagam, A. D. and Runnalts, Halle, M., "Environment and Globalisation Understanding the Linkages", Available online at www.google/environmental-2bisuses.band.com,

Nagam, A.D. and Runnalts, Halle M, "Environment and Globalisation Understanding the Linkages" www.google/environmental-2bisuses.band.com

Nicholson, Michael, *International Relations*, www.ebafQuestion.com,

Ogunade, Raymond, "Environmental Issues in Yoruba Religion: Implications for Leadership and Society in Nigeria", a paper prepared for "Science and Religion: Global Perspectives", June 4-8, 2005, in Philadelphia, PA, USA, a program of the Metanexus Institute. Available online at www.metanexus.net

Omatseye, Sam, "Redeeming Ibadan", available online at http://thenationonlineng.net/new/sam-omatseye/redeeming-ibadan/

Omoleke, Ishaq Ishola, "Management of Environmental Pollution in Ibadan, An African City: The Challenges of Health Hazard Facing Government and the People", available online at www.krepublishers.com/.../JHE-15-4-265-275-2004-omoleke.pdf

Onibokun, A.G., and Kumuyi, A.J. "Ibadan, Nigeria", in Onibokun, A.G. (ed.) *Managing the Monster: Urban Waste and Governance in Africa*, available online at http://web.idrc.ca/en/ev-9402-201-1-DO_TOPIC.html

Oosthoek, K. Jan, "Environmental History - Between Science and Philosophy" http://science.jrank.org/pages/7662/Environmental-History.html#ixzz0eOxOibWY,

Owen, Greene, "Environmental Issues," www.google.com/owengreen,

Oyeranmi, Soji "Globalisation as a source of Environmental Tragedy in Sub-Saharan Africa: The Role Multinational Oil Corporations in Nigeria", *Global South*, SEPHIS e-Magazine, Volume 7, No. 3rd July 2011, www.sephisemagazine.org

Oyesiku, Kayode and Alade, Wale, "Historical Development of Urban and Regional Planning in Nigeria" in *State of Planning Report*, The Nigerian Institute of Town Planners, www.nitpng.com.

Roever, Sally, "The People's Economy and the Future of Cities," www.wiego.org

Ryan, O. "25 Big companies that Are Going Green," www.businesspundit.com/25-big-companies

Special Report by The United Nations' Agency for Human Habitat: The State of European cities in transition: taking stock after 20 years of reform" 2013, availa *www.unhabitat.org*,

Taiwo, David, (SIP Project Manager), "Ibadan: Mobilising Resources through a Technical Coordinating Committee", http://ww2.unhabitat.org/programmes/uef/cities/summary/ibadan.html,

UNDP *Governance for Sustainable Human Development,* New York, 2000; Asian Development Bank, *Elements of Governance,* http://www.adb.org/Governance/gov_elements.asp,

UNEP 2007 Reports on Climate Change, http://www.unep.org

UNEP's Report on "Africa's Vulnerability to Climate Change, UNEP Climate Change Resources." http://www.unep.org/themes/climatechange/

UN-HABITAT *Global Campaign on Urban Governance: Concept Paper, 2nd Edition.* Nairobi. 2002. 5. www.unhabitat.org/issue-papers-and-policy-units

UN-HABITAT (United Nations Human Settlements Programme), "I'm a City Changer", available online at www.worldurbancampaign.org and www.imacitychanger.org

UN-HABITAT (United Nations Human Settlements Programme), "I'm a City Changer", www.worldurbancampaign.org and www.imacitychanger.org.

United Cities and Local Governments' Global Charter-Agenda for Human Rights in the City, www.cities-localgovernments.org.

United Nations Conference on Environment and Development (UNCED) *Agenda 21* 1992 http://www.un.org/esa/sustdev/agenda21text.htm

United Nations Environmental Programme (UNEP) *Global Environment Outlook 3: Past, Present and the Future Perspectives,* 2002, http://www.unep.org/geo/geo3/

United Nations Environmental Programme (UNEP) *Global Environment Outlook 3: Past, Present and the Future Perspectives,* 2002, http://www.unep.org/geo/geo3/

United Nations Environmental Programme (UNEP) *Global Environment Outlook 3: Past, Present and the Future Perspectives,* 2002, available at: http://www.unep.org/geo/geo3/ "Oil for Nothing: Multinational Corporations, Environmental Destruction, and Death".

United Nations Environmental Programme (UNEP) *Global Environment Outlook 3: Past, Present and the Future Perspectives,* [On-line], 2002, available at: http://www.unep.org/geo/geo3/

United Nations Millennium Declaration. 2000. Available at, www.un.org/millennium/declaration/ares552e.pdf

UNU-wider conference on urbanisation, 2007, www.wider.unu.edu.

Vaughan, O. "Globalisation and Marginalisation," http://goggle.csnencarta.com/nigeria

Wall, Tim, "2011 Natural Disasters Worst Ever", available at
   http://news.discovery.com/earth/natural-disasters-in-2011-costliest-ever-
   110712.html

Wikipedia article of 9th May, 2007 on the Maldives available online at
   http://en.wikipedia.org/wiki/Maldives

Woods, Ngaire, *International Political Economy in the Age of Globalisation*. Available
   on www.questia.com

World urban forum 2007, www.unhabitat.org

Unpublished Materials

Adesola, Bisi, "Application of Environmental Planning and Management (EPM)
   Process to Sustainable Urban Planning and Development in Nigeria", (An
   Opening Address by the Special Guest of Honour and Secretary to the Oyo
   State Government, at a conference Organised by the Institute for Human
   Settlement and Environment IHSE, December, 1997 .

Akinyele, T.A. "Economic Relevance of Ibadanland: Past, Present and Future",
   Annual Lecture of the Oluyole Club of Lagos delivered at Kakanfo Inn, Ibadan
   on 9th January, 2010.

Buchenrieder, G. and Göltenboth, A.R., "Sustainable Freshwater Resource
   Management in the Tropics: The Myth of Effective Indicators", A Paper
   Presented at the 25th International Conference of Agricultural Economists
   (IAAE) on "Reshaping Agriculture's Contributions to Society" in Durban,
   South Africa, 2003.

Global Deforestation, Global Change Curriculum, University of Michigan, Global
   Change Program, January 2006.

Kjaer, A. M., "Central Government Intervention as Obstacle to Local Participatory
   Governance: The Case of Uganda." Paper prepared for the ILO Conference on
   Governance, 2005.

Makinde, Abayomi, "Local Administration in Ibadan City, 1893-1934",
   Undergraduate Long Essay, Department of History, University of Ibadan,
   Ibadan, Nigeria,1974.

Okunfulire, J.O., "The Challenges of Managing The Nigerian Urban
   Environment", (Paper Presented by the Directorate of Lands, Environment,
   Urban and Regional Development, Federal Ministry of Works and Housing, at
   the Conference Center, University of Ibadan, Ibadan, 5-17 December,1997.

Oluyitan, J. A., "Sanitation in Ibadan, 1942-1999," M.A. Dissertation, Department of History," University of Ibadan, Ibadan, Nigeria, 2005.

Raifu, Isiaka, "Urbanisation in Ibadan, 1951-2000", M.A. Dissertation, Department of History," University of Ibadan, Nigeria, 2003.

Ramkumar, T.K.,"Community Initiative for Environmental Management: The Work of Exnora International, an NGO", Paper Presented at the International Conference on Environmental Strategies for Asian Cities, Madras, India, February 14-17,1996

www.ingramcontent.com/pod-product-compliance
Lightning Source LLC
Chambersburg PA
CBHW030236230326
41458CB00091B/310